N

3

1

Computer-Aided
Heat Transfer
Analysis

Computer-Aided Heat Transfer Analysis

J. ALAN ADAMS
Mechanical Engineering Department
United States Naval Academy

DAVID F. ROGERS
Aerospace Engineering Department
United States Naval Academy

McGRAW-HILL BOOK COMPANY
New York   St. Louis   San Francisco
Düsseldorf   Johannesburg   Kuala Lumpur
London   Mexico   Montreal
New Delhi   Panama   Rio de Janeiro
Singapore   Sydney   Toronto

Computer-Aided
HEAT TRANSFER ANALYSIS

4 5 6 7 8 9 0 KPKP 7 9 8

This book was set in Times New Roman. The
editors were B. J. Clark and M. E. Margolies; and
the production supervisor was John A. Sabella. The
drawings were done by Oxford Illustrated Limited.
The printer and binder was Kingsport Press, Inc.

Library of Congress Cataloging in Publication Data

Adams, James Alan, 1936–
   Computer-aided heat transfer analysis.

   Includes bibliographies.
    1. Electronic data processing—Heat—Transmission.
I. Rogers, David F., 1937–   joint author.
II. Title.
TJ265.A24   621.4′022′02854   72–1366
ISBN 0–07–000285–1

"There is no natural wine.
Bread and wine are a product
of man's genius."

Jaurés

Contents

Preface

The objective of this book, which serves as an introductory text in heat transfer, is to present material which leads to physical insight concerning heat transfer processes. This objective is achieved by maintaining a proper balance between theory and analysis The variety of topics is limited to allow more concentration on basic principles and on a more generalized method of analysis. The prerequisites for this book are introductory courses in thermodynamics and fluid mechanics, although they are not absolutely necessary.

The approach utilized in this book is made possible by the digital computer. A computer can be described as a powerful source of information. The plethora of information readily produced by computers must have a significant effect on both the teaching and practice of any technical profession. Traditionally, a student is expected to apply his knowledge to obtain a single solution to a given problem. The student using this book is shown how to thoroughly analyze a given problem by varying parameters, boundary conditions, and mathematical assumptions. Experience has shown that this approach produces greater physical insight and understanding.

A thorough development of the methods for formulating mathematical models in terms of nondimensional variables is presented. The importance of understanding the physical meaning of these nondimensional parameters is stressed. The depth of the subject matter is increased by supplementing the classical analyses with computer-aided analysis programs. These contribute to the understanding of many problems that are not amenable to analytical solutions. The sample programs are written in the BASIC[1] language. Experience has shown BASIC to be the most useful language for students who desire to use the digital computer but have limited programming ability. Experienced programmers can make their own improvements and modifications to the program logic to improve machine efficiency.

The particular pedagogic approach presented in this book was developed while using remote time-sharing computer terminals for both classroom demonstration and analysis, as well as independent student analysis. However, the approach is useful for introductory courses in heat transfer even if a computer is not available. The methodology presented provides a sound foundation for a generalized approach to heat transfer analysis.

The three modes of heat transfer are defined in the Introduction. Concepts associated with radiation heat transfer are introduced in the first chapter. Subsequently a study of steady-state and transient heat transfer by radiation and convection using a lumped thermal analysis is made. In Chapter 2, steady-state, one-dimensional conduction in fins with various profiles and boundary conditions is analyzed. Both linear and nonlinear problems are considered. Steady-state, two-dimensional conduction, which requires solutions to the Laplace or Poisson equation, is treated in Chapter 3. Several combinations of geometry and boundary conditions are considered. The next chapter deals with one- and two-dimensional, transient conduction with various initial conditions and boundary conditions. Chapter 5 contains a differential analysis of laminar, two-dimensional boundary layers in forced and free convection. Chapter 6 is a study of convective heat transfer by the use of approximate integral techniques. These techniques are applied to external and internal convection and to laminar and turbulent flow. Empirical correlations and heat exchanger concepts are included in the final chapter.

The problems at the end of each chapter are generally of two types: numerical and parametric. The numerical problems give specific information and require the calculation of specific results; the parametric problems require the investigation of the effects of the variation of a parameter on the numerical solution. These problems generally are amenable to solution by using analysis programs similar to those presented in the text. They allow the student to ask the question "What if . . .?" and lead to a physical understanding of the mathematical model. The authors have found this type of problem very useful in teaching the concepts of heat transfer.

Because of the availability of many printed textbooks and papers, the formal lecture for the purpose of teaching has been an anachronism for many years. However, many educators continue to use it for lack of a better alternative. The authors have found an effective alternative in the approach to teaching which is compatible with this text. Although some lectures are necessary, the most effective use of classroom time is

---

[1] BASIC was originally developed at Dartmouth College.

to stimulate concentration and application of the subject matter by use of parametric studies and a generalized analysis.

The statements of the suggested problems at the end of each chapter use either the SI (System International) units or the British system of units. It is the accepted opinion that the SI system will soon become the preferred system. However, it will be necessary to be familiar with both systems during the next decade. The classification below can be used when choosing problems to analyze. The availability of a computer will vary from nonexistent to adequate. Problems listed in the first two columns may be analyzed without the need of a computer. Problems listed in the third column are designed to be analyzed with the aid of a digital computer, either in the batch or time-sharing mode.

PROBLEM CLASSIFICATION

Chapter	Problems which illustrate basic principles	Problems which use computer results presented in the text	Problems which require original computer solutions
1	1, 2, 3, 4, 5, 6, 7, 8, 9, 10, 11, 12, 13, 14		15, 16, 17, 18, 19, 20, 21, 22
2	1, 13	2, 3, 4, 5, 6, 7, 8	9, 10, 11, 12, 14, 15, 16, 17
3	1, 2, 3, 4, 5, 6, 7, 9, 10, 11, 12, 15, 16, 17, 19	8, 13, 14, 18	3c, 20, 21, 22, 23
4	1, 2, 3, 7, 10, 11, 12, 13, 14, 15, 16, 18a, 19, 21c, 23a	4, 5, 6, 8, 9	16c, 17, 18b, 18c, 19c, 20, 21a, 21b, 22, 23b, 24
5	1, 2, 3, 6, 7, 9, 10	4, 5, 8	11, 12, 13, 14, 15, 16
6	1, 2, 3, 4, 5, 9, 10, 14, 15, 16, 17, 18	11	6, 7, 8, 12, 13
7	1, 2, 3, 10, 11, 12, 13, 14, 15, 16, 17		4, 5, 6, 7, 8, 9

ACKNOWLEDGMENTS

The authors gratefully acknowledge the encouragement and support of the U.S. Naval Academy. The academic environment provided by the administration, faculty, and midshipmen was conducive to the development of the educational philosophy expressed in this book. Finally, we wish to acknowledge the patience and understanding of our wives and families during the creation of this book.

J. Alan Adams
David F. Rogers

I-1 Introduction to Heat Transfer

A study of heat transfer is founded on the concepts of energy, mass, and momentum. These physical concepts have meaning because they can be related to other measurable properties, such as temperature and velocity, by use of the physical laws and empirical relationships presented in this book. The property *energy* is used in mechanics and thermodynamics to help specify the state of a system. On the other hand, energy *transfer* across the boundaries of a thermodynamic system is considered to be in the form of either work or heat. *Heat transfer* is the expression used to indicate the transfer of energy due to a temperature difference. A *heat transfer rate* is the expression of thermal energy transfer per unit time, and *heat flux* is a heat transfer rate per unit area. The calculation of local heat transfer rates requires a knowledge of the local temperature distributions which provide the potential for heat transfer.

   When one is concerned with heat transfer, interest is usually focused on the transfer of energy across a surface. The interface which forms the surface of interest can take many forms. Perhaps most common is the interface between a solid and a gas, or a solid and a liquid. A solid-solid interface is associated with contact resistance which affects heat transfer between the two solids. The interface between a liquid and a gas is important in two-phase behavior, such as that associated with evaporation, boiling, and cavitation. Liquid-liquid and gas-gas interfaces can also be important in the study of transport phenomena.

A proper analysis of heat transfer across an interface must be based upon an awareness of all three modes of heat transfer—radiation, conduction, and convection. In general, the temperature of a surface and the temperature gradients at the interface are controlled by the combined effect of these three modes of energy transfer.

Thermal radiation is energy in the form of electromagnetic radiation emitted because of the absolute temperature of a body. The wavelength range of electro-magnetic radiation which is important in the analysis of thermal radiation is $0.1 \ \mu m \leq \lambda \leq 100 \ \mu m$, where λ is the wavelength measured in microns ($1 \mu m = 10^{-4}$ cm). This range includes the visible band (specified approximately by $0.38 \ \mu m \leq \lambda \leq 0.76 \ \mu m$) and most of the infrared portion of the electromagnetic spectrum.

Heat transfer by *conduction* refers to kinetic and internal energy exchange between molecules. Knowledge of this complex microscopic mechanism can be replaced by phenomenological laws which are based upon a macroscopic model. This change allows heat conduction rates to be calculated as a function of the temperature gradients. Conduction occurs in liquids and gases where energy exchange occurs between molecules in motion, as well as in solids where the molecules are held relatively stationary within a lattice.

When fluids are subjected to macroscopic motions due to surface or body forces, the thermal energy transferred to or from the flowing fluid can be classified as heat transfer by *convection*. The motion of the fluid over a surface depends upon the geometry of the surface, the properties of the fluid, and the velocity of the flow. This convection motion in the presence of a temperature gradient causes heat transfer by convection.

We now consider how the three modes of heat transfer occur at an interface under steady-state conditions. (For steady-state conditions, the temperature profile is not a function of time.) Consider an isothermal surface with temperature T_s which forms an interface between a solid and its surroundings as shown in Fig. I-1. This isothermal interface can be treated as a closed thermodynamic system in thermal equilibrium. (A closed system requires that no mass cross the system boundary. Thermal equilibrium requires that the system have uniform temperature throughout.) The first law of thermodynamics applied to this closed system under steady-state conditions gives (Ref. I-1)

$$\delta Q = 0 \qquad \text{(I-1)}$$

That is, the net heat transfer crossing the interface equals zero.

Possible heat transfer paths at an interface are indicated in Fig. I-1. The single

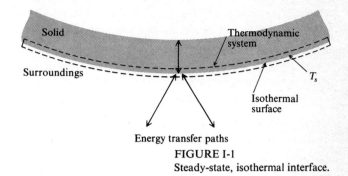

Energy transfer paths

FIGURE I-1

Steady-state, isothermal interface.

path in the solid indicates that conduction is the dominant mode of energy transfer. At the surface of the solid the energy transfer to or from the interface may occur by conduction, convection, radiation, or a combination of these modes. If the surrounding medium is a fluid, the conductive and convective modes are considered in series rather than in parallel, and together they constitute the convective heat transfer path. This is necessary because the velocity of the fluid relative to the surface at the interface is zero in most situations. Thus, molecular conduction at the interface must provide the mechanism for energy transfer to the fluid. Once the energy is conducted away from the solid surface, the convective motion of the fluid provides a second mechanism for energy transfer.

A detailed study of the conductive and convective modes of heat transfer based upon the laws of continuum mechanics is presented later in this book. Once the average convective heat transfer rate q across an interface is found from the controlling velocity and temperature fields, it is possible to define an average convection heat transfer coefficient $\bar{h}$ based upon a reference temperature difference $T_s - T_{ref}$; it is given by

$$\bar{h} = \frac{q}{A_s(T_s - T_{ref})} \qquad \text{(I-2)}$$

This expression is referred to as *Newton's law of cooling*. The area A_s is the surface area exposed to the convective heat transfer, and T_s is the surface temperature. The dimensions of q in terms of mass m, length l, temperature T, and time t are ml^2/t^3. Typical units of q are Btu/h or watts. The dimensions of $\bar{h}$ are m/t^3T, and typical units are Btu/h-ft^2-°F or W/m^2-°C. The relationships between units are given in Appendix C.

In the first four chapters the average convection heat transfer coefficient is treated as a known constant. The rate of convective heat transfer is then expressed by writing $q = \bar{h}A_s(T_s - T_\infty)$, where the environmental temperature T_∞ is used as the reference temperature in Eq. (I-2). In later chapters the variation of the local convection heat transfer coefficient h_x as a function of distance x measured along a surface will be calculated from a knowledge of the surface geometry, fluid properties, and temperature profiles in the fluid adjacent to the solid interface. The relationship between local and average convection heat transfer coefficients for a flat surface of length L and unit width is based upon the mean value theorem of calculus and is given by

$$\bar{h}A_s = \int_0^L h_x \, dx \, (1) \qquad (\text{I-3})$$

where $A_s = L(1)$.

Possible values for average convection heat transfer coefficients for various situations are indicated in Table I-1. Values are given for both free convection and forced convection. In *free* convection the energy causing fluid motion is produced by a thermally induced density difference in the fluid which arises because of the presence of a heated or cooled surface. In *forced* convection the fluid motion is caused by an external source. The largest values of $\bar{h}$ occur under conditions of boiling or condensation.

There is an assumption which underlies the entire analysis presented in this book which the reader should understand. In a first course in thermodynamics, one studies systems in equilibrium. Properties such as temperature, pressure, and enthalpy are defined only for an equilibrium state. For example, a system in thermal equilibrium must have uniform temperature. On the other hand, heat transfer requires a temperature gradient to provide the necessary potential. The assumption which allows one to apply equilibrium thermodynamic concepts to a nonequilibrium rate process such as heat transfer is local thermodynamic equilibrium.

Table I-1   ORDER OF MAGNITUDE FOR AVERAGE CONVECTION HEAT TRANSFER COEFFICIENTS* (1 Btu/h-ft²-°F = 5.67826 W/m²-°C)

Fluid	Free convection		Forced convection	
	$\bar{h}$ (Btu/h-ft²-°F)	$\bar{h}$ (W/m²-°C)	$\bar{h}$ (Btu/h-ft²-°F)	$\bar{h}$ (W/m²-°C)
Gases	1–5	5–30	5–50	30–300
Oils	1–20	5–100	10–500	50–3,000
Water (single phase)	5–50	30–300	50–2,000	300–10,000
Liquid metals	10–100	50–500	100–3,000	500–20,000
Water (boiling)	300–3,000	2,000–20,000	500–20,000	3,000–100,000
Steam (condensing)	500–5,000	3,000–30,000	500–30,000	3,000–200,000

* See Chaps. 5 to 7 for the determination of accurate values of $\bar{h}$ for specific situations.

The concept of a local thermodynamic state is coupled to the continuum model of matter. A *continuum model* can be visualized as a large number of differentially small subsystems. These make up the finite system which may or may not be in thermodynamic equilibrium. All thermodynamic properties are defined at each subsystem "point" within the finite system at each instant of time. Each subsystem is treated as a system in thermodynamic equilibrium. Although each subsystem must be small enough to be approximated by a system in equilibrium, it must be large enough to be accurately described by macroscopic thermodynamic concepts. For example, if the finite system is a gas, each subsystem must contain enough molecules to produce statistically meaningful average values for its properties, such as density. For air at 0°C, the number of molecules in a cubic subsystem 10^{-4} cm on each side is of the order of 10^7 molecules. Thus, subsystems of this size have meaningful macroscopic properties when there are slight departures from true equilibrium within the finite system.

In most engineering problems the departure from thermodynamic equilibrium can be considered slight, and the continuum model of matter and the assumption of local thermodynamic equilibrium are appropriate. Proof of this statement is the fact that the continuum model predicts the actual measured behavior of the real physical system. Certain situations, such as detonations and shock waves, do not allow the continuum approach subject to conditions of local thermodynamic equilibrium, and a more advanced analysis is required. These processes are characterized by the fact that the process time is of the same order as the relaxation time. (The *relaxation time* is the time required for a system to resume equilibrium after a sudden change in one of its properties.) The continuum model of matter and the assumption of local thermodynamic equilibrium are the basis for all analyses appearing in this book.

Reference

I-1 REYNOLDS, W. C.: "Thermodynamics," 2d ed., McGraw-Hill, New York, 1968.

1

RADIATION HEAT TRANSFER

1-1 Blackbody Radiation

Thermal radiation due to the absolute temperature of a body is generally emitted from a surface in all directions. The emitted electromagnetic radiation is a result of the thermal vibration of the particles, atoms, and molecules in the body. The total thermal radiation per unit time is called *radiation power*. Radiation power per unit area is defined as the *emissive power E*. Total emissive power includes thermal radiation emitted in all directions over all wavelengths between 0.1 μm $\leq \lambda \leq$ 100 μm. The *monochromatic* (spectral) emissive power E_λ is the radiation power per unit area within a wavelength range $d\lambda$. It follows that

$$E = \int_0^\infty E_\lambda \, d\lambda \approx \int_{0.1}^{100} E_\lambda \, d\lambda \qquad (1\text{-}1)$$

Because of the electromagnetic nature of thermal radiation, the emission and absorption of radiation are governed by the laws of wave mechanics. The Stefan-Boltzmann law for blackbody emission, as derived in statistical thermodynamics, and Planck's law of radiation distribution, obtained from the principles of quantum mechanics, provide the basic descriptions of thermal radiation emission from a surface. Thermodynamic limitations impose the maximum amount of thermal radiation that can be emitted at a given temperature and wavelength. This maximum emission is called *blackbody radiation*. The maximum amount of monochromatic emissive power $E_{b,\lambda}$ is given by *Planck's law* (Ref. 1-2):

$$E_{b,\lambda} = \frac{C_1}{\lambda^5} \frac{1}{\exp(C_2/\lambda T) - 1} \tag{1-2}$$

where
$$C_1 = 1.187 \times 10^8 \text{ Btu-}\mu\text{m}^4/\text{h-ft}^2$$
$$= 3.740 \times 10^4 \text{ W-}\mu\text{m}^4/\text{cm}^2$$
$$C_2 = 2.5896 \times 10^4 \ \mu\text{m-}°\text{R}$$
$$= 1.4387 \times 10^4 \ \mu\text{m-}°\text{K}$$

The constants C_1 and C_2 contain classical constants that arise in statistical mechanics. These are Boltzmann's constant, Planck's constant, and the velocity of light in a vacuum. It can be seen that monochromatic emissive power has units of Btu/h-ft^2-μm or W/cm^2-μm. The total emissive power for a blackbody can be expressed by (Ref. 1-4)

$$E_b = \sigma T^4 = \int_0^\infty E_{b,\lambda} \, d\lambda \tag{1-3}$$

where
$$\sigma = 0.1714 \times 10^{-8} \text{ Btu/h-ft}^2\text{-}°\text{R}^4$$
$$= 5.6699 \times 10^{-12} \text{ W/cm}^2\text{-}°\text{K}^4$$

Equation (1-3) is called the *Stefan-Boltzmann law*.

The fraction of blackbody radiation emitted by a nonblack specimen at the same temperature is defined as the *emittance ε*. Emittance is a nondimensional ratio of emissive powers and can be defined for both monochromatic and total emissive power. The monochromatic emittance is given by

$$\varepsilon_\lambda = \frac{E_\lambda}{E_{b,\lambda}} \tag{1-4}$$

and the total emittance is

$$\varepsilon = \frac{E}{E_b} = \frac{E}{\sigma T^4} \tag{1-5}$$

Since the values of emittance for various surfaces are normally determined by experimental measurements on test specimens, a certain nomenclature has developed. The term *emittance* is used to specify the property of a specimen. On the other hand, *emissivity* is a material property measured by using an opaque, optically smooth test specimen. Thus, emissivity is a special case of emittance. The suffix "ance" is used in this chapter to specify the properties of engineering materials since optically smooth materials are seldom encountered.

From Eqs. (1-4) and (1-5) it is obvious that $\varepsilon_\lambda = \varepsilon = 1.0$ for a blackbody. Using Eq. (1-4) we can write the total emissive power as

$$E = \int_0^\infty E_\lambda \, d\lambda = \int_0^\infty \varepsilon_\lambda E_{b,\lambda} \, d\lambda \qquad (1-6)$$

Differentiating Eq. (1-2) to obtain $(\partial E_{b,\lambda}/\partial \lambda)_T$ and setting the results equal to zero gives the value of $(E_{b,\lambda})_{max}$. It follows that the product $\lambda_{max} T = C$, where C is a constant equal to 5,215.5 μm-°R or 2,897.6 μm-K, and λ_{max} is the wavelength where $E_{b,\lambda}$ is a maximum for a given absolute surface temperature T. The relation $\lambda_{max} T = C$ is called *Wien's displacement law*. This feature of blackbody radiation is clear in Fig. 1-1 which gives a plot of Planck's law.

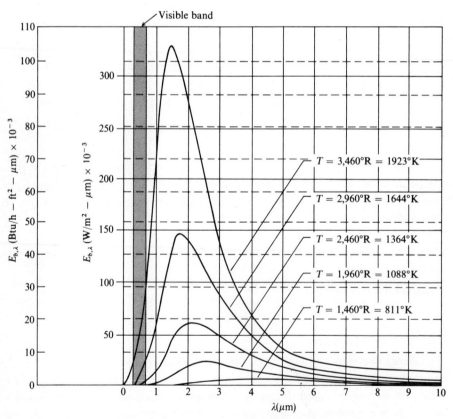

FIGURE 1-1
Blackbody monochromatic emissive power.

The characteristics of blackbody radiation are more efficiently represented if we form the nondimensional ratio $E_{b,\lambda}/(E_{b,\lambda})_{max}$. By use of Eq. (1-2) and the relation $\lambda_{max} T = 5,215.6$ μm-$^\circ$R, it follows that

$$\frac{E_{b,\lambda}}{(E_{b,\lambda})_{max}} = \left(\frac{5.215.6}{\lambda T}\right)^5 \frac{\exp{(4.965)} - 1}{\exp{(25,896/\lambda T)} - 1}$$

or

$$\frac{E_{b,\lambda}}{(E_{b,\lambda})_{max}} = \left(\frac{\lambda_{max}}{\lambda}\right)^5 \frac{\exp{(4.965)} - 1}{\exp{(4.965\lambda_{max}/\lambda)} - 1} \qquad (1\text{-}7)$$

where

$$(E_{b,\lambda})_{max} = \frac{C_1 T^5}{(5,215.6)^5[\exp{(C_2/5,215.6)} - 1]}$$

$$= 2.161 \times 10^{-13} T^5$$

for T expressed in degrees Rankine. Equation (1-7) then gives

$$\frac{E_{b,\lambda}}{\sigma T^5} = f\left(\frac{\lambda_{max}}{\lambda}\right) = F(\lambda T) \qquad (1\text{-}8)$$

Dunkle (Ref. 1-9) has tabulated radiation functions which are based upon Eqs. (1-7) and (1-8). These functions appear in Table 1-1.

Many interesting properties of thermal radiation can be observed by using Table 1-1 along with the basic radiation principles. Consider the following example:

EXAMPLE 1-1   (a) The sun may be treated as a blackbody with $T = 10,400^\circ$F. From Wien's displacement law it follows that

$$\lambda_{max} = 0.480 \ \mu\text{m}$$

The wavelength at which $(E_{b,\lambda})_{max}$ occurs falls within the narrow visible band of radiation, 0.38 μm $\leq \lambda \leq 0.76$ μm.

(b) The corresponding solar monochromatic emissive power at $\lambda = \lambda_{max}$ is

$$(E_{b,\lambda})_{max} = \left[\frac{C_1 \lambda^{-5}}{\exp{(C_2/\lambda T)} - 1}\right]_{\lambda = \lambda_{max}} = 3.26 \times 10^7 \ \text{Btu/h-ft}^2\text{-}\mu\text{m}$$

or from Table 1-1 at $\lambda_{max} T = 5,216$, we read

$$\frac{E_{b,\lambda}}{\sigma T^5} = 12.6 \times 10^{-5}$$

and then

$$E_{b,\lambda} = 3.26 \times 10^7 \ \text{Btu/h-ft}^2\text{-}\mu\text{m}$$

Table 1-1 PLANCK RADIATION FUNCTIONS*

λT, μm-°R	$\dfrac{E_{b\lambda} \times 10^5}{\sigma T^5}$	$\dfrac{E_b(0 - \lambda T)}{\sigma T^4}$	λT, μm-°R	$\dfrac{E_{b\lambda} \times 10^5}{\sigma T^5}$	$\dfrac{E_b(0 - \lambda T)}{\sigma T^4}$
1,000.0	0.000039	0.0000	10,400.0	5.142725	0.7183
1,200.0	0.001191	0.0000	10,600.0	4.921745	0.7284
1,400.0	0.012008	0.0000	10,800.0	4.710716	0.7380
1,600.0	0.062118	0.0000	11,000.0	4.509291	0.7472
1,800.0	0.208018	0.0003	11,200.0	4.317109	0.7561
2,000.0	0.517405	0.0010	11,400.0	4.133804	0.7645
2,200.0	1.041926	0.0025	11,600.0	3.959010	0.7726
2,400.0	1.797651	0.0053	11,800.0	3.792363	0.7803
2,600.0	2.761875	0.0098	12,000.0	3.633505	0.7878
2,800.0	3.882650	0.0164	12,200.0	3.482084	0.7949
3,000.0	5.093279	0.0254	12,400.0	3.337758	0.8017
3,200.0	6.325614	0.0368	12,600.0	3.200195	0.8082
3,400.0	7.519353	0.0507	12,800.0	3.069073	0.8145
3,600.0	8.626936	0.0668	13,000.0	2.944084	0.8205
3,800.0	9.614973	0.0851	13,200.0	2.824930	0.8263
4,000.0	10.463377	0.1052	13,400.0	2.711325	0.8318
4,200.0	11.163315	0.1269	13,600.0	2.602997	0.8371
4,400.0	11.714711	0.1498	13,800.0	2.499685	0.8422
4,600.0	12.123821	0.1736	14,000.0	2.401139	0.8471
4,800.0	12.401105	0.1982	14,200.0	2.307123	0.8518
5,000.0	12.559492	0.2232	14,400.0	2.217411	0.8564
5,200.0	12.613057	0.2483	14,600.0	2.131788	0.8607
5,400.0	12.576066	0.2735	14,800.0	2.050049	0.8649
5,600.0	12.462308	0.2986	15,000.0	1.972000	0.8689
5,800.0	12.284687	0.3234	16,000.0	1.630989	0.8869
6,000.0	12.054971	0.3477	17,000.0	1.358304	0.9018
6,200.0	11.783688	0.3715	18,000.0	1.138794	0.9142
6,400.0	11.480102	0.3948	19,000.0	0.960883	0.9247
6,600.0	11.152254	0.4174	20,000.0	0.815714	0.9335
6,800.0	10.807041	0.4394	21,000.0	0.696480	0.9411
7,000.0	10.450309	0.4607	22,000.0	0.597925	0.9475
7,200.0	10.086964	0.4812	23,000.0	0.515964	0.9531
7,400.0	9.721078	0.5010	24,000.0	0.447405	0.9579
7,600.0	9.355994	0.5201	25,000.0	0.389739	0.9621
7,800.0	8.994419	0.5384	26,000.0	0.340978	0.9657
8,000.0	8.638524	0.5561	27,000.0	0.299540	0.9689
8,200.0	8.290014	0.5730	28,000.0	0.264157	0.9717
8,400.0	7.950202	0.5892	29,000.0	0.233807	0.9742
8,600.0	7.620072	0.6048	30,000.0	0.207663	0.9764
8,800.0	7.300336	0.6197	40,000.0	0.074178	0.9891
9,000.0	6.991475	0.6340	50,000.0	0.032617	0.9941
9,200.0	6.693786	0.6477	60,000.0	0.016479	0.9965
9,400.0	6.407408	0.6608	70,000.0	0.009192	0.9977
9,600.0	6.132361	0.6733	80,000.0	0.005521	0.9984
9,800.0	5.868560	0.6853	90,000.0	0.003512	0.9989
10,000.0	5.615844	0.6968	100,000.0	0.002339	0.9991
10,200.0	5.373989	0.7078			

* From J. WIEBELT: "Engineering Radiation Heat Transfer," Holt, New York, 1965, by permission.

(c) The fraction of emitted solar thermal radiation between $0 \ \mu m < \lambda < 1.0 \ \mu m$ can be read from Table 1-1. At $\lambda T = (1.0)(10,860) = 10,860$, the ratio $E_{b,(0-\lambda T)}/\sigma T^4$ is approximately 0.74. Thus 74 percent of the blackbody radiation is emitted as electromagnetic radiation with a wavelength less than 1.0 μm. Over 99 percent is emitted with a wavelength less than 10 μm.

(d) When the short wavelength solar radiation falls on a surface, an equilibrium temperature of the surface may be reached. At equilibrium, the temperature of the surface is such that the net surface heat transfer by radiation, convection, and conduction is zero. If the equilibrium surface temperature of a body is 100°F, the wavelength of the maximum monochromatic emissive power from the body is given by Wien's law as $\lambda_{\text{max}} = 9.3 \ \mu m$.

(e) The corresponding monochromatic emissive power at $\lambda = 9.3 \ \mu m$ is $E_{b,\lambda} = 11.9$ Btu/h-ft²-μm.

(f) The fraction of thermal radiation emitted from the surface at $T = 560°R$ between $0 \ \mu m < \lambda < 10 \ \mu m$ is read from Table 1-1 at $\lambda T = 5,600$. As seen, 29.860 percent of the radiation falls within this range. None falls within the range $0 \ \mu m < \lambda < 1.0 \ \mu m$. The majority of the radiation emitted is at wavelengths in the far infrared, $\lambda > 10 \ \mu m$.

(g) The monochromatic emissive power at $\lambda = 100 \ \mu m$, the extreme end of the thermal radiation band, is found from the value $E_{b,\lambda}/\sigma T^5 = 0.022 \times 10^{-5}$ at $\lambda T = 56,000$. For $T = 560°R$ this gives $E_{b,\lambda} = 0.0208$ Btu/h-ft²-μm.   ////

The difference in the wavelength between incoming and emitted radiation must be carefully considered in analyzing radiation. The surface properties of radiation to be discussed later in this chapter are quite dependent upon these wavelength characteristics. Only in an isothermal enclosure will the wavelength range of incoming radiation be exactly equal to the wavelength range of emitted radiation. In the previous example it was shown that the wavelengths of radiation emitted from a high-temperature source are much shorter than the wavelengths of electromagnetic radiation emitted from a low-temperature source.

1-2   Radiation between Black Surfaces

The exchange of energy between two or more surfaces by radiation heat transfer depends upon two effects—a surface effect and a space effect. The amount of radiation emitted by a surface is a function of the radiation properties of the surface. For a black surface the total emission is the maximum possible as given by Eq. (1-3). The transfer of radiation between surfaces also depends upon the fraction of energy leaving a given surface which falls on the surrounding surfaces. This space effect is a function of geometry.

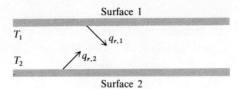

FIGURE 1-2
Infinite black surfaces.

The surface and space effects are easy to determine for two parallel, infinite black surfaces as shown in Fig. 1-2. By treating the actual surfaces as infinite we can say that all radiation emitted by surface 1 falls on surface 2, and vice versa; that is, we neglect radiation escaping out the ends of the parallel-plate assembly. For this simple system, the radiation emitted by surface 1 and falling on surface 2 is given by

$$q_{r,1} = AE_{b,1} = A_1 \sigma T_1{}^4 \qquad (1\text{-}9)$$

In a similar manner

$$q_{r,2} = AE_{b,2} = A_2 \sigma T_2{}^4 \qquad (1\text{-}10)$$

If one treats surface 1 as a thermodynamic system in thermal equilibrium, the first law of thermodynamics requires that

$$\delta Q = q_{r,2} - q_{r,1} + q_1 = 0 \qquad (1\text{-}11)$$

Here q_1 is the additional energy that must be supplied to the interface, such as by conduction or convection, to replace the net energy lost by radiation if T_1 is to remain constant. The net energy lost by radiation from surface 1, if one assumes that $T_1 > T_2$, is

$$q_{r,1} - q_{r,2} = \sigma A(T_1{}^4 - T_2{}^4) \qquad (1\text{-}12)$$

A similar analysis for surface 2 gives

$$\delta Q = q_{r,1} - q_{r,2} - q_2 = 0 \qquad (1\text{-}13)$$

Thus the energy $q_2 = q_{r,1} - q_{r,2}$ must be removed from surface 2 to maintain steady-state conditions.

When the geometry requires that surfaces be treated as finite, the determination of the space effect can be the major problem in calculating radiation heat transfer. Radiation leaving a surface must now be considered as a "pencil" of rays as shown in Fig. 1-3 in order to describe the directional properties. This concept leads to the definition of radiation intensity. *Thermal radiation intensity* is defined as the radiant energy leaving a surface per unit area *normal* to the pencil of rays per unit solid angle per unit

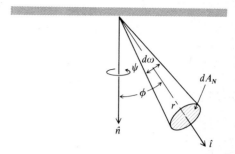

FIGURE 1-3
Pencil of rays.

time. A *solid angle* is defined as the ratio of the area on a sphere cut by a conical pencil of rays issuing from the center of the sphere to the square of the radius. Because of the geometric definition of a sphere, this area will always be normal to the radius vector that defines the direction of the pencil rays. Using the nomenclature in Fig. 1-3, the solid angle is

$$d\omega = \frac{dA_N}{r^2} \qquad (1\text{-}14)$$

Note that the dimensionless solid angle $d\omega$, *steradian*, is formed by a ratio of terms with dimensions of length squared, as opposed to a plane angle which is a ratio of lengths.

We now return to the definition of radiation intensity I. Let dq'' be the radiant energy per unit area per unit time (radiant heat flux) contained within the solid angle $d\omega$ which leaves a unit surface in the direction ϕ. The corresponding radiant energy per unit *normal* area is then $dq''/\cos\phi$. Then by definition we write

$$I = \frac{dq''/\cos\phi}{d\omega} \qquad (1\text{-}15)$$

The total radiation leaving the surface per unit time and area is then

$$q'' = \iint dq'' = \iint I \cos\phi \, d\omega \qquad (1\text{-}16)$$

The integration with respect to the solid angle is carried out over the hemisphere which covers the surface.

The most convenient way to integrate Eq. (1-16) is to express the solid angle $d\omega$ in terms of the spherical coordinates shown in Fig. 1-4. The differential area dA_N on a hemisphere cut out by the pencil of rays is $(r \, d\phi)(r \sin\phi \, d\psi) = r^2 \sin\phi \, d\psi \, d\phi$. Thus

$$d\omega = \frac{r^2 \sin\phi \, d\psi \, d\phi}{r^2} = \sin d\phi \, d\psi \qquad (1\text{-}17)$$

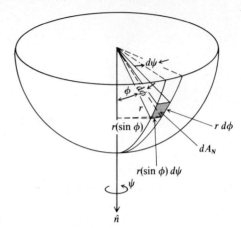

FIGURE 1-4
Spherical coordinate system.

Equation (1-16) is now written

$$q'' = \int_0^{2\pi} \int_0^{\pi/2} I \cos \phi \sin \phi \, d\phi \, d\psi \qquad (1\text{-}18)$$

Note that the limits on ϕ and ψ are defined such that integration occurs over the entire hemisphere which covers the emitting surface.

We now use another conclusion obtained by Planck for blackbody radiation. It is shown in Ref. 1-2 that the intensity of blackbody radiation is not a function of direction. This type of radiation behavior is called *isotropic* and allows the intensity in Eq. (1-18) to be treated as a constant. Then, integration with respect to ϕ yields

$$q'' = I \int_0^{2\pi} \left(\frac{\sin^2 \phi}{2} \right)_0^{\pi/2} d\psi = \frac{I}{2} \int_0^{2\pi} d\psi \qquad (1\text{-}19)$$

A second integration gives

$$q'' = I\pi \qquad (1\text{-}20)$$

This simple result relates the radiation flux leaving a black surface to the intensity of radiation.

Up to this point the discussion has been related to thermal radiation *leaving* a surface due to the emission of electromagnetic waves in the wavelength region $0.1 \ \mu m \le \lambda \le 100 \ \mu m$. The character of emitted radiation is a function of the surface temperature and surface emittance. We now consider thermal radiation *falling* on a surface.

Total *irradiation G* is defined as the total thermal radiation per unit area per unit time falling on a surface. The irradiation confined within a small wavelength band $d\lambda$

is the monochromatic irradiation G_λ. The interaction at a solid interface between incoming radiation and the surface is a strong function of the wavelength characteristics of the irradiation, which in turn depends upon the temperature at the original source of the incoming radiation.

1-3 Surface Characteristics

When radiation falls on a surface of a body, it may be reflected, absorbed within the body, or transmitted through the body. This leads to the definitions of *reflectance*, *absorptance*, and *transmittance*, respectively. As before, the suffix "ance" indicates the property of a specimen, and "ivity" is used for the property of an opaque, optically smooth surface. The monochromatic reflectance is defined as

$$\rho_\lambda = \frac{\text{reflected radiation/time-area-wavelength}}{G_\lambda}$$

The total reflectance is then

$$\rho = \frac{\int_0^\infty \rho_\lambda G_\lambda \, d\lambda}{\int_0^\infty G_\lambda \, d\lambda} \qquad (1\text{-}21)$$

In a similar manner the monochromatic absorptance is defined as

$$\alpha_\lambda = \frac{\text{absorbed radiation/time-area-wavelength}}{G_\lambda}$$

The total absorptance is given by

$$\alpha = \frac{\int_0^\infty \alpha_\lambda G_\lambda \, d\lambda}{\int_0^\infty G_\lambda \, d\lambda} \qquad (1\text{-}22)$$

Total transmittance τ is defined in a similar manner. Conservation of energy requires that $\alpha + \rho + \tau = 1.0$. Many engineering materials are opaque to thermal radiation, and $\tau = 0$. For opaque materials we can write

$$\alpha_\lambda + \rho_\lambda = 1.0 \qquad (1\text{-}23)$$

and also
$$\alpha + \rho = 1.0 \qquad (1\text{-}24)$$

In more advanced analyses of thermal radiation one considers the directional and polarization effects of electromagnetic radiation. Such an analysis leads to important laws, called *Kirchhoff's laws*, that relate surface radiation properties such as α_λ, ε_λ, and E. Kirchhoff's laws are summarized and discussed in Ref. 1-3. The important restriction that must be realized when using Kirchhoff's laws is that they are derived under the assumption of an isothermal enclosure. Thus the surface is in thermal equilibrium with its surroundings. The importance of this limitation becomes clear when it is realized that the surface emittance depends on the temperature of the surface emitting radiation; whereas the surface absorptance depends on the wavelength of the irradiation, and thus on the temperature of the surrounding sources of irradiation.

The following three equations are based on Kirchhoff's laws, neglecting the effects of polarization and the angular variation of properties. Equation (1-25) states that in an isothermal enclosure the monochromatic absorptance is equal to the monochromatic emittance.

$$\alpha_\lambda = \varepsilon_\lambda \qquad (1\text{-}25)$$

Experimental measurements have shown that this relationship between *monochromatic* properties is also valid in nonisothermal enclosures.

Equation (1-26) states that the ratio of total surface emissive power to total surface absorptance is a constant when that surface is in thermal equilibrium with its surroundings.

$$\frac{E}{\alpha} = \text{constant} \qquad (1\text{-}26)$$

This conclusion is based on the realization that bodies continue to emit radiation even when in thermal equilibrium with their surroundings.

Equation (1-27) states that the irradiation falling on a black surface within an isothermal enclosure is equal to the total emissive power of the black surface within the enclosure, since the blackbody temperature equals the enclosure temperature.

$$G = E_b \qquad (1\text{-}27)$$

For a black surface, $\varepsilon_\lambda = 1.0 = \varepsilon$ as required by the definition, Eq. (1-4). It follows from Eq. (1-25) that $\alpha_\lambda = 1.0$. Since $\alpha_\lambda = 1.0$, a black surface can be defined as a perfect absorber of radiant energy. It is common to define an ideal black surface as one which absorbs *all* incoming radiation, regardless of direction, wavelength, or other characteristics of thermal radiation. This definition requires that α_λ be independent of wavelength and $\alpha_\lambda = \alpha = 1.0$. An ideal black surface is not necessarily in thermal equilibrium with its surroundings.

Since $\alpha = 1.0$ for an ideal black surface, $\rho = \tau = 0$, and the only thermal radiation leaving the surface is due to the emissive power of the body. Recall that the radiant heat flux q'' used in the definition of intensity [cf. Eq. (1-16)] is the total radiation leaving the surface. For a blackbody it follows that $q'' = E_b = \sigma T^4$, and from Eq. (1-20) we can write

$$E_b = \pi I \qquad (1\text{-}28)$$

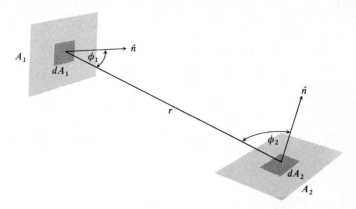

FIGURE 1-5
Finite black surfaces.

1-4   Shape Factors

Consider two finite black surfaces as shown in Fig. 1-5. We focus our attention on the two differential areas dA_1 and dA_2. Using Eq. (1-15) we can express the radiation per unit area per unit time which leaves dA_1 and directly falls on dA_2 as

$$dq''_{1-2} = I_1 \cos \phi_1 \, d\omega_{12} \qquad (1\text{-}29)$$

where the solid angle $d\omega_{12} = dA_2 \cos \phi_2/r^2$. The total radiation from dA_1 to dA_2 is then

$$dq_{1-2} = dq''_{1-2} \, dA_1 = \frac{I_1 \cos \phi_1 \cos \phi_2 \, dA_1 \, dA_2}{r^2} \qquad (1\text{-}30)$$

Combining Eqs. (1-30) and (1-28) and integrating (for $I_1 = \text{constant} = E_{b,1}/\pi$) gives

$$q_{1-2} = E_{b,1} \int_{A_2} \int_{A_1} \frac{\cos \phi_1 \cos \phi_2 \, dA_1 \, dA_2}{\pi r^2} \qquad (1\text{-}31)$$

By similar reasoning it can be shown that

$$q_{2-1} = E_{b,2} \int_{A_1} \int_{A_2} \frac{\cos \phi_2 \cos \phi_1 \, dA_2 \, dA_1}{\pi r^2} \qquad (1\text{-}32)$$

The net radiation heat transfer between A_1 and A_2 is $q_{1-2} - q_{2-1} = q_{1 \rightleftarrows 2}$. If one uses Eqs. (1-31) and (1-32), $q_{1 \rightleftarrows 2}$ becomes

$$q_{1 \rightleftarrows 2} = (E_{b,1} - E_{b,2}) \int_{A_1} \int_{A_2} \frac{\cos \phi_1 \cos \phi_2 \, dA_1 \, dA_2}{\pi r^2} \qquad (1\text{-}33)$$

A shape factor is now defined such that Eqs. (1-31) and (1-32) can be written

$$q_{1-2} = E_{b,1}(A_1 F_{1-2}) \tag{1-34a}$$

where

$$A_1 F_{1-2} = \int_{A_2} \int_{A_1} \frac{\cos \phi_1 \cos \phi_2 \, dA_1 \, dA_2}{\pi r^2} \tag{1-34b}$$

and

$$q_{2-1} = E_{b,2}(A_2 F_{2-1}) \tag{1-35a}$$

where

$$A_2 F_{2-1} = \int_{A_1} \int_{A_2} \frac{\cos \phi_2 \cos \phi_1 \, dA_2 \, dA_1}{\pi r^2} \tag{1-35b}$$

If one compares Eqs. (1-34b) and (1-35b), it is obvious that $A_1 F_{1-2} = A_2 F_{2-1}$. This relationship is called *reciprocity* and is limited to surfaces with constant radiation intensity. The net radiation between finite black surfaces may now be written

$$q_{1 \rightleftarrows 2} = A_1 F_{1-2}(E_{b,1} - E_{b,2}) = A_1 F_{1-2} \sigma(T_1^4 - T_2^4) \tag{1-36}$$

Physically, F_{1-2} is the fraction of energy emitted from surface 1 which falls directly on surface 2. It is also called a configuration factor, view factor, or angle factor in the literature.

When isothermal, finite black surfaces exchange thermal radiation, the major task in calculating radiation heat transfer rates is in determining the values of the appropriate shape factors F_{i-j}. The evaluation of the integrals in Eq. (1-33) can become difficult, even for fairly simple geometry. Several methods, including analytical, numerical, and experimental techniques, have been used to determine shape factors. Useful results can be found in Refs. (1-3) through (1-8). Table 1-2 gives shape factors for some common configurations that can be found in the above references.

Sometimes it is possible to determine the values of a shape factor without detailed calculations. This can be done by using the physical definition of F_{i-j} along with the reciprocity relationships. Consider the two examples shown in Fig. 1-6, page 23.

Figure 1-6a shows a small sphere enclosed in a large spherical cavity. By inspection, $F_{1-2} = 1.0$, and then by reciprocity $F_{2-1} = (A_1/A_2)F_{1-2}$. Figure 1-6b shows a long, flat surface covered by a long, semicircular duct. Neglecting end radiation we see that $F_{1-2} = 1.0$, and then $F_{2-1} = A_1/A_2$. We can also write $F_{2-1} + F_{2-2} = 1.0$, and it follows that $F_{2-2} = 1.0 - A_1/A_2$. Physically F_{2-2} is the fraction of radiation emitted by surface 2 which falls directly on surface 2. In general, for n surfaces it follows that the total radiation leaving surface 1 is accounted for by writing

$$F_{1-1} + F_{1-2} + F_{1-3} + \cdots + F_{1-n} = 1.0$$

If surface 1 is flat, it cannot "see" itself, and $F_{1-1} = 0$. One can also write reciprocity relations between any pair of surfaces.

Table 1-2   SHAPE FACTORS

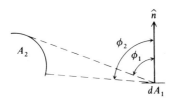

dA_1 = infinite plane of differential width

A_2 = infinite surface parallel to itself and the plane of dA_1

$F_{1-2} = 0.5(\sin \phi_2 - \sin \phi_1)$

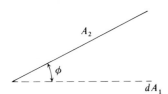

dA_1 = infinite plane of differential width

A_2 = infinite plane intersecting plane of dA_1 at an angle

$F_{1-2} = 0.5(1 + \cos \phi)$

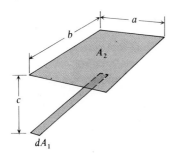

dA_1 = differential strip parallel to a finite, plane, rectangular surface A_2

A_2 = rectangle with one edge opposite to dA_1

$$\eta = \frac{b}{c} \qquad \xi = \frac{a}{c}$$

$$F_{1-2} = \frac{1}{\pi\eta} \left[(1 + \eta^2)^{1/2} \tan^{-1} \frac{\xi}{(1 + \eta^2)^{1/2}} - \tan^{-1} \xi \right.$$

$$\left. + \frac{\eta\xi}{(1 + \xi^2)^{1/2}} \tan^{-1} \frac{\eta}{(1 + \xi^2)^{1/2}} \right]$$

Table 1-2 (*continued*)

d

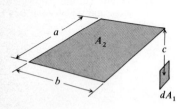

$dA_1 =$ differential area in a plane perpendicular to A_2

$A_2 =$ finite rectangle with one corner directly above dA_1

$$\eta = \frac{b}{c} \quad \xi = \frac{a}{c}$$

$$F_{1-2} = \frac{1}{2\pi} \left[\frac{\xi}{(1+\xi^2)^{1/2}} \tan^{-1} \frac{\eta}{(1+\xi^2)^{1/2}} + \frac{\eta}{(1+\eta^2)^{1/2}} \tan^{-1} \frac{\xi}{(1+\eta^2)^{1/2}} \right]$$

$$\lim_{\xi \to \infty} F_{1-2} = \frac{\eta/4}{(1+\eta^2)^{1/2}} \qquad \lim_{\eta \to \infty} F_{1-2} = \frac{\xi/4}{(1+\xi^2)^{1/2}}$$

e

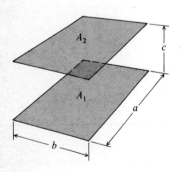

$A_1 =$ finite rectangle parallel to A_2

$A_2 =$ finite rectangle identical to A_1 and directly opposed to A_1

$$\eta = \frac{b}{c} \quad \xi = \frac{a}{c}$$

$$F_{1-2} = \frac{2}{\pi\eta\xi} \left\{ \ln \left[\frac{(1+\eta^2)(1+\xi^2)}{1+\eta^2+\xi^2} \right]^{1/2} + \xi(1+\eta^2)^{1/2} \tan^{-1} \frac{\xi}{(1+\eta^2)^{1/2}} \right.$$

$$\left. + \eta(1+\xi^2)^{1/2} \tan^{-1} \frac{\eta}{(1+\xi^2)^{1/2}} - \xi \tan^{-1} \xi - \eta \tan^{-1} \eta \right\}$$

$$\lim_{\eta \to \infty} F_{1-2} = \left(1 + \frac{1}{\xi^2} \right)^{1/2} - \frac{1}{\xi} \qquad \lim_{\xi \to \infty} F_{1-2} = \left(1 + \frac{1}{\eta^2} \right)^{1/2} - \frac{1}{\eta}$$

Table 1-2 (*continued*)

f

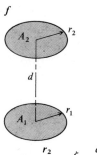

$A_1 =$ disk of radius r_1 parallel to disk A_2 of radius r_2

$A_2 =$ disk of radius r_2 directly opposed to A_1 separated by distance d

$$\eta = \frac{r_2}{d} \quad \xi = \frac{d}{r_1} \quad \gamma = 1 + (1 + \eta^2)\xi^2$$

$$F_{1-2} = 0.5[\gamma - (\gamma^2 - 4\eta^2\xi^2)^{1/2}]$$

g

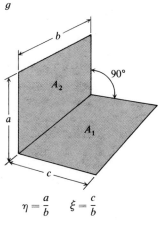

$A_1 =$ finite rectangle with an edge common to A_2

$A_2 =$ finite rectangle perpendicular to rectangle A_1

$$\eta = \frac{a}{b} \quad \xi = \frac{c}{b}$$

$$F_{1-2} = \frac{1}{\pi\xi} \left\{ \xi \tan^{-1}\frac{1}{\xi} + \eta \tan^{-1}\frac{1}{\eta} - (\eta^2 + \xi^2)^{1/2} \tan^{-}\xi \frac{1}{(\eta^2 + \xi^2)^{1/2}} \right.$$
$$\left. + \frac{1}{4} \ln \left[\frac{(1 + \xi^2)(1 + \eta^2)}{1 + \eta^2 + \xi^2} \xi^2 \frac{1 + \xi^2 + \eta^2}{(1 + \xi^2)(\xi^2 + \eta^2)} \eta^2 \frac{1 + \xi^2 + \eta^2}{(1 + \eta^2)(\xi^2 + \eta^2)} \right] \right\}$$

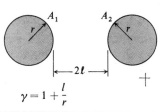

$A_1 =$ portion of the infinite cylinder parallel to cylinder A_2 and seen by A_2

$A_2 =$ portion of the infinite cylinder identical to cylinder A_1 and seen by A_1

$$\gamma = 1 + \frac{l}{r}$$

$$F_{1-2} = \frac{2}{\pi} \left[(\gamma^2 - 1)^{1/2} - \gamma + \frac{\pi}{2} - \cos^{-1}\frac{1}{\gamma} \right]$$

Table 1-2 (*continued*)

i

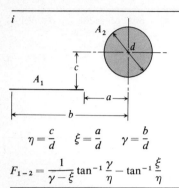

$A_1 =$ horizontal plane of finite width and infinite extent
$A_2 =$ portion of infinite cylinder seen by A_1 when the cylinder axis is parallel to the plane of A_1

$$\eta = \frac{c}{d} \qquad \xi = \frac{a}{d} \qquad \gamma = \frac{b}{d}$$

$$F_{1-2} = \frac{1}{\gamma - \xi} \tan^{-1} \frac{\gamma}{\eta} - \tan^{-1} \frac{\xi}{\eta}$$

j

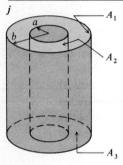

$A_1 =$ inner surface of outer cylinder
$A_2 =$ outer surface of inner cylinder
$A_3 =$ cross-sectional area of annulus at end of cylinders

$$\eta = \frac{b}{a} \qquad \xi = \frac{c}{a}$$

$$A = \xi^2 + \eta^2 - 1 \qquad B = \xi^2 - \eta^2 + 1$$

$$F_{1-2} = \frac{1}{\eta} - \frac{1}{\pi\eta}\left(\cos^{-1}\frac{B}{A} - \frac{1}{2\xi}\left\{[(A+2)^2 - (2\eta)^2]^{1/2}\cos^{-1}\frac{B}{\eta A} + B\sin^{-1}\frac{1}{\eta} - \frac{\pi A}{2}\right\}\right)$$

$$F_{1-1} = 1 - \frac{1}{\eta} + \frac{2}{\pi\eta}\tan^{-1}\frac{2(\eta^2-1)^{1/2}}{\xi} - \frac{\xi}{2\pi\eta}\left\{\frac{(4\eta^2+\xi^2)^{1/2}}{\xi}\sin^{-1}\left[4(\eta^2-1)+\frac{(\xi^2/\eta^2)(\eta^2-2)}{\xi^2+4(\eta^2-1)}\right]\right.$$

$$\left. -\sin^{-1}\frac{\eta^2-2}{\eta^2} + \frac{\pi}{2}\left[\frac{(4\eta^2+\xi^2)^{1/2}}{\xi} - 1\right]\right\}$$

1-5 Gray Surfaces

Most surfaces do not behave as ideal black surfaces. Therefore, $\rho > 0$, and radiation leaves the surface by both reflection and emission. To account for the total radiant energy leaving a nonblack surface, we introduce the term radiosity J. *Radiosity* is the total radiation leaving a surface per unit area per unit time. For an opaque surface this radiation can consist of emission and reflection, and we write

$$J = E + \rho G \qquad (1\text{-}37)$$

For a black surface, $\rho = 0$ and the radiosity is equal to the emissive power.

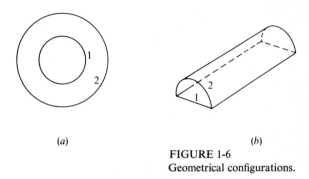

(a)   (b)

FIGURE 1-6
Geometrical configurations.

By use of Eq. (1-4) the monochromatic radiation emitted from a nonblack surface can be written $E_\lambda = \varepsilon_\lambda E_{b,\lambda}$, and the total emitted radiation is then

$$E = \int_0^\infty \varepsilon_\lambda E_{b,\lambda}\, d\lambda \qquad (1\text{-}38)$$

For real surfaces, ε_λ is a function of λ. Typical variations of ε_λ for both conductors and nonconductors for short wavelengths, $0\ \mu\text{m} \le \lambda \le 10\ \mu\text{m}$, are shown in Fig. 1-7.[1]

If we assume that $\varepsilon_\lambda = \text{constant} = \varepsilon$, the emissive power may be easily determined from Eqs. (1-38) and (1-3). The approximation that ε_λ and α_λ are constant and therefore independent of wavelength defines a *gray body*. Thus, for a gray body $\varepsilon_\lambda = \varepsilon$ and $\alpha_\lambda = \alpha$. It follows from Kirchhoff's law, Eq. (1-25), that $\varepsilon = \alpha$ for a gray body.

The gray body approximation may only be appropriate over a limited wavelength range. Also, one constant value of ε_λ may apply for short wavelengths, and another value may approximate the emittance for long wavelength radiation. The appendices of Refs. 1-3 and 1-5 contain total and spectral values for many common engineering materials. For a gray body, Eq. (1-38) gives the following expression for the emissive power:

$$E = \varepsilon \int_0^\infty E_{b,\lambda}\, d\lambda = \varepsilon\sigma T^4 \qquad (1\text{-}39)$$

The approximation of a gray body is often used to calculate the total absorptance of a surface from a knowledge of the emittance. There is difficulty in calculating α from Eq. (1-22), which requires knowledge of the irradiation G_λ as a function of wavelength. If the temperature difference between the surfaces exchanging radiation is not too great, the wavelength characteristics of the emitted and absorbed radiation are similar, and the approximation $\alpha = \varepsilon$ may be useful. On the other hand, when the emitted radiation and irradiation occur in different wavelength bands, this approximation is usually not correct. For example, for most materials it is not correct to assume $\alpha = \varepsilon$ for a low-temperature surface exposed to solar radiation.

---

[1] Values for specific materials are given in Appendix D.

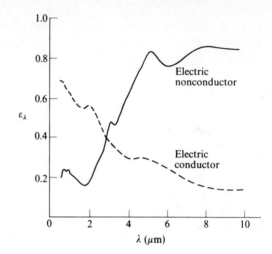

FIGURE 1-7
Typical emittance variation for conductors and nonconductors.

Sometimes it is possible to also use the simple result given by Eq. (1-20) for gray bodies. This equation was based on the fact that blackbody radiation emitted from a surface is isotropic, i.e., independent of direction. If the radiation leaving an opaque gray surface is to be independent of direction, it must be true for both the emitted and reflected components.

Reflected energy can be idealized as either specular (mirrorlike) or diffuse, depending upon the surface roughness and wavelength of irradiation. A diffusely reflecting surface reflects irradiation with a uniform intensity, regardless of the incoming direction. The assumption of diffuse reflection, along with diffuse emission, leads to the concept of diffuse radiosity. If the radiosity is treated as diffuse, the reflected and emitted components need not be treated separately, the radiant intensity is independent of direction, and $q'' = J = \pi I$ from Eq. (1-20).

The shape factor defined in the previous section can also be applied to gray surfaces if a few more restricting assumptions are made. If the shape factor is to be independent of surface variation of radiant energy, we must assume a constant radiosity along the surface. For an isothermal gray surface the emission is constant since ε_λ and T are constants. It is also necessary that the irradiation be the same at every point on the surface if the shape factor is to be a constant. The physical interpretation of the shape factor for gray surfaces is then the fraction of diffusely emitted and reflected energy from surface 1 which falls directly on surface 2.

The equation for the net radiation between gray surfaces, subject to the above assumptions, is written

$$q_{1 \rightleftarrows 2} = A_1 F_{1-2}(J_1 - J_2) \qquad (1\text{-}40)$$

For a black surface, $J = E_b$ and Eq. (1-40) reduces to Eq. (1-36). Although all the limitations employed to obtain Eq. (1-40) will seldom be met in practice, the approximate analysis is useful from an engineering point of view. A more exact analysis leads to integral equations rather than algebraic equations. These are discussed in books on radiation heat transfer, such as Refs. 1-3, 1-4, and 1-5.

1-6   Surface and Space Resistance

As mentioned in Sec. 1-3, radiation heat transfer is determined by a surface effect and a space effect. The space effect is expressed in terms of a shape factor. Since surface temperatures are usually of interest in heat transfer calculations, it is convenient to express the radiation heat transfer between surfaces in terms of temperature differences rather than radiosity differences as in Eq. (1-40). This formulation leads to the concept of surface resistance to thermal radiation.

From Eqs. (1-37) and (1-39), the radiosity from a gray surface is given by

$$J = \varepsilon \sigma T^4 + \rho G \qquad (1\text{-}41)$$

For an opaque surface the total reflectance is given by $\rho = 1 - \alpha$. Using the gray body approximation $\varepsilon_\lambda = $ constant and $\alpha = \varepsilon$, we write

$$\rho = 1 - \varepsilon \qquad (1\text{-}42)$$

Application of the first law of thermodynamics to an isothermal gray surface under steady-state conditions requires that

$$\delta Q = -J + G + q'' = 0 \qquad (1\text{-}43)$$

where q'' is the combined energy transfer per unit area per unit time at the surface by conduction and convection. Combining Eqs. (1-41), (1-42), and (1-43) and using $E_b = \sigma T^4$ gives

$$q = q'' A_1 = \frac{A_1 \varepsilon_1}{1 - \varepsilon_1} (E_{b,1} - J_1) \qquad (1\text{-}44)$$

A similar equation applies for each gray surface emitting and receiving radiation.

For a steady state the combined heat transfer by conduction and convection equals the net radiation to or from the surface, as given by Eq. (1-44). The net radiation is also given by Eq. (1-40). The meaning of Eqs. (1-40) and (1-44) can best be explained by use of an electrical analogy. Current flow through a resistance is given by an equation of the form

$$\text{Current flow} = \frac{\text{potential difference}}{\text{resistance}}$$

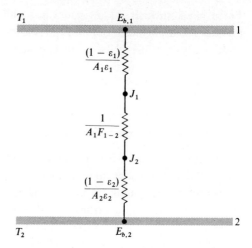

FIGURE 1-8
Electrical analogy.

Equating Eqs. (1-40) and (1-44) and writing them in the above form gives

$$q_{1 \rightleftarrows 2} = \frac{E_{b,1} - J_1}{(1 - \varepsilon_1)/A_1\varepsilon_1} = \frac{J_1 - J_2}{1/A_1 F_{1-2}} \qquad (1\text{-}45)$$

We can identify a surface resistance between nodes $E_{b,1}$ and J_1 given by $(1 - \varepsilon_1)/A_1\varepsilon_1$ and a space resistance between nodes J_1 and J_2 given by $1/A_1 F_{1-2}$. The heat transfer between two gray surfaces can be considered as a current flowing through these resistances in series, with nodal potentials as indicated in Fig. 1-8. A surface resistance is associated with each surface, and a space resistance is associated with the pair of surfaces. By use of the electrical analogy we can express the net radiation heat transfer between two gray surfaces:

$$\text{Net radiation} = \frac{\text{total potential difference}}{\text{total resistance}}$$

The mathematical expression is

$$q_{1 \rightleftarrows 2} = \frac{E_{b,1} - E_{b,2}}{\dfrac{1 - \varepsilon_1}{A_1\varepsilon_1} + \dfrac{1}{A_1 F_{1-2}} + \dfrac{1 - \varepsilon_2}{A_2\varepsilon_2}} \qquad (1\text{-}46)$$

where $E_{b,1} = \sigma T_1^4$ and $E_{b,2} = \sigma T_2^4$.

1-7 Radiation between Gray Surfaces

A gray body shape factor can be defined by writing the net radiation heat transfer between two gray surfaces as

$$q_{1 \rightleftarrows 2} = A_1 \mathscr{F}_{1-2}(E_{b,1} - E_{b,2}) \qquad (1\text{-}47)$$

By comparison with Eq. (1-46) it must follow that

$$A_1 \mathscr{F}_{1-2} = \frac{1}{\dfrac{1 - \varepsilon_1}{A_1 \varepsilon_1} + \dfrac{1}{A_1 F_{1-2}} + \dfrac{1 - \varepsilon_2}{A_2 \varepsilon_2}} \qquad (1\text{-}48)$$

where $\mathscr{F}_{1-2}$ is the shape factor for diffuse radiation between two gray surfaces.

Two special cases are of interest. For infinite, parallel plates, $A_1 = A_2$ and $F_{1-2} = 1.0$. It follows that $\mathscr{F}_{1-2}$ may be written

$$\mathscr{F}_{1-2} = \frac{1}{1/\varepsilon_1 + 1/\varepsilon_2 - 1} \qquad (1\text{-}49)$$

For a small body with surface area A_1 completely enclosed by a large area A_2 we can write $F_{1-2} = 1.0$ and $A_1/A_2 \ll 1$. Then $\mathscr{F}_{1-2}$ can be approximated by

$$\mathscr{F}_{1-2} = \frac{1}{(1 - \varepsilon_1)/\varepsilon_1 + 1} = \varepsilon_1 \qquad (1\text{-}50)$$

Remember that these results are based on the approximation that $\alpha = \varepsilon$.

For two finite gray surfaces the values of ε_1, ε_2, and F_{1-2} are used in Eq. (1-48) to determine $\mathscr{F}_{1-2}$. Note that F_{1-2} is the same blackbody shape factor that was defined in Sec. 1-4. If several surfaces are exchanging radiation, equivalent electrical circuits such as shown in Fig. 1-8 are useful in calculating the net radiant energy exchange. The equivalent circuit for three infinite gray surfaces is shown in Fig. 1-9a, and the circuit for a three-sided enclosure is shown in Fig. 1-9b. Notice that a surface resistance is associated with each surface, and a space resistance with each pair of surfaces that "see" each other. The equation for the net radiation heat transfer between any two gray surfaces can be expressed as a ratio of the total potential difference between the surface nodes $E_i - E_j$ to the equivalent resistance between the surface nodes in question. For example, the expression for $q_{1 \rightleftarrows 3}$ in Fig. 1-9a is

$$q_{1 \rightleftarrows 3} = \frac{\sigma(T_1^4 - T_2^4)}{\dfrac{1 - \varepsilon_1}{A_1 \varepsilon_1} + \dfrac{1}{A_1 F_{1-2}} + \dfrac{2(1 - \varepsilon_2)}{A_2 \varepsilon_2} + \dfrac{1}{A_2 F_{2-3}} + \dfrac{1 - \varepsilon_3}{A_3 \varepsilon_3}} \qquad (1\text{-}51)$$

A radiant energy flow balance can be written for each of the three nodes shown in Fig. 1-9b. Since we have assumed uniform radiosity over each surface, we write

$$q_{\text{net}, 1} = A_1 J_1 - A_1 F_{1-1} J_1 - A_2 F_{2-1} J_2 - A_3 F_{3-1} J_3$$

$$q_{\text{net}, 2} = A_2 J_2 - A_1 F_{1-2} J_1 - A_2 F_{2-2} J_2 - A_3 F_{3-2} J_3 \qquad (1\text{-}52)$$

$$q_{\text{net}, 3} = A_3 J_3 - A_1 F_{1-3} J_1 - A_2 F_{2-3} J_2 - A_3 F_{3-3} J_3$$

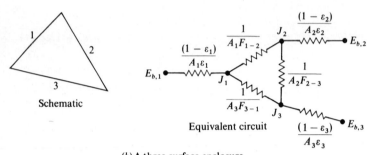

(a) Three parallel surfaces

(b) A three surface enclosure

FIGURE 1-9
Equivalent circuits.

When all surfaces are flat, $F_{1-1} = F_{2-2} = F_{3-3} = 0$. Using Eq. (1-45) for each surface gives

$$q_{net, 1} = \frac{E_{b, 1} - J_1}{(1 - \varepsilon_1)/A_1 \varepsilon_1}$$

$$q_{net, 2} = \frac{E_{b, 2} - J_2}{(1 - \varepsilon_2)/A_2 \varepsilon_2} \qquad (1\text{-}53)$$

$$q_{net, 3} = \frac{E_{b, 3} - J_3}{(1 - \varepsilon_3)/A_3 \varepsilon_3}$$

Combining Eqs. (1-52) and (1-53) gives three equations for three unknown radiosities J_1, J_2, and J_3, assuming that the shape factors, surface emittances, and surface temperatures are known. Knowledge of the three radiosities allows the net radiation from each surface to be calculated by use of Eqs. (1-53). For n surfaces of uniform radiosity which form an enclosure, Eqs. (1-52) can be generalized for each surface i by writing

$$q_{net, i} = A_i J_i - \sum_{j=1}^{n} A_j F_{j-i} J_j \qquad (1\text{-}54)$$

1-8   Combined Radiation and Convection

Heat transfer from a solid-gas interface usually occurs simultaneously by radiation and convection. This convective path can be represented by an equivalent circuit as was done in the previous section for radiation. We first write Newton's law of cooling, Eq. (I-2), in the form

$$q_c = \frac{T_s - T_{\text{ref}}}{1/\bar{h}A_s} = \frac{\Delta T}{R_c} \qquad (1\text{-}55)$$

where the convective resistance $R_c = 1/\bar{h}A_s$.

When the gas does not participate in the radiation by scattering, absorption, and reemission, the heat transfer by convection can be determined independently of the radiation. Then the total thermal energy leaving a surface is the sum of the convection and radiation. If the heat transfer process is steady, energy transfer from the interface by radiation and convection must be equal to the energy transfer to the interface, such as by conduction from within the solid and by radiation from the surroundings.

For a small gray body completely enclosed by a large area, $\mathscr{F}_{1-2} = \varepsilon_1$ by Eq. (1-50), and Eq. (1-47) gives

$$q_{1 \rightleftarrows \infty} = A_1 \varepsilon_1 \sigma (T_1{}^4 - T_\infty{}^4) \qquad (1\text{-}56)$$

To base the radiation heat transfer on a linear temperature difference, one can write Eq. (1-56) as

$$q_{1 \rightleftarrows \infty} = \frac{A_1 \varepsilon_1 \sigma (T_1{}^4 - T_\infty{}^4)(T_1 - T_\infty)}{T_1 - T_\infty}$$

If one uses the equality $T_1{}^4 - T_\infty{}^4 = (T_1{}^3 + T_1{}^2 T_\infty + T_1 T_\infty{}^2 + T_\infty{}^3)(T_1 - T_\infty)$, this equation becomes

$$q_{1 \rightleftarrows \infty} = A_1 \varepsilon_1 \sigma (T_1{}^3 + T_1{}^2 T_\infty + T_1 T_\infty{}^2 + T_\infty{}^3)(T_1 - T_\infty) = \frac{T_1 - T_\infty}{1/\bar{h}_r A_1} \qquad (1\text{-}57)$$

where the denominator has been written in the form of a resistance to radiation heat transfer. This equation defines a radiation coefficient $\bar{h}_r$ which is similar to the convection coefficient defined by Newton's law of cooling. When the difference between T_1 and T_∞ is not great,

$$T_1{}^3 + T_1{}^2 T_\infty + T_1 T_\infty{}^2 + T_\infty{}^3 \approx 4\bar{T}^3$$

where $\bar{T} = (T_1 + T_\infty)/2$. If one uses this approximation, the expression for $\bar{h}_r$ becomes

$$\bar{h}_r = 4\varepsilon_1 \sigma \bar{T}^3 \qquad (1\text{-}58)$$

Many times it is convenient to define $\bar{h}_r$ as in Eq. (1-57) even if the approximation that leads to Eq. (1-58) is not used. This allows the radiation heat transfer to be expressed in terms of the linear temperature difference $T_1 - T_\infty$.

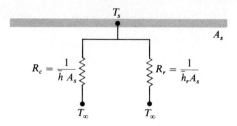

FIGURE 1-10
Parallel circuit for convection and radiation.

We can now use the electrical analogy to represent a parallel path for convection and radiation. This consists of two parallel resistances between the same temperature nodes T_1 and T_∞ as shown in Fig. 1-10.

The more general expression for $\bar{h}_r$ can be obtained from Eq. (1-47) by writing

$$q_{1 \rightleftarrows 2} = \frac{A_1 \mathscr{F}_{1-2} \, \sigma (T_1{}^4 - T_2{}^4)(T_1 - T_{\text{ref}})}{T_1 - T_{\text{ref}}} = \frac{T_1 - T_{\text{ref}}}{1/\bar{h}_r A_1}$$

It follows that

$$\bar{h}_r = \frac{\mathscr{F}_{1-2} \, \sigma (T_1{}^4 - T_2{}^4)}{T_1 - T_{\text{ref}}} \qquad (1\text{-}59)$$

The total heat transfer leaving the surface in Fig. 1-10 can be expressed by

$$q = q_c + q_r = \frac{T_1 - T_\infty}{1/\bar{h}A} + \frac{T_1 - T_{\text{ref}}}{1/\bar{h}_r A} \qquad (1\text{-}60)$$

Choosing $T_{\text{ref}} = T_\infty$ gives

$$q = \bar{h}_T A(T_1 - T_\infty) \qquad (1\text{-}61)$$

where the total heat transfer coefficient $\bar{h}_T = \bar{h} + \bar{h}_r$.

1-9   Transient Radiation and Convection

Consider two parallel, rectangular, thin surfaces as shown in Fig. 1-11. Assume that the geometric shape factor $F_{2-3} = F_{3-2} = 1.0$. Physically this requires that the plates be very large or the spacing l be very small, so that the radiation escaping to the environment out the ends of the plates is negligible. A gas flows between the plates and over the top plate at a free-stream velocity V_∞ and a temperature T_∞ equal to the temperature of the surroundings. The methods for determining the convection heat transfer coefficients on the surfaces of the plates are presented in later chapters. For now we consider that the average values of $\bar{h}$ are known constant quantities and are equal on each exposed surface.

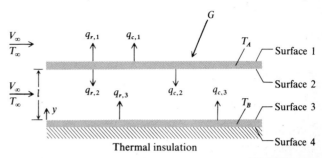

FIGURE 1-11
Parallel plates.

Initially the system is in thermal equilibrium with $T_A = T_B = T_\infty$ where T_A and T_B are the temperatures of the two plates. At time $t = 0$, additional energy from a source of thermal radiation falls on the upper surface of the upper plate. The radiation absorbed is given by $\alpha_1 G$. As the temperature of the upper surface increases, convection occurs from surface 1 to the environment, conduction occurs through the upper plate, and thermal radiation emitted by the surfaces increases because of the temperature increase. Conduction increases the temperature at the lower surface of the upper plate, and this temperature increase affects the convection and radiation from surface 2. The radiation emitted by surface 2 which falls on surface 3 has a similar effect on the convection and radiation from surface 3 as well as on the temperature of the lower plate.

A lumped thermal analysis can be made by assuming that the temperature within each plate is uniform throughout the plate at any instant. We specify this condition mathematically by writing $T_1 = T_2 = T_A$ and $T_3 = T_4 = T_B$ at any time t. Physically this is reasonable for thin plates with good thermal conductivity. The exact thermal conditions necessary for this approximation are discussed in detail in Chap. 4.

Since the temperature in each plate is assumed to be uniform, each plate can be treated as a closed thermodynamic system. The first law of thermodynamics for a plate during the transient heating process can be written (Ref. 1-1)

$$\delta Q = q_T \, dt = dU \qquad (1\text{-}62)$$

where q_T is the total heat transfer rate crossing the surfaces of the plate during the time increment dt, and dU is the change in total internal energy within the plate. If one uses the thermodynamic definition of the solid specific heat at constant volume, $c =$

$(\partial u/\partial T)_v \approx du/dT$, and the relation $U = mu$, where u is the internal energy per unit mass, Eq. (1-62) becomes

$$q_T = mc\,\frac{dT}{dt} \qquad (1\text{-}63)$$

The net radiation between the two plates is $q_{r,2} - q_{r,3}$. This can be written as

$$q_{1 \rightleftharpoons 2} = q_{r,2} - q_{r,3} = \sigma A_2 \mathscr{F}_{2-3}(T_2{}^4 - T_3{}^4) \qquad (1\text{-}64a)$$

for diffuse surfaces with constant emittances. The net radiation between surface 1 and the environment is written as

$$q_{1 \rightleftharpoons \infty} = \sigma A_1 \mathscr{F}_{1 \to \infty}(T_1{}^4 - T_\infty{}^4) \qquad (1\text{-}64b)$$

The application of Eq. (1-63) to the upper plate, if one uses the nomenclature in Fig. 1-11, gives

Total heat transfer, plate A	Absorbed radiation	Convection from surface 1	Convection from surface 2			
↓	↓	↓	↓			
$q_{T,A}$	$= A_1 \alpha_1 G$	$-q_{c,1}$	$-q_{c,2}$			

	Net radiation from surface 1		Net radiation between plates		Change in internal energy, plate A	
	↓		↓		↓	
	$-q_{r,1}$		$-\sigma A_2 \mathscr{F}_{2-3}(T_2{}^4 - T_3{}^4)$		$= m_A c_A \dfrac{dT_A}{dt}$	$(1\text{-}65)$

If one assumes that the upper surface of plate A can be treated as a small surface in large surroundings, i.e., $\mathscr{F}_{1-\infty} = \varepsilon_1$, it follows that $q_{r,1} = \sigma A_1 \varepsilon_1(T_1{}^4 - T_\infty{}^4)$.

The corresponding equation for the lower plate, if one assumes no heat transfer from the lower surface due to thermal insulation, is

Total heat transfer, plate B	Net radiation between plates	Convection from surface 3	Change in internal energy, plate B	
↓	↓	↓	↓	
$q_{T,B}$	$= \sigma A_3 \mathscr{F}_{3-2}(T_2{}^4 - T_3{}^4)$	$-q_{c,3}$	$= m_B c_B \dfrac{dT_B}{dt}$	$(1\text{-}66)$

The convective heat transfer can be expressed in terms of a convection heat transfer coefficient by using Newton's law of cooling with $T_{\text{ref}} = T_\infty$ in Eq. (I-2). The environmental temperature T_∞ is treated as a constant. The two governing equations then become

$$A_1 \alpha_1 G - 2\bar{h} A_1 (T_A - T_\infty) - \sigma A_1 \varepsilon_1 (T_A{}^4 - T_\infty{}^4) - \sigma A_2 \mathscr{F}_{2-3}(T_A{}^4 - T_B{}^4)$$

$$= m_A c_A \frac{dT_A}{dt} \qquad (1\text{-}67)$$

and
$$\sigma A_3 \mathscr{F}_{3-2}(T_A{}^4 - T_B{}^4) - \bar{h} A_3 (T_B - T_\infty) = m_B c_B \frac{dT_B}{dt} \qquad (1\text{-}68)$$

where the equalities $T_1 = T_2 = T_A$, $A_1 = A_2$, and $T_3 = T_B$ have been used. To find T_A and T_B as a function of time, we must simultaneously solve the above two first-order nonlinear ordinary differential equations.

We first define two nondimensional dependent variables as

$$\theta_A = \frac{T_A}{T_\infty}$$

$$\theta_B = \frac{T_B}{T_\infty}$$

A nondimensional independent variable is defined as $\tau = t/t_c$, where t_c is a reference time yet to be chosen. Since $dT/dt = (d\theta/d\tau)(d\tau/dt)(dT/d\theta) = (T_\infty/t_c)(d\theta/d\tau)$ and mass m equals density ρ times volume V, the resulting nondimensional equations are

$$\frac{d\theta_A}{d\tau} = \frac{A_1 \alpha_1 G t_c}{\rho_A V_A T_\infty c_A} - \frac{2\bar{h} A_1 t_c}{\rho_A V_A c_A}(\theta_A - 1)$$

$$- \frac{A_1 \sigma \varepsilon_1 T_\infty{}^3 t_c (\theta_A{}^4 - 1)}{\rho_A V_A c_A} - \frac{A_2 \sigma T_\infty{}^3 t_c \mathscr{F}_{2-3}}{\rho_A V_A c_A}(\theta_A{}^4 - \theta_B{}^4) \qquad (1\text{-}69)$$

and
$$\frac{d\theta_B}{d\tau} = \frac{A_3 \sigma T_\infty{}^3 t_c}{\rho_B V_B c_B} \mathscr{F}_{3-2}(\theta_A{}^4 - \theta_B{}^4) - \frac{\bar{h} A_3 t_c}{\rho_B V_B c_B}(\theta_B - 1) \qquad (1\text{-}70)$$

The initial conditions are $\theta_A = \theta_B = 1.0$ at $\tau = 0$.

We assume that the plates have identical geometrical and physical properties. The equations are simplified if we choose $t_c = \rho V c / \bar{h} A$. This ratio can be interpreted as a time constant for the system under study. The equations are then

$$\frac{d\theta_A}{d\tau} = \frac{\alpha_1 G}{\bar{h} T_\infty} - 2(\theta_A - 1) - \frac{\sigma \varepsilon_1 T_\infty{}^3}{\bar{h}}(\theta_A{}^4 - 1) - \frac{\sigma T_\infty{}^3}{\bar{h}} \mathscr{F}_{2-3}(\theta_A{}^4 - \theta_B{}^4) \qquad (1\text{-}71)$$

and $$\frac{d\theta_B}{d\tau} = \frac{\sigma T_\infty{}^3}{\bar{h}} \mathscr{F}_{3-2}(\theta_A{}^4 - \theta_B{}^4) - (\theta_B - 1) \qquad (1\text{-}72)$$

We can now identify important nondimensional parameters in the governing equations. The physical significance of the nondimensional parameters α_1, ε_1, and $\mathscr{F}_{2-3}$ has been discussed previously. We can also define

$$N1 = \frac{\alpha_1 G}{h T_\infty} \qquad (1\text{-}73)$$

and

$$N2 = \frac{\sigma T_\infty^3}{h} \qquad (1\text{-}74)$$

The ratio N1/N2 is the ratio of the radiation absorbed to the radiation emitted by a blackbody at T_∞. Individually, N1 may be interpreted as the ratio of absorbed radiation to convection, and N2 as the ratio of emitted blackbody radiation to convection.

Finally, to analyze the problem under consideration, we need to solve the following two coupled first-order nondimensional ordinary differential equations:

$$\frac{d\theta_A}{d\tau} = N1 - 2(\theta_A - 1) - (N2)\varepsilon_1(\theta_A^4 - 1) - (N2)\mathscr{F}_{2-3}(\theta_A^4 - \theta_B^4) \qquad (1\text{-}75)$$

and

$$\frac{d\theta_B}{d\tau} = (N2)\mathscr{F}_{3-2}(\theta_A^4 - \theta_B^4) - (\theta_B - 1) \qquad (1\text{-}76)$$

with initial conditions at $\tau = 0$, given by $\theta_A = \theta_B = 1.0$.

Here

$$\tau = \frac{h A t}{\rho V c} \qquad \theta_A = \frac{T_A}{T_\infty} \qquad \theta_B = \frac{T_B}{T_\infty}$$

For infinite, parallel plates the value of $\mathscr{F}_{2-3}$ is given by Eq. (1-48) with $F_{2-3} = 1.0$. If the plates are finite rectangles of width a and length b, the value of F_{2-3} can be calculated from the appropriate equation in Table 1-2. When finite surfaces are involved, one must consider two radiation transfer terms from each of the surfaces which see each other. One is the net radiation between surfaces as discussed above. The second is the radiation leaving each surface that does not strike the other surface. One can write $F_{2-3} + F_{2-\infty} = 1.0$ and $F_{3-2} + F_{3-\infty} = 1.0$. Once the shape factors F_{2-3} and F_{3-2} are known, the values of $F_{2-\infty}$ and $F_{3-\infty}$ follow directly. Equations of the form of Eq. (1-48) can then be used to determine $\mathscr{F}_{2-\infty}$ and $\mathscr{F}_{3-\infty}$.

1-10   Numerical Technique

The radiation and convection problem in the previous section was formulated in terms of two first-order nonlinear ordinary differential equations (1-75) and (1-76). Since the value of the dependent variable θ in each equation was known at the initial value of the independent variable ($\tau = 0$), the problem is termed an *initial-value* problem. In general, to formulate higher-order differential equations of the form

$$\frac{d^n y}{dx^n} = f\left(x, y, \frac{dy}{dx}, \ldots, \frac{d^{n-1} y}{dx^{n-1}}\right) \qquad (1\text{-}77)$$

in terms of an initial-value problem, one must not only know y at $x = 0$, but also $dy/dx, \ldots, d^{n-1}y/dx^{n-1}$ at $x = 0$.

In this book the numerical technique used to obtain solutions to ordinary differential equations is the fourth-order Runge-Kutta integration scheme. The theoretical basis for this method is discussed in Appendix A.

We now apply this method to the solution of a first-order differential equation. Mathematically we represent this type of equation by writing

$$\frac{d\theta}{d\tau} = \theta' = f(\tau, \theta) \qquad (1\text{-}78)$$

The fourth-order Runge-Kutta technique utilizes a recurrence formula of the form

$$\theta(\tau + \Delta\tau) = \theta(\tau) + (w_1 k_1 + w_2 k_2 + w_3 k_3 + w_4 k_4) \qquad (1\text{-}79)$$

to calculate successive values of θ. The weighting functions w's are determined by matching the coefficients in Eq. (1-79) to a fourth-order Taylor series expansion of θ about τ. If we use the results indicated in Appendix A, Eq. (1-79) becomes

$$\theta(\tau + \Delta\tau) = \theta(\tau) + \tfrac{1}{6}(k_1 + 2k_2 + 2k_3 + k_4) \qquad (1\text{-}80)$$

where

$$k_1 = f(\tau, \theta)\,\Delta\tau \qquad (1\text{-}81)$$

$$k_2 = f\left(\tau + \frac{\Delta\tau}{2}, \theta + \frac{k_1}{2}\right)\Delta\tau \qquad (1\text{-}82)$$

$$k_3 = f\left(\tau + \frac{\Delta\tau}{2}, \theta + \frac{k_2}{2}\right)\Delta\tau \qquad (1\text{-}83)$$

$$k_4 = f(\tau + \Delta\tau, \theta + k_3)\,\Delta\tau \qquad (1\text{-}84)$$

By use of Eqs. (1-78) and (1-81), $k_1 = f(\tau, \theta)\,\Delta\tau = \theta'\,\Delta\tau$. The physical significance of k_1 is best understood by a graphical interpretation. As shown in Fig. 1-12a, k_1 is equal to the cross-hatched rectangular area with height θ'_1 and width $\Delta\tau$. Note that $\Delta\tau$ has been greatly enlarged to clearly show the areas. Also, since

$$\theta = \int_0^\tau \frac{d\theta}{d\tau}\,d\tau$$

a point on the curve in Fig. 1-12b at $\tau = c_1$ (e.g., point A) has a value equal to the area under the curve in Fig. 1-12a between 0 and $\tau = c_1$ plus the initial value of θ.

A geometrical interpretation may also be applied to the other three k values. Point B in Fig. 1-12b is formed by the intersection of the lines $\tau + \Delta\tau/2$ and $\theta(\tau) + k_1/2$. Point B must also correspond to an area in Fig. 1-12a, according to the mathematical equation for k_2. The additional area is equal to the increase in θ between point A and B in Fig. 1-12b. It must also have width $\Delta\tau$. The height of the rectangle which produces the required area is the value $\theta'_2 = f(\tau + \Delta\tau/2, \theta + k_1/2)$. The area in Fig. 1-12a which represents k_2 is the lightly shaded area.

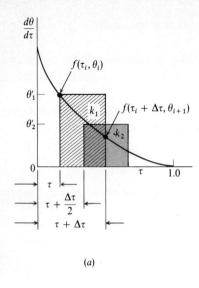

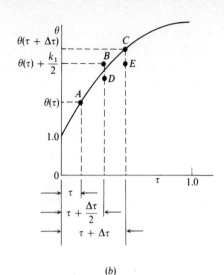

(a)   (b)

FIGURE 1-12
Graphs of $d\theta/d\tau$ and θ versus τ.

The geometrical interpretation of k_3 and k_4 is similar. We first identify a point such as D in Fig. 1-12a formed by the intersection of $\tau + \Delta\tau/2$ and $\theta + k_2/2$. See Eq. (1-83). A third area equal to k_3 can then be identified in Fig. 1-12a with width $\Delta\tau$ and height $\theta'_3 = f(\tau + \Delta\tau/2, \theta + k_2/2)$. Finally, point E formed by $\tau + \Delta\tau$ and $\theta + k_3$ is identified. See Eq. (1-84). A final area equal to k_4 can then be identified in Fig. 1-12a with width $\Delta\tau$ and height $\theta'_4 = f(\tau + \Delta\tau, \theta + k_3)$.

The geometrical interpretation of the Runge-Kutta recurrence formula, Eq. (1-80), is now clear. The actual area under the curve of $d\theta/d\tau$ versus τ between τ and $\tau + \Delta\tau$ is equal to $\theta(\tau + \Delta\tau) - \theta(\tau)$. This area A is approximated by four rectangular areas, weighted according to the equation $A \approx (k_1 + 2k_2 + 2k_3 + k_4)/6$.

1-11   Numerical Solutions

We first obtain a numerical solution to a single first-order differential equation by means of the fourth-order Runge-Kutta integration scheme. Consider the problem depicted in Fig. 1-11, but without the lower plate. The governing equation can be obtained directly from Eq. (1-75) by treating surface 2 as a small gray surface in a large enclosure and letting $\theta_B = T_\infty/T_\infty = 1.0$. This method gives

$$\frac{d\theta_A}{d\tau} = N1 - 2(\theta_A - 1) - N2(\varepsilon_1)(\theta_A{}^4 - 1) - N2(\varepsilon_2)(\theta_A{}^4 - 1) \qquad (1\text{-}85)$$

The equality $\mathscr{F}_{2-3} = \varepsilon_2$ [Eq. (1-50)] now applies to surface 2.

A flow chart (Flow Chart 1-1) and a listing of the BASIC program used to numerically integrate Eq. (1-85) are given below. Table 1-3 gives a listing of nomenclature

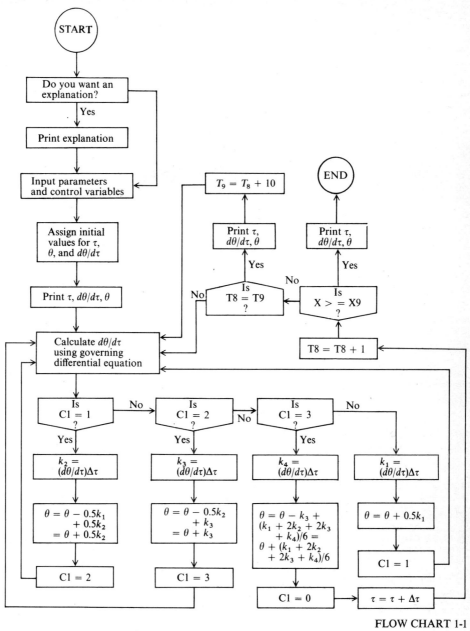

FLOW CHART 1-1
RKINT

Listing 1-1 RKINT

```
10   PRINT"TYPE 1. IF YOU WANT A PROGRAM EXPLANATION."
12   PRINT
13   PRINT"OTHERWISE, TYPE 2."
14   INPUT L
16   IF L=1 THEN 20
18   IF L=2 THEN 39
20   PRINT
22   PRINT"THIS PROGRAM USES A FOURTH ORDER RUNGE-KUTTA SCHEME"
24   PRINT"TO INTEGRATE THE GOVERNING DIFFERENTIAL EQUATION"
26   PRINT"DESCRIBING THE TRANSIENT RESPONSE OF A THIN PLATE"
27   PRINT"TO THERMAL RADIATION, GIVEN BY:"
28   PRINT
30   PRINT"   T*=N1-2(T-1)-N2(E1)(T^4-1)-N2(E2)(T^4-1)"
32   PRINT
34   PRINT" WHERE T IS A NON-DIMENSIONAL TEMPERATURE AND"
36   PRINT" T* IS ITS TIME DERIVATIVE."
38   PRINT
39   REM INPUT NON-DIMENSIONAL PARAMETERS
40   PRINT
50   PRINT"TYPE THE VALUE OF N1."
55   INPUT N1
60   PRINT"TYPE THE VALUE OF N2."
65   INPUT N2
70   PRINT"TYPE THE VALUES OF THE TWO SURFACE EMISSIVITIES E1,E2"
75   INPUT E1,E2
80   PRINT
85   PRINT"THE SOLUTION IS"
100  PRINT
200  REM INITIALIZE CONTROL CONSTANTS
210  LET C1=0
220  LET H=0.01
240  LET T8=0
250  LET T9=10
260  LET X9=1
1000 REM SET PROGRAM VALUES AND INITIAL CONDITIONS
1030 REM INITIALIZE INDEPENDENT VARIABLE
1040 LET X=0
1050 REM SPECIFY INITIAL CONDITION
1060 LET Y0=1
1070 LET Y=Y0
1075 REM CALCULATE INITIAL GRADIENT USING GOVERNING EQN.
1080 LET F=N1-2*(Y-1)-N2*E1*(Y^4-1)-N2*E2*(Y^4-1)
1100 PRINT"TIME", "T*", "T"
1110 PRINT
1120 PRINT X,F,Y
1130 REM
1140 REM FOURTH ORDER RUNGE-KUTTA INTEGRATION SCHEME
1150 REM
1160 LET F=N1-2*(Y-1)-N2*E1*(Y^4-1)-N2*E2*(Y^4-1)
1190 IF C1=1 THEN 1300
1200 IF C1=2 THEN 1380
1210 IF C1=3 THEN 1460
1220 LET K1=H*F
1270 LET Y=Y+0.5*K1
1280 LET C1=1
1290 GOTO 1160
1300 LET K2=H*F
1350 LET Y=Y-0.5*K1+0.5*K2
1360 LET C1=2
1370 GOTO 1160
1380 LET K3=H*F
1430 LET Y=Y-0.5*K2+K3
1440 LET C1=3
1450 GOTO 1160
```

```
1460 LET K4=H*F
1510 LET Y=Y-K3+(K1+2*K2+2*K3+K4)/6
1520 LET C1=0
1530 LET X=X+H
1540 LET T8=T8+1
1550 REM
1560 REM PRINT SEQUENCE
1570 REM
1580 IF X>=X9 THEN 2000
1590 IF T8=T9 THEN 1610
1600 GOTO 1160
1610 PRINT X,F,Y
1620 LET T9=T8+10
1660 GOTO 1160
2000 PRINT
2005 PRINT"THE PARAMETERS USED IN THIS SOLUTION ARE"
2006 PRINT
2010 PRINT"N1="N1,"N2="N2,"E1="E1,"E2="E2
3000 PRINT"DO YOU WANT TO OBTAIN AN ADDITIONAL SOLUTION?"
3010 PRINT
3020 PRINT"IF YES TYPE 1, IF NO TYPE 2."
3030 INPUT L1
3040 IF L1=1 THEN 39
4000 END
```

Computer Results 1-1   RKINT

```
TYPE 1. IF YOU WANT A PROGRAM EXPLANATION.

OTHERWISE, TYPE 2.
? 2

TYPE THE VALUE OF N1.
? 1
TYPE THE VALUE OF N2.
? 1
TYPE THE VALUES OF THE TWO SURFACE EMISSIVITIES E1,E2
? 0.8,0.8

THE SOLUTION IS
```

TIME	T*	T
0	1	1
0.1	0.399817	1.06621
0.2	0.145844	1.09151
0.3	5.13176 E-2	1.10058
0.4	1.78232 E-2	1.10375
0.5	6.16208 E-3	1.10485
0.6	2.12712 E-3	1.10523
0.7	7.34011 E-4	1.10536
0.8	2.53109 E-4	1.1054
0.9	8.71935 E-5	1.10542
1.	3.02479 E-5	1.10543

```
THE PARAMETERS USED IN THIS SOLUTION ARE

N1= 1          N2= 1          E1= 0.8          E2= 0.8
DO YOU WANT TO OBTAIN AN ADDITIONAL SOLUTION?

IF YES TYPE 1, IF NO TYPE 2.
? 1
```

Computer Results 1-1   RKINT (*continued*)

```
TYPE THE VALUE OF N1.
? 2
TYPE THE VALUE OF N2.
? 1
TYPE THE VALUES OF THE TWO SURFACE EMISSIVITIES E1,E2
? 0.8,0.8

THE SOLUTION IS

TIME            T*              T

0               2               1
0.1             0.736703        1.1295
0.2             0.224641        1.17302
0.3             6.41311 E-2     1.18586
0.4             1.79521 E-2     1.18949
0.5             4.9972 E-3      1.1905
0.6             1.38911 E-3     1.19079
0.7             3.86398 E-4     1.19087
0.8             1.06969 E-4     1.19089
0.9             2.94746 E-5     1.19089
1.              8.09671 E-6     1.19089

THE PARAMETERS USED IN THIS SOLUTION ARE

N1= 2           N2= 1           E1= 0.8         E2= 0.8
DO YOU WANT TO OBTAIN AN ADDITIONAL SOLUTION?

IF YES TYPE 1, IF NO TYPE 2.
? 2
```

used in the program. Since the initial condition θ_A is known at $\tau = 0$, the initial value of $d\theta_A/d\tau$ can be calculated from the governing equation.

Since the program is written for use on a time-sharing computer terminal, the user can obtain a program explanation if desired by properly responding to the initial question. The user must then supply the values of N1, N2, ε_1, and ε_2 from the keyboard as required in Statements 50 to 75. The value of $\Delta\tau$ is specified as H = 0.01 in Statement 220 and can be changed if desired. The output τ, $d\theta/d\tau$, and θ is specified in Statement 1120. Since C1 is initially zero (Statement 210), the first pass through the Runge-Kutta scheme beginning at Statement 1160 executes State-

Table 1-3   VARIABLES USED IN RKINT PROGRAM

Variable	Definition
C1	Logic control variable
E1	Surface emittance ε_1
E2	Surface emittance ε_2
F	Derivative $d\theta/d\tau$
H	Increment $\Delta\tau$
K1, K2, K3, K4	Coefficients required for the Runge-Kutta algorithm
N1	Parameter $\alpha_1 G/\hbar T_\infty$
N2	Parameter $\sigma T_\infty{}^3/\hbar$
T8, T9	Print control variables
X	Nondimensional independent variable τ
X9	Maximum value of τ
Y	Nondimensional dependent variable θ_A
Y0	Initial value of dependent variable

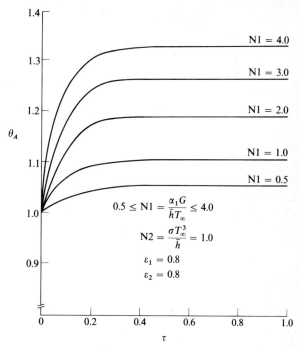

θ_A

$$0.5 \le \text{N1} = \frac{\alpha_1 G}{\hbar T_\infty} \le 4.0$$

$$\text{N2} = \frac{\sigma T_\infty^3}{\hbar} = 1.0$$

$$\varepsilon_1 = 0.8$$

$$\varepsilon_2 = 0.8$$

FIGURE 1-13
Transient response of an irradiated plate.

ments 1220, 1270, 1280, and 1290. This computes k_1 and assigns C1 = 1. The second pass with C1 = 1 computes k_2 (Statement 1300) and increments C1. The third pass with C1 = 2 computes k_3 (Statement 1380), and the final pass with C1 = 3 computes k_4 (Statement 1460). The Runge-Kutta algorithm [Eq. (1-80)] is executed in Statement 1510. Then the independent variable is increased by $\Delta\tau$ (Statement 1530), and the process repeated until the value of the independent variable is greater or equal to the value of X9 as specified in Statement 260. After one solution is obtained, an additional solution for different values of the nondimensional, governing parameters may be obtained by responding in the affirmative to the question asked in Statement 3000.

The variation of the nondimensional temperature gradient and nondimensional temperature as a function of nondimensional time for a typical computer solution obtained by using the program RKINT is given following the program listing. The effect of varying the parameter N1 is shown in the results given in Fig. 1-13, where θ_A is given as a function of τ. Both the transient and steady-state solutions can be seen in Fig. 1-13. The transient behavior dies out at $\tau \approx 0.4$. The corresponding dimensional time is given by $t = 0.4\rho Vc/\hbar A$. The maximum temperature T_A reached by the plate for N1 = 4.0 is given by $T_A = \theta_A T_\infty = 1.326 T_\infty$, where $T_\infty = \alpha_1 G/\hbar\text{N1} = \frac{1}{4}\alpha_1 G/\hbar$. Note that T_∞ and $\hbar$ appear in both N1 and N2. However, N1 can be varied

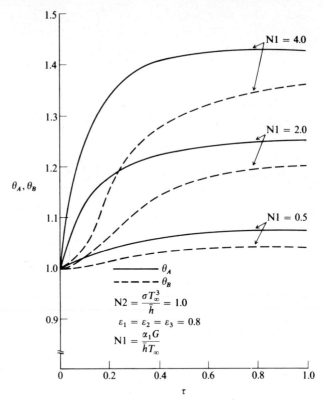

FIGURE 1.14
Transient response of two plates.

independently of N2 by varying α_1 or G. Often T_∞ and $\bar{h}$ can be treated as known constants, and thus N2 is fixed for each specified condition of convective heat transfer.

The steady-state analytical solution for the nondimensional temperature may be obtained from the governing equation with $d\theta_A/d\tau = 0$. This gives $0 = \mathrm{N1} - 2(\theta_A - 1) - 2(\mathrm{N2})\varepsilon(\theta_A^4 - 1)$. For example, if $\mathrm{N1} = 3.0$, $\mathrm{N2} = 1.0$, and $\varepsilon = 0.8$, it can be shown that $\theta_A = 1.263$ is a root of this equation. The heat transfer by convection from each surface for steady-state conditions is then

$$\bar{h}A(T_A - T_\infty) = \bar{h}AT_\infty(\theta_A - 1)$$

The net radiation crossing the top surface is

$$A[\alpha_1 G - \sigma\varepsilon_1(T_A^4 - T_\infty^4)] = \bar{h}AT_\infty[\mathrm{N1} - \varepsilon_1(\mathrm{N2})(\theta_A^4 - 1)]$$

and the radiation leaving the lower surface is

$$A\sigma\varepsilon_2(T_A^4 - T_\infty^4) = \bar{h}AT_\infty[\varepsilon_2(\mathrm{N2})(\theta_A^4 - 1)]$$

Now consider the numerical solution for the thermal response of two infinite, parallel plates as shown in Fig. 1-11. The equations governing θ_A and θ_B as a function of τ are Eqs. (1-75) and (1-76). The gray body shape factor between the two infinite surfaces is given by Eq. (1-49) as

$$\mathscr{F}_{2-3} = \frac{1}{1/\varepsilon_2 + 1/\varepsilon_3 - 1} = \mathscr{F}_{3-2}$$

The simultaneous solution to the two coupled first-order nonlinear governing differential equations can be obtained numerically by using the program PLATES given below. Here we let $Y = \theta_A$ and $Z = \theta_B$, with derivatives $F = d\theta_A/d\tau$ and $G = d\theta_B/d\tau$. The shape factor is F2. The algorithm given in Appendix A-3 for the Runge-Kutta technique applied to simultaneous equations is employed. The program logic is similar to that used in the previous program. By comparing the two programs it can be seen how the Runge-Kutta technique is expanded to simultaneously obtain solutions to more than one equation. The set of coefficients l_1, l_2, l_3, l_4 is obtained along with the values of k_1, k_2, k_3, k_4. The fourth-order Runge-Kutta algorithm is used twice: once to increment Y using the k values (Statement 1510) and once to increment Z using the l values (Statement 1515). A printout for a typical solution obtained from the program PLATES is given following the program (see Computer Results 1-2). These results show the variation of θ_A and θ_B as a function of τ. The derivatives $d\theta_A/d\tau$ and $d\theta_B/d\tau$ can also be obtained by printing out F and G in Statements 1120 and 1610.

The results given in Fig. 1-14 show the transient behavior of the two surfaces as a

Listing 1-2   PLATES

```
10    PRINT"TYPE 1 IF YOU WANT A PROGRAM EXPLANATION."
12    PRINT
13    PRINT"OTHERWISE, TYPE 2."
14    INPUTL
16    IF L=1 THEN 20
18    IF L=2 THEN 39
20    PRINT
22    PRINT"THIS PROGRAM USES A FOURTH ORDER RUNGE-KUTTA SCHEME"
24    PRINT"TO INTEGRATE THE GOVERNING DIFFERENTIAL EQUATIONS"
26    PRINT"DESCRIBING THE TRANSIENT RESPONSE OF TWO THIN PLATES,"
27    PRINT"THE UPPER PLATE BEING SUBJECTED TO RADIATION, GIVEN BY:"
28    PRINT
30    PRINT"   T1*=N1-2(T1-1)-N2(E1)(T1^4-1)-N2(F2)(T1^4-T2^4)"
31    PRINT
32    PRINT"    T2*=N2(F2)(T1^4-T2^4)-(T2-1)"
33    PRINT
34    PRINT"WHERE T1 AND T2 ARE NON-DIMENSIONAL TEMPERATURES AND"
36    PRINT"T1* AND T2* ARE TIME DERIVATIVES."
38    PRINT
39    REM INPUT NON-DIMENSIONAL PARAMETERS
40    PRINT
50    PRINT"TYPE VALUES OF N1,N2."
55    INPUTN1,N2
60    PRINT"TYPE THE VALUES OF SURFACE EMISSIVITIES E1, E2, E3."
65    INPUTE1,E2,E3
70    PRINT
80    PRINT"THE SOLUTION IS"
```

Listing 1-2   PLATES (*continued*)

```
100    PRINT
110    LETF2=1/(1/E2+1/E3-1)
200    REM INITIALIZE CONTROL CONSTANTS
210    LETC1=0
220    LETH=0.01
240    LETT8=0
250    LETT9=10
260    LETX9=1
1000   REM SET PROGRAM VALUES AND INITIAL CONDITIONS
1030   REM INITIALIZE INDEPENDENT VARIABLE
1040   LETX=0
1050   REM SPECIFY INITIAL CONDITION
1060   LETY0=1
1065   LETZ0=1
1070   LETY=Y0
1075   LETZ=Z0
1080   LETF=N1-2*(Y-1)-N2*E1*(Y^4-1)-N2*F2*(Y^4-Z^4)
1085   LETG=N2*F2*(Y^4-Z^4)-(Z-1)
1100   PRINT"TIME","T1","T2"
1110   PRINT
1120   PRINTX,Y,Z
1130   REM
1140   REM FOURTH ORDER RUNGE-KUTTA INTEGRATION SCHEME
1150   REM
1160   LETF=N1-2*(Y-1)-N2*E1*(Y^4-1)-N2*F2*(Y^4-Z^4)
1170   LETG=N2*F2*(Y^4-Z^4)-(Z-1)
1190   IF C1=1 THEN 1300
1200   IF C1=2 THEN 1380
1210   IF C1=3 THEN 1460
1220   LETK1=H*F
1230   LETL1=H*G
1270   LETY=Y+0.5*K1
1275   LETZ=Z+0.5*L1
1280   LETC1=1
1290   GOTO 1160
1300   LETK2=H*F
1310   LETL2=H*G
1350   LETY=Y-0.5*K1+0.5*K2
1355   LETZ=Z-0.5*L1+0.5*L2
1360   LETC1=2
1370   GOTO 1160
1380   LETK3=H*F
1390   LETL3=H*G
1430   LETY=Y-0.5*K2+K3
1435   LETZ=Z-0.5*L2+L3
1440   LETC1=3
1450   GOTO 1160
1460   LETK4=H*F
1470   LETL4=H*G
1510   LETY=Y-K3+(K1+2*K2+2*K3+K4)/6
1515   LETZ=Z-L3+(L1+2*L2+2*L3+L4)/6
1520   LETC1=0
1530   LETX=X+H
1540   LETT8=T8+1
1550   REM
1560   REM PRINT SEQUENCE
1570   REM
1580   IF X>=X9 THEN 2000
1590   IF T8=T9 THEN 1610
1600   GOTO 1160
1610   PRINTX,Y,Z
1620   LETT9=T8+10
1660   GOTO 1160
2000   PRINT
```

Listing 1-2   PLATES (*continued*)

```
2005 PRINT"THE PARAMETERS USED IN THIS SOLUTION ARE"
2006 PRINT
2010 PRINT"NI="N1,"N2="N2
2020 PRINT
2030 PRINT"EI="EI,"E2="E2,"E3="E3
2040 PRINT
2050 PRINT"SHAPE FACTOR="F2
3000 PRINT
3010 PRINT"DO YOU WANT TO OBTAIN AN ADDITIONAL SOLUTION?"
3020 PRINT
3030 PRINT"IF YES TYPE 1, IF NO TYPE 2."
3040 INPUTLI
3050 IF LI=I THEN 39
4000 END
```

Computer Results 1-2   PLATES

```
TYPE I IF YOU WANT A PROGRAM EXPLANATION.

OTHERWISE, TYPE 2.
? 2

TYPE VALUES OF NI,N2.
? 4,1
TYPE THE VALUES OF SURFACE EMISSIVITIES EI, E2, E3.
? 0.8,0.8,0.8

THE SOLUTION IS

TIME            TI              T2

0               I               I
0.1             1.25768         1.04685
0.2             1.34187         1.13781
0.3             1.37621         1.21888
0.4             1.39786         1.27802
0.5             1.41352         1.31844
0.6             1.42467         1.34536
0.7             1.43236         1.36306
0.8             1.43753         1.37459
0.9             1.44095         1.38207
1.              1.44318         1.3869

THE PARAMETERS USED IN THIS SOLUTION ARE

NI= 4           N2= 1

EI= 0.8         E2= 0.8         E3= 0.8

SHAPE FACTOR= 0.666667

DO YOU WANT TO OBTAIN AN ADDITIONAL SOLUTION?

IF YES TYPE 1, IF NO TYPE 2.
? 2
```

function of the parameter N1. The effect of the second surface on the variation of θ_A with τ can be seen by comparing Figs. 1-13 and 1-14. The presence of the second surface has caused the transient response of the system to be more sluggish; i.e., it takes longer for the transients to die out. Also, the steady-state temperature of surface *A* is significantly increased for every value of N1 because of the reflected radiation from the second plate. It is also interesting to observe that the maximum thermal lag of surface B occurs at $\tau = 0.2$ for all values of N1 shown.

By treating the radiating surfaces as lumped thermal elements we have neglected the effects of a temperature drop through the solid materials. To determine when this assumption is valid and to learn how to include the effects of temperature gradients in solids, we must rely on the theory of heat conduction that is presented in the next three chapters.

References

1-1   REYNOLDS, W. C.: "Thermodynamics," McGraw-Hill, New York, 1968.

1-2   PLANCK, M.: "The Theory of Heat Radiation," Dover, New York, 1959.

1-3   WIEBELT, J. A.: "Engineering Radiation Heat Transfer," Holt, New York, 1966.

1-4   SPARROW, E. M., and R. D. CESS: "Radiation Heat Transfer," Brooks/Cole, Belmont, Calif., 1966.

1-5   LOVE, T. J.: "Radiative Heat Transfer," Merrill, Columbus, Ohio, 1968.

1-6   HAMILTON, D. C., and W. R. MORGAN: Radiant Interchange Configuration Factors, NACA TN 2836, 1952.

1-7   HOTTEL, H. C., and A. F. SAROFIM: "Radiative Transfer," McGraw-Hill, New York, 1967.

1-8   KRIETH, F.: "Radiation Heat Transfer," International Textbook, Scranton, Pa., 1962.

1-9   DUNKLE, R. F.: Thermal Radiation Tables and Applications, *Trans. ASME*, vol. 76, p. 549, May 1954.

1-10   CARNAHAN, B., H. A. LUTHER, and J. O. WILKES: "Applied Numerical Methods," Wiley, New York, 1969.

1-11   KEMENY, J. G., and T. E. KURTY: "Basic Programming," Wiley, 1968.

1-12   WATTS, R. G.: Radiant Heat Transfer to Earth Satellites, *Heat Transfer*, vol. 87, ser. C, p. 369, August 1965.

Problems[1]

1-1   The monochromatic emissive power for blackbody radiation is given by Planck's law, Eq. (1-2).

(*a*) Evaluate the derivative $(\partial E_{b,\lambda}/\partial\lambda)_T$, and show that setting $(\partial E_{b,\lambda}/\partial\lambda)_T = 0$ leads to the transcendental equation $A/5 + 1/\exp A = 1$, where $A = C_2/\lambda T$.

(*b*) Show that the transcendental equation in part (*a*) has a root $A = 4.9651$.

(*c*) From the results in part (*b*) calculate the value of the product $\lambda_{\max} T$.

---

[1] Problems whose numbers are followed by a superscript italic *c* should be analyzed by obtaining solutions on a digital computer.

1-2 A pane of window glass transmits 60 percent of the incident thermal radiation which falls within the wavelength band $0.1 \ \mu m \leq \lambda \leq 3.0 \ \mu m$ and 2 percent of the thermal radiation with wavelength above $3.0 \ \mu m$.

(*a*) Solar radiation falls on the outside of the glass at a rate of 400 Btu/h-ft². How much thermal radiation is transmitted through the glass per unit area? Treat the sun as a blackbody at $T = 10,000°F$.

(*b*) Thermal radiation from a surface at 40°F falls on the inside of the glass at a rate of 400 Btu/h-ft². How much of this radiation is transmitted through the glass?

(*c*) What is the monochromatic emissive power at $\lambda = 100 \ \mu m$ for a blackbody at $T = 40°F$?

(*d*) What is the maximum monochromatic emissive power for a blackbody at $T = 40°F$?

1-3 Two partially transparent materials are available for use in construction. One material transmits 80 percent of radiation between $0.3 \ \mu m \leq \lambda \leq 0.6 \ \mu m$. The other transmits 20 percent of radiation between $0.5 \ \mu m \leq \lambda \leq 0.8 \ \mu m$. Which material will transmit the most solar radiation? Assume that the sun is a blackbody at $T = 10,000°F$.

1-4 The temperature of an isothermal enclosure is 1,063°C. Calculate the following quantities:

(*a*) the emissive power of a small opening in the enclosure

(*b*) the intensity of the opening

(*c*) the monochromatic intensity at 0.655 μm

(*d*) the wavelength at which the maximum intensity occurs

1-5 Quartz transmits 90 percent of incident thermal radiation between $0.2 \ \mu m \leq \lambda \leq 4.0 \ \mu m$. Consider blackbody radiation incident on the quartz. Calculate the percent of the radiation transmitted through the quartz if the radiation is emitted from a source at the following temperatures:

(*a*) 500°R

(*b*) 1,000°R

(*c*) 2,000°R

(*d*) 6,000°R

1-6 A polished aluminum plate is exposed to irradiation from the sun ($G = 1,000 \ \text{W/m}^2$) and loses energy to the atmosphere by both convection and radiation. If the convection heat transfer coefficient is 6.0 W/m²-°C and the environmental temperature is 15°C, calculate the equilibrium temperature of the plate ($\alpha_1 = 0.3$, $\varepsilon_1 = 0.04$).

1-7 By definition, the apparent monochromatic temperature T_λ is the temperature of a blackbody which has the same radiation intensity at a wavelength of $\lambda = 0.665 \ \mu m$ (a red wavelength) as the surface under consideration. From Planck's law the relation between T_λ and the true surface temperature is given by

$$\frac{\lambda^{-5}}{\exp(C_2/\lambda T_\lambda) - 1} = \varepsilon_\lambda \frac{\lambda^{-5}}{\exp(C_2/\lambda T) - 1}$$

(a) Show that for commonly encountered temperatures the above equation can be approximated by the following:

$$\frac{1}{T} = \frac{1}{T_\lambda} + \frac{\lambda}{C_2} \ln \varepsilon_\lambda$$

Note: $C_2 = hc/k$, where h is Planck's constant, c is the speed of light in a vacuum, and k is Boltzmann's constant.

(b) The apparent monochromatic temperature T_λ can be measured with an optical pyrometer and the true surface temperature estimated by use of the results in part (a). Estimate the accuracy of this procedure if the actual surface temperature is 100, 1,000, 4,000, and 10,000 K.

1-8 A large area A_2 is situated directly above a small area A_1. Physically, we expect the shape factor $F_{dA_1 - A_2}$ to approach unity as A_2 becomes infinite. Treating A_1 as a differential area dA_1, one can write [see Eq. (1-34b)]

$$F_{1-2} = \int_{A_2} \frac{\cos \phi_1 \cos \phi_2 \, dA_2}{\pi r^2}$$

where $dA_2 = 2\pi l \, dx$ and $\phi_1 = \phi_2$, as shown in the figure.

(a) Evaluate the above integral between the limits $r = a$ and $r = \infty$ to determine F_{1-2}.

(b) Since $r = (l^2 + a^2)^{1/2}$, repeat the above integration between $r = a$ and $r = r$ to obtain an expression for F_{1-2}. Explain the physical significance of your answer.

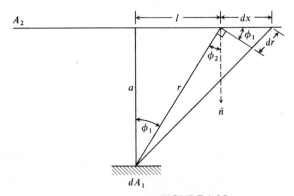

FIGURE 1-8P
Shape factor configuration.

1-9 Consider the configuration of an annulus formed by two cylinders as shown in Table 1.2j. What expression can be used to calculate the fraction of thermal radiation emitted by the inner surface of the outer cylinder A_1 which escapes out one end of the annulus without striking either cylindrical surface? The values of F_{1-2} and F_{1-1} are given in Table 1-2. What expression in terms of F_{1-1} and F_{1-2} gives the fraction of radiation emitted by the outer surface of the inner cylinder A_2 which directly escapes out one end?

1-10 A cup of hot coffee is losing energy to the environment by convection and radiation. The outside temperature of the cup is 80°C.

(*a*) Calculate the ratio of heat transfer by radiation to heat transfer by convection for an environmental temperature of 30°C. Treat the cup as a small gray surface in a large enclosure. The cup surface emittance is 0.72, and the average convection heat transfer coefficient is $\bar{h} = 6.0$ W/m²-°C.

(*b*) Repeat part (*a*) for a horizontal iron pipe ($\varepsilon = 0.85$) with a surface temperature $T = 210°F$ and $T_\infty = 70°F$. The average convection heat transfer coefficient $\bar{h} = 1.10$ Btu/h-ft²-°F.

1-11 A thermocouple used to measure gas temperature can be represented by a small gray sphere enclosed by a black environment which is a large duct at a wall temperature $T_w = 330°C$. The emittance of the thermocouple surface is 0.9 and the thermocouple indicates a temperature of $T_s = 500°C$.

(*a*) If $\bar{h} = 200$ W/m²-°C, calculate the temperature T_∞ of the hot gas flowing through the duct by equating the convective heat transfer to the radiation heat transfer, treating the thermocouple as a closed thermodynamic system in thermal equilibrium. Explain why there is a difference between T_∞ and T_s.

(*b*) If T_∞ is known to be 200°C, what temperature T_s will the thermocouple indicate when the duct walls are at a temperature of 330°C?

1-12 (*a*) Show that the values of θ_A and θ_B at $\tau = 1.0$ in Fig. 1-14 are approximately equal to the steady-state values by direct substitution into the governing equations.

(*h*) For steady-state conditions, calculate the net radiation heat transfer per unit area between the two plates if $T_\infty = 500°R$.

1-13 (*a*) The reflectance of an opaque surface is a function of wavelength as shown in the solid curve in Fig. 1-13P. Approximate this variation by using the dashed lines, and plot α_λ versus λ.

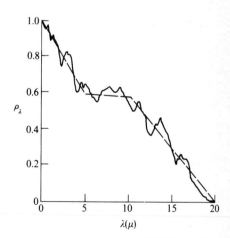

FIGURE 1-13*Pa*
Reflectance.

(b) Irradiation G_λ falls on the opaque surface. The spectral distribution of the irradiation is shown in Fig. 1-13P. Calculate the total irradiation between $0\mu m \leq \lambda \leq 20\ \mu m$.

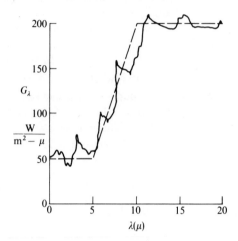

FIGURE 1-13Pb
Irradiation.

(c) Determine the average absorptance α of the surface, where

$$\alpha = \frac{\text{energy absorbed}}{\text{energy incident}}$$

1-14 A spherical space satellite receives incident solar radiation $G = 443$ Btu/h-ft² (based upon the projected area perpendicular to the sun's rays). Neglect other sources of radiation, and assume that the satellite surface is isothermal. Three surface coatings are available for use. Calculate the steady-state surface temperature for each coating based upon the absorptance values indicated in Fig. 1-14P. Treat the sun as a blackbody at $T = 10,000°$F.

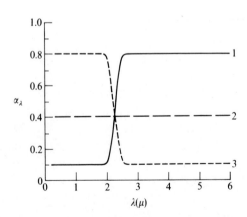

FIGURE 1-14P
Absorptance vs. wavelength.

1-15ᶜ It is desired to study the effect of increasing the cooling rate by convective heat transfer on the results presented in Fig. 1-14. Assume that the average convection heat transfer coefficient $\bar{h}$ is increased by a factor of 10. Physically, this could be accomplished by increasing the velocity of the air flowing over the surfaces. All other conditions and properties remain the same. Obtain numerical solutions corresponding to those given in Fig. 1-14, and discuss the results. How would you proceed if $\alpha \neq \varepsilon$?

1-16ᶜ Three thin surfaces make up a radiation shielding arrangement as shown. At time $t = 0$ radiation falls on the top surface. Obtain numerical solutions for the non-dimensional transient response of surface C. Treat the surfaces as infinite gray surfaces. Use a lumped analysis by assuming that each plate is in thermal equilibrium at any instant of time.

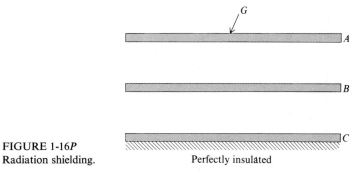

FIGURE 1-16*P*
Radiation shielding.   Perfectly insulated

1-17ᶜ Obtain solutions similar to those presented in Fig. 1-14 except treat the surfaces as finite, rather than infinite. Account for radiation to and from the environment at T_∞ for all three exposed surfaces. The expression for F_{1-2} is given in Table 1-2. Note that $\eta = b/c$ and $\xi = a/c$ are two additional nondimensional parameters that now enter into the problem. Study two special configurations specified by:

$$\eta = 1.0 \qquad \xi = 1.0$$
$$\eta = 10 \qquad \xi = 20$$

Discuss the physical significance of the results. How would you proceed if $\alpha \neq \varepsilon$?

1-18ᶜ Consider a system of two plates similar to that shown in Fig. 1-11 except that the second plate is oriented as shown in Fig. 1-18*P*. The governing equations for the temperature response of the two plates are given by Eqs. (1-75) and (1-76). The expression for F_{1-2} is given in Table 1-2. Let $\eta = a/b = 1.0$ and $\xi = c/b = 1.0$. Obtain the transient solution for θ_A and θ_B for different values of $N1 = \alpha_1 G / \bar{h} T_\infty$.

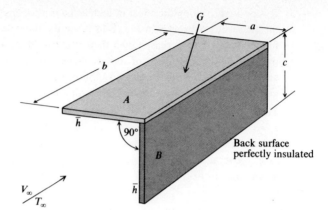

FIGURE 1-18P
Orthogonal plates.

1-19^c Obtain steady-state and transient solutions for a radiation shielding system consisting of three surfaces as shown in Fig. 1-19P. The third surface is maintained at constant temperature by an evaporating cryogenic fluid on its inside surface. Treat the thin plates as infinite, parallel gray surfaces. Initially the plates are in thermal equilibrium at $T = T_0$. At time $t = 0$ the outer surface of plate 1 is exposed to an environment $T_\infty = 3T_0$ and $\bar{h}_1 = 10\bar{h}_2$. Study the effect of changing the surface emittance values.

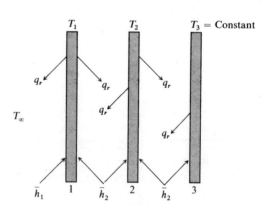

FIGURE 1-19P
Shielding for cryogenic fluid.

1-20^c A long pipe, surface 1, is to be enclosed by an outer concentric cylinder, surface 2. The annulus is evacuated so convection from surfaces 1 and 2 may be neglected. The surfaces are gray, $\varepsilon_1 = \varepsilon_2 = 0.2$. The system is initially at uniform temperature equal to T_∞. Under certain conditions surface 1 suddenly increases in temperature to $T_1 = 2T_\infty$. To detect this change, a thin coating of wax is to be put on the outside surface 3 which will melt in a certain minimum time. The wax gives the outer surface an effective emittance of 0.5. To choose a wax with a proper melting point, it is necessary to obtain solutions for the temperature of surface 3 as a function of time and the ratio d_2/d_1.

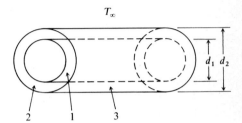

FIGURE 1-20P
Evacuated annulus.

1-21ᶜ A single pipe at a temperature T_s carries hot liquid metal. Initially $T_s = T_\infty$ at time $t = 0$. When the liquid metal begins to flow through the pipe, energy is convected to the inside pipe-wall surface. Calculate the transient response of T_s. Assume that the pipe is a small gray surface in a large enclosure. Study the effect of placing a second identical pipe adjacent to the first as shown in the figure. Obtain results as a function of various internal convective heat flux rates and the ratios L/D. For internal convection, $q = \bar{h}_i A_i (T_L - T_s)$, where $T_L = 800°C$ and $1{,}000 \text{ W/m}^2\text{-}°C \leq \bar{h}_i \leq 10{,}000 \text{ W/m}^2\text{-}°C$. For external convection, $q = \bar{h}_o A_o (T_s - T_\infty)$, where $T_\infty = 40°C$ and $\bar{h}_o = 30 \text{ W/m}^2\text{-}°C$.

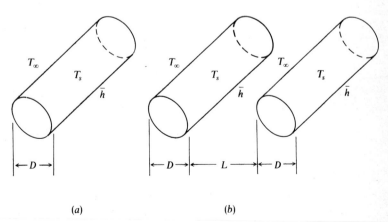

(a)   (b)

FIGURE 1-21P
Piping systems.

1-22^c The energy equation applied to the thin skin of a spherical earth satellite of radius r leads to the following equation (see Ref. 1-12)

$$q_s + q_t + q_r = \varepsilon\sigma4\pi r^2 T^4 + \rho VC \frac{dT}{dt}$$

where q_s = absorbed solar radiation directly incident on surface

q_t = absorbed terrestrial radiation

q_r = absorbed solar radiation reflected from the earth's atmosphere (*albedo* radiation)

The incoming radiation sees the projected satellite area $A_p = \pi r^2$. The irradiation falling on the satellite directly from the sun is G_s. The effective blackbody emissive powers of terrestrial and reflected radiation are denoted by E_t and E_r, respectively. The corresponding absorptances for the three sources of incoming radiation are α_s, α_t, and α_r, where $\alpha_r = \alpha_s$ and $\alpha_t = \varepsilon$.

(a) Define $\tau = E_t A_p t/\rho VCT_s$ and $\theta = T/T_s$, where T_s is the steady-state temperature of the satellite. Express the governing equation in nondimensional form, and identify the important nondimensional parameters.

(b) Derive an expression for the steady-state temperature T_s, and calculate T_s when $\alpha_s = 0.1$, $\varepsilon = 0.01$, $G_s/E_t = 6.0$, $E_r/E_t = 2.0$, and $G_s = 420$ Btu/h-ft^2. The shape factor for reflected radiation $F_{r-s} = 0.05$, and the shape factor for terrestrial radiation $F_{t-s} = 0.8$. (See Ref. 1-12.) Repeat the calculation exchanging the values of α_s and ε.

(c) When the satellite enters the shadow of the earth, G_s and E_r suddenly become zero. If the satellite is initially at steady-state conditions, calculate the transient behavior of the surface temperature due to the earth's shadow. Study the problem for various values of ε.

(d) Assume that the satellite reaches a steady-state condition in the earth's shadow. Calculate the transient response as the satellite reenters the sunlit portion of the earth.

ONE-DIMENSIONAL CONDUCTION

2-1   Introduction to One-dimensional Analysis

The objective of this chapter is to form a mathematical model for one-dimensional heat conduction. This is done by applying the conservation of energy principle to a properly chosen differential element. In a one-dimensional, steady-state model we assume that the temperature is only a function of one independent variable, say x. When the temperature varies only with x, an appropriate differential element has a differential thickness dx and parallel faces normal to the x direction. The other dimensions of the element may have any size and shape.

Consider a differential element in the form of a thin parallelepiped of differential thickness dx and finite y and z dimensions. What are the physical conditions which allow only one-dimensional conduction in the x direction through this differential element? The most obvious situation is when the thin edges of the element are adiabatic. This condition prohibits any heat conduction normal to the x direction. In most practical situations, however, some heat transfer will occur from exposed edges of the element by either conduction, convection, radiation, or a combination of these modes. Even when this occurs, there are certain conditions where a one-dimensional model will give accurate results for the temperature distribution within the solid material. We shall learn what these conditions are after analyzing the results of the model. Our approach is to assume a one-dimensional temperature distribution and derive the mathematical model based upon this assumption. After we use this model and compare the predicted results with the corresponding results of a two-dimensional model, the limitations of the model can be specified.

Steady-state, one-dimensional, constant-property heat conduction can usually be treated analytically. Several examples of one-dimensional analytical solutions are discussed in later chapters as special cases of multidimensional problems. One class of problems in which an analytical analysis becomes more involved, and sometimes impractical due to nonlinear terms or difficult boundary conditions, is the analysis of fins. This chapter is concerned with the analytical and numerical analysis of problems broadly classified as fin problems.

A *fin* is a surface connected to a wall at its base. It transfers thermal energy to its surroundings by convection and radiation. The geometry of fins can take many forms. Longitudinal fins and radial fins are commonly used in heat transfer applications. Once the temperature distribution within the fin is known, the steady-state heat transfer through the fin can be calculated by using Fourier's law, which gives the relationship between the heat transfer per unit area and the temperature gradient. In vector form it is written

$$\mathbf{q}'' = -k \, \nabla T \qquad (2\text{-}1)$$

where $\mathbf{q}''$ is called the *heat flux vector*. The proportionality constant k between the heat flux and the temperature gradient is called the *thermal conductivity*. The negative sign gives a positive value for the heat transfer caused by a negative temperature gradient.

Table 2-1 ORDER OF MAGNITUDE FOR THERMAL CONDUCTIVITIES
(1 Btu/h-ft-°F = 1.73073 W/m-°C)

Material	Thermal conductivities*	
	k(Btu/h-ft-°F)	k(W/m-°C)
Gases	0.001–0.1	0.002–0.2
Oils	0.05–0.5	0.1–1.0
Water	0.3–0.4	0.5–0.7
Liquid metals	5–50	10–100
Solids (nonmetals)	0.02–2.0	0.03–3.0
Solids (alloys)	10–100	20–200
Solids (pure metals)	20–200	40–400

* See Appendix D for values of k for specific materials.

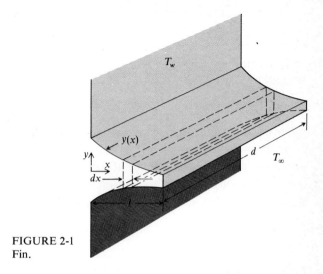

FIGURE 2-1
Fin.

Equation (2-1) applies to a stationary, heterogeneous, isotropic material. In a heterogeneous material the thermal conductivity can vary from point to point within the material. In an isotropic material the thermal conductivity at a point is independent of direction. Fourier's law is a phenomenological law determined by experiment. It is the relationship that allows the complicated microscopic, thermal interaction of electrons to be analyzed from a macroscopic point of view which disregards the molecular structure. The numerical value of the thermal conductivity is usually determined experimentally. The highest values occur in metals such as copper, and the lowest values occur in gases. Typical values are indicated in Table 2-1.

Due to the symmetry of conductivity in an isotropic material, the heat transfer per unit area at a point will be normal to the isothermal surface which passes through that point. For one-dimensional conduction expressed in terms of a Cartesian coordinate system, the total heat transfer in the x direction as given by Fourier's law is

$$q_x = -kA \frac{\partial T}{\partial x} \qquad (2\text{-}2)$$

The area A is the cross-sectional area of the isothermal surface normal to the x axis at the point in question, and $\partial T/\partial x$ is the temperature gradient at that point.

To illustrate the analysis of steady-state heat transfer in a fin, consider a longitudinal fin of arbitrary profile as shown in Fig. 2-1. The fin is symmetrical about the x axis. We assume unit fin depth ($d = 1$), constant thermal conductivity, and steady, one-dimensional heat conduction in the x direction. The differential element shown in dashed lines is chosen as the thermodynamic system. Since the temperature is assumed to vary in the x direction, a differential thickness dx is necessary to form a closed thermodynamic system in thermal equilibrium.

Writing the conservation of energy principle for the defined system yields (Ref. 2-2)

$$\delta Q = dE + \delta W \qquad (2\text{-}3)$$

where δQ is the heat transfer across the system boundaries, δW is the work done on or by the system due to an interaction with the surroundings, and dE is the change in stored energy within the system. If we assume that the energy associated with magnetism, electricity, and surface tension is negligible, then the stored energy is the sum of internal, kinetic, and potential energy:

$$E = U + \frac{mV^2}{2} + gmZ$$

For a stationary closed system there is no change in the system's kinetic and potential energy. The change in total stored energy is then equal to the change in internal energy:

$$dE = dU$$

The internal energy can be related to molecular energies and the internal structure of the molecules. In heat transfer studies we express a change in internal energy as a function of the change of the system temperature. A steady-state system is one in which the temperature does not change with time, and thus $dU = 0$. For the stationary, steady-state differential system shown in Fig. 2-1, no work crosses the boundary and Eq. (2-3) reduces to

$$\delta Q = 0 \qquad (2\text{-}4)$$

Thus, the net heat transfer into and out of the system equals zero. This result does not imply that there is no heat transfer into or out of the system. In general, energy may be conducted into the system at x and may also be generated internally within the system, e.g., by nuclear fission, electrical dissipation, chemical reaction, or other sources of energy. Energy also can cross the system boundary by conduction at $x + dx$ and by convection and radiation through the top and bottom surfaces of the system. The conduction and convection terms are shown in Fig. 2-2. Since fins are usually very thin, energy transfer from the edges of the fin may often be neglected.

Equation (2-4) may be satisfied by equating the energy which enters the system with the energy leaving the system. Using the nomenclature shown in Fig. 2-2 to express Eq. (2-4) and neglecting radiation to or from the fin surface, we obtain

$$q_x + q'''A \, dx = q_{x+dx} + 2q_c \qquad (2\text{-}5)$$

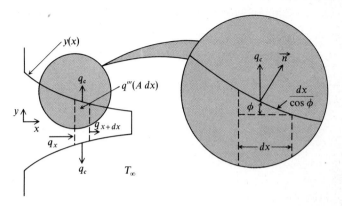

FIGURE 2-2
Fin cross section.

The term q''' represents a uniform rate of internal generation of energy per unit volume. The volume of the differential system is $A\,dx$, where A is the cross-sectional area. The rate of convection from the fin surface q_c is a function of the temperature gradient in the fluid adjacent to the fin surface, which in turn depends upon the flow field around the fin.

Using Newton's law of cooling, q_c can be expressed as

$$q_c = h_x A_s (T - T_{ref}) \qquad (2\text{-}6)$$

where A_s is the surface area of the differential element across which convection occurs, and T_{ref} is chosen as the environmental temperature T_∞. For the fin element, the convective heat transfer per unit depth is expressed by

$$q_c = h_x \frac{dx}{\cos \phi}(1)(T - T_\infty) \qquad (2\text{-}7)$$

Fourier's law of conduction and Newton's law of cooling can be used to obtain expressions for the heat transfer terms in Eq. (2-5). The cross-sectional area per unit depth normal to the one-dimensional heat conduction within the fin is $A = 2y(x)(1)$. The term for heat conduction at the right face

$$q_{x+dx} = \left\{ -k[2y(x)]\frac{dT}{dx} \right\}_{x+dx} \qquad (2\text{-}8)$$

can be expressed by using the first two terms of a Taylor series expansion about x.

The resulting form of Eq. (2-5) is

$$-k[2y(x)]\frac{dT}{dx} + q'''[2y(x)]\,dx = -k[2y(x)]\frac{dT}{dx}$$

$$+\frac{d}{dx}\left\{-k[2y(x)]\frac{dT}{dx}\right\}dx + \frac{2h_x\,dx}{\cos\phi}(T - T_\infty) \qquad (2\text{-}9)$$

Assuming constant thermal conductivity reduces this equation to

$$k[2y(x)]\frac{d^2T}{dx^2} + 2k\frac{dy(x)}{dx}\frac{dT}{dx} + [2y(x)]q''' - \frac{2h_x(T - T_\infty)}{\cos\phi} = 0 \qquad (2\text{-}10)$$

For convenience we define a nondimensional temperature

$$\theta = \frac{T - T_\infty}{T_w - T_\infty} \qquad (2\text{-}11)$$

where T_w is the fin temperature at $x = 0$. Equation (2-11) can be written as $\theta(T_w - T_\infty) = T - T_\infty$. Differentiating each side with respect to x yields

$$\theta\left(-\frac{dT_\infty}{dx}\right) + (T_w - T_\infty)\frac{d\theta}{dx} = \frac{dT}{dx} - \frac{dT_\infty}{dx} \qquad (2\text{-}12)$$

Since it is usually appropriate to treat T_∞ as a constant, the first and last terms in the above equation become zero. In like manner

$$\frac{d^2\theta}{dx^2}(T_w - T_\infty) = \frac{d^2T}{dx^2}$$

In terms of the nondimensional temperature θ, Eq. (2-10) becomes

$$k[2y(x)]\frac{d^2\theta}{dx^2} + 2k\frac{dy(x)}{dx}\frac{d\theta}{dx} + \frac{2y(x)}{T_w - T_\infty}q''' - \frac{2h_x\theta}{\cos\phi} = 0 \qquad (2\text{-}13)$$

This second-order differential equation, along with two boundary conditions, forms the mathematical model which governs the steady-state temperature distribution in the fin.

2-2   Mathematical Boundary Conditions

To specify boundary conditions, something must be known about the dependent variable at specific locations. One end boundary condition that is easy to treat mathematically is to specify the end temperature T_e. However, this condition is usually unknown at a solid-gas interface. To attain a temperature T_∞ at the end surface, it is necessary to have a large heat sink in contact with the surface and a small convective resistance at the surface. The resistance to convection is inversely proportional to the value of the convection heat transfer coefficient. Small values of the convective resistance often occur when the end surface is submerged in a liquid. The temperature difference between a surface and its surroundings is proportional to the product of the heat transfer and the convective resistance. [See Eq. (1-55).]

A second type of boundary condition specifies the value of the derivative of the dependent variable at an interface. The energy conducted to a solid-fluid interface from within the solid is given by Fourier's law. Energy may leave the interface by convection and/or radiation. At steady-state conditions the energy conducted to the interface must equal the sum of the convection and radiation. If the heat transfer crossing the surface can be determined from independent measurements or calculations, the temperature gradient within the solid at the interface can be calculated and used as a boundary condition. A common example is when a surface is assumed to be perfectly insulated and the value of the temperature gradient can be specified as zero. This adiabatic condition can be approached physically by use of insulating materials such as asbestos and glass wool. When a symmetrical temperature profile occurs within a body, the adiabatic condition also exists at the point of zero temperature gradient.

When the steady-state heat transfer from a surface is by convection only, the energy conducted to the interface equals the energy convected from the interface, or vice versa, and we write

$$\frac{q}{A} = \pm k\left(\frac{\partial T}{\partial n}\right)_s = h_x(T - T_\infty) \qquad (2\text{-}14)$$

Here the thermal conductivity and temperature gradient are values in the solid material. Equation (2-14) allows the temperature gradient at the surface to be expressed in terms of the convection heat transfer coefficient. Across a solid-solid interface between materials 1 and 2, the proper boundary condition for steady-state conduction is

$$k_1\left(\frac{\partial T_1}{\partial n}\right)_i = k_2\left(\frac{\partial T_2}{\partial n}\right)_i \qquad (2\text{-}15)$$

If contact resistance between the two solid surfaces is negligible, then $T_1 = T_2$.

If the heat transfer from a surface with constant emittance (gray surface) is by radiation only, one can use Eq. (1-47) and write

$$\frac{q}{A} = \pm k\left(\frac{\partial T}{\partial n}\right)_s = \sigma \mathscr{F}_{1-2}(T_1^4 - T_2^4) \qquad (2\text{-}16)$$

The shape factor $\mathscr{F}_{1-2}$ accounts for the surface resistance to radiation due to surface emittance and the space resistance to radiation due to the geometric configuration of the surfaces exchanging radiation. The use of a radiation boundary condition

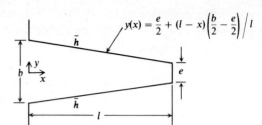

FIGURE 2-3
Trapezoidal fin.

introduces nonlinear terms into the mathematical model. Two other boundary conditions which introduce nonlinear terms are a change of phase and free convection at large temperature differences. Other sources of energy which can enter into the formulation of boundary conditions include work done by friction, mass flow through a porous solid, and energy released by chemical reaction. In each situation the proper boundary condition results from an application of the principle of conservation of energy at the interface.

Based upon the known physical conditions for a fin, it is usually necessary to specify the value of the boundary conditions at two points, unless both the temperature and temperature gradient at $x = 0$ are known. For example, in a fin with an adiabatic end, one can specify the value of θ at $x = 0$ and the value of $d\theta/dx$ at $x = l$. It is then necessary to find a value of $d\theta/dx$ at $x = 0$ which allows the resulting temperature profile to satisfy the end boundary condition at $x = l$. Methods for solving this type of two-point boundary-value problem will be treated later on in this chapter.

2-3 Tapered Fins

A tapered fin of trapezoidal profile is now analyzed. A proper specification of the surface variation of the local convection heat transfer coefficient h_x is based upon the analysis presented in later chapters. For now we simplify the mathematical model by replacing h_x with $\bar{h}$, an average convection heat transfer coefficient. Using the nomenclature shown in Fig. 2-3, Eq. (2-13) then becomes

$$k\left[e + \frac{(l-x)(b-e)}{l}\right]\frac{d^2\theta}{dx^2} - \frac{k(b-e)}{l}\frac{d\theta}{dx}$$

$$+ \frac{e + (l-x)(b-e)/l}{T_w - T_\infty}q''' - \frac{2\bar{h}\theta}{\cos\phi} = 0$$

or

$$\frac{d^2\theta}{dx^2} = \frac{2\bar{h}\theta/k\cos\phi + [(b-e)/l](d\theta/dx)}{e + (l-x)(b-e)/l} - \frac{q'''}{k(T_w - T_\infty)} \qquad (2\text{-}17)$$

An alternate choice of a coordinate system is to choose x measured from the end of the fin toward the base of the fin with $x = 0$ at the fin end. The expression for $y(x)$ would then be

$$y(x) = \frac{b}{2} - (l - x)\left(\frac{b}{2} - \frac{e}{2}\right) \Big/ l$$

and

$$\frac{d^2\theta}{dx^2} = \frac{2\bar{h}\theta/k \cos \phi - [(b - e)/l](d\theta/dx)}{b - (l - x)(b - e)/l} - \frac{q'''}{k(T_w - T_\infty)} \quad (2\text{-}18)$$

This choice of coordinate systems alters the boundary-value problem since at $x = 0$ the value of θ is now unknown and the value of $d\theta/dx = 0$. Hence, it is necessary to find the value of θ at $x = 0$ which satisfies the value $\theta = 1.0$ at $x = l$, the fin base where $T = T_w$.

For a rectangular fin with no internal generation, Eq. (2-17) reduces to

$$\frac{d^2\theta}{dx^2} - \frac{2\bar{h}\theta}{ke} = 0 \quad (2\text{-}19)$$

Neglecting edge effects, the perimeter for a fin of unit depth is $p = 2.0$. The area normal to the heat conduction is $A = 1(e)$. The equation can then be written

$$\frac{d^2\theta}{dx^2} - m^2\theta = 0 \quad (2\text{-}20)$$

where $m = (\bar{h}p/kA)^{1/2}$.

The solution to this differential equation is $\theta = C_1 \exp(mx) + C_2 \exp(-mx)$. For an adiabatic end the boundary conditions are

$$\text{at} \quad x = 0 \quad \theta = 1.0$$

$$\text{at} \quad x = l \quad \frac{d\theta}{dx} = 0$$

These two boundary conditions lead to the following solution

$$\frac{T - T_\infty}{T_w - T_\infty} = \theta = \frac{\cosh m(l - x)}{\cosh ml} \quad (2\text{-}21)$$

Using Fourier's law and Eq. (2-21), the one-dimensional heat transfer at the fin base predicted by the analytical model is

$$q = (\bar{h}pkA)^{1/2}(\tanh ml)(T_w - T_\infty) \quad (2\text{-}22)$$

The analytical solution to a triangular fin with no internal generation can be obtained by choosing $x = 0$ at the tip of a fin of length l. The governing differential equation is Eq. (2-18). For $q''' = 0$ and $\cos \phi \approx 1$, we obtain

$$x\frac{d^2\theta}{dx^2} + \frac{d\theta}{dx} - N^2 l\theta = 0 \qquad (2\text{-}23)$$

where $N = (2\bar{h}/kb)^{1/2}$.

The general solution to this equation is expressed in terms of zero-order Bessel functions of the first and second kind

$$\theta = C_1 J_0\left(2iN\sqrt{lx}\right) + C_2 Y_0\left(2iN\sqrt{lx}\right) \qquad (2\text{-}24)$$

where J_0 and Y_0 are the Bessel functions of the first and second kind. The boundary conditions based on the chosen coordinate system are

$$\text{at } x = l \qquad T = T_w \qquad \theta = 1.0$$

$$\text{at } x = 0 \qquad \frac{dT}{dx} = 0 \qquad \frac{d\theta}{dx} = 0$$

The second boundary condition requires that $C_2 = 0$, and the first boundary condition leads to the solution

$$\theta = \frac{J_0\left(2iN\sqrt{lx}\right)}{J_0(2iNl)} \qquad (2\text{-}25)$$

According to the mathematical theory of Bessel functions, for integer values of n the equality

$$I_n(x) = i^{-n}J_n(ix) \qquad (2\text{-}26)$$

is valid, and for $n = 0$

$$I_0(x) = J_0(ix)$$

where $I_n(x)$ is a modified Bessel function. Tabulated values of Bessel functions and modified Bessel functions are available. (See Ref. 2-4.)

2-4   Nondimensional Equations

As pointed out in Chap. 1, when a parametric study is to be made, the most logical and efficient analysis requires that the mathematical model be put in nondimensional form. This move in turn leads to the identification of the nondimensional parameters which control the behavior of the solutions. We have already chosen a nondimensional temperature θ as a dependent variable for the analysis of heat conduction in fins. In addition, let us define a nondimensional distance $\eta = x/l$ as the independent variable in the governing equation. The value of the nondimensional fin length then varies from $0 \leq \eta \leq 1$ for all fins. The dimensional independent variable derivatives may be written in terms of the nondimensional independent variable η as

$$\frac{d\theta}{dx} = \frac{d\theta}{d\eta}\frac{d\eta}{dx} = \frac{1}{l}\frac{d\theta}{d\eta}$$

$$\frac{d^2\theta}{dx^2} = \frac{d}{d\eta}\left(\frac{d\theta}{dx}\right)\frac{d\eta}{dx} = \frac{d}{d\eta}\left(\frac{1}{l}\frac{d\theta}{d\eta}\right)\left(\frac{1}{l}\right) = \frac{1}{l^2}\frac{d^2\theta}{d\eta^2}$$

The governing equation for a tapered fin of trapezoidal profile is then

$$kl\left[\frac{e}{l} + \left(\frac{b}{l} - \frac{e}{l}\right)(1-\eta)\right]\frac{1}{l^2}\frac{d^2\theta}{d\eta^2} - \left(\frac{b}{l} - \frac{e}{l}\right)\frac{1}{l}\frac{d\theta}{d\eta}$$

$$+ l\left[\frac{e}{l} + \left(\frac{b}{l} - \frac{e}{l}\right)(1-\eta)\right]\frac{q'''}{T_w - T_\infty} - \frac{2\bar{h}\theta}{\cos\phi} = 0$$

Solving for the second derivative of the dependent variable yields

$$\frac{d^2\theta}{d\eta^2} = \frac{2\bar{h}l\theta/k\cos\phi + (b/l - e/l)(d\theta/d\eta)}{e/l + (b/l - e/l)(1-\eta)} - \frac{q'''l^2}{k(T_w - T_\infty)} \qquad (2\text{-}27)$$

We see that the solution to the nondimensional temperature field is governed by four nondimensional parameters. The ratio $\bar{h}l/k$, called a *Biot number,* is an indication of the ratio of the internal resistance to heat transfer by conduction to that of the external resistance to heat transfer by convection. As we shall see, the Biot number value indicates whether a one-dimensional or a two-dimensional analysis is appropriate for a given problem.

The physical meaning of the Biot number can be clearly shown by considering steady-state, one-dimensional heat conduction through a wall of thickness l and a constant cross-sectional area A. For constant thermal properties of the wall material, the temperature profile is linear, and $dT/dx = \Delta T/l$. Fourier's law can then be written in a form that makes the electrical analogy clear:

$$q = kA\frac{\Delta T}{l} = \frac{\Delta T}{l/kA} \qquad (2\text{-}28)$$

where ΔT is the temperature drop across the wall. The term in the denominator is interpreted as the internal resistance to heat conduction within the wall. At the surface, heat transfer occurs to the surroundings by convection, and Newton's law can be written

$$q = \bar{h}A_s(T_w - T_\infty) = \frac{T_w - T_\infty}{1/\bar{h}A_s} \qquad (2\text{-}29)$$

The term in the denominator is interpreted as the external resistance to convective heat transfer at the surface. For a plane wall with one-dimensional conduction, $A = A_s$, and the ratio of internal resistance to heat transfer by conduction to that of the external resistance to heat transfer by convection is

$$\frac{l/kA}{1/\bar{h}A} = \frac{\bar{h}l}{k} = \text{Bi} \qquad (2\text{-}30)$$

For this simple problem it is seen that the Biot number Bi is exactly equal to the ratio of the internal resistance to conduction to the external resistance to convection.

The other three nondimensional parameters are e/l, b/l, and $q'''l^2/k(T_w - T_\infty)$. For the purpose of analysis, the four nondimensional parameters are defined as $N1 = q'''l^2/k(T_w - T_\infty)$, $N2 = e/l$, $N3 = b/l$, and $N4 = \bar{h}l/k$. It is important to keep the physical significance of these parameters in mind when making an analysis. The significance of N4 has just been discussed. The physical meaning of N1 may be more easily interpreted by writing

$$N1 = \frac{q'''l^3}{kl^2(T_w - T_\infty)/l}$$

The numerator is a heat transfer rate resulting from internal generation in a reference volume l^3. The denominator is a reference heat conduction term across a reference area l^2 and along a reference temperature gradient $(T_w - T_\infty)/l$. Thus N1 can be thought of as a ratio of energy internally generated to energy conducted. The values of N2 and N3 are related to the geometry of the fin. We can now obtain parametric solutions by holding three of the four nondimensional parameters constant and varying the value of the fourth parameter within the range of interest. Another major advantage of the nondimensional results is that they are valid for variables expressed in any system of units.

2-5 Numerical Techniques

Equation (2-27) can be written in terms of two equivalent first-order differential equations by writing

$$Y = \theta \qquad (2\text{-}31)$$

$$Z = \frac{d\theta}{d\eta} \qquad (2\text{-}32)$$

This technique is discussed in Appendix A-3. The two first-order equations are then

$$F = \frac{dY}{d\eta} = Z \tag{2-33}$$

$$G = \frac{dZ}{d\eta} = \frac{2(N4)\,Y\,/\cos\phi + (N3 - N2)Z}{N2 + (N3 - N2)(1 - \eta)} - N1 \tag{2-34}$$

where $F = d\theta/d\eta$ and $G = d^2\theta/d\eta^2$.

These two equations can be solved by means of the fourth-order Runge-Kutta technique used in Chap. 1. The primary difference is that now the value of $Z = d\theta/d\eta$ at $\eta = 0$ is unknown. The two known boundary conditions for a fin with an adiabatic end are $\theta = 1.0$ at $\eta = 0$ and $d\theta/d\eta = 0$ at $\eta = 1.0$. Since these two boundary conditions are specified at two different points on the fin, the problem is called a *two-point boundary-value problem.*

One must determine input values for the geometric and physical parameters appearing in Eq. (2-27), as well as the increment in the independent variable $\Delta\eta$ to be used in the Runge-Kutta integration scheme. To start the integration, an assumed value of $d\theta/d\eta$ at $\eta = 0$ is also required. The integration is then carried out by using the Runge-Kutta algorithm between $\eta = 0$ and the end of the fin at $\eta = 1.0$, in incremental steps of $\Delta\eta$.

The initial results based upon an assumed temperature gradient at $\eta = 0$ cannot be expected to satisfy the adiabatic boundary condition at $\eta = 1.0$. However, by comparing the results obtained to the desired results, it should be obvious whether the value of $d\theta/d\eta$ at $\eta = 0$ is too large or too small. It is necessary to repeat the integration using a second value of $d\theta/d\eta$ at $\eta = 0$ to determine how sensitive the value of $d\theta/d\eta$ at $\eta = 1.0$ is to a change in $d\theta/d\eta$ at $\eta = 0$. Once the results of two trial integrations have been obtained, it is possible to calculate a more accurate value of $d\theta/d\eta$ at $\eta = 0$ for the next integration. This is done by comparing the required boundary condition at $\eta = 1.0$ to the calculated value.

There is some unknown functional relationship between $(d\theta/d\eta)_{\eta=0}$ and $(d\theta/d\eta)_{\eta=1}$. This relationship is represented by the solid curve in Fig. 2-4. After two integrations we can plot two points (A and B) on this curve. If we require $(d\theta/d\eta)_{\eta=1} = 0$, we seek the value of $(d\theta/d\eta)_{\eta=0}$ indicated by point S. The results of the second integration are represented by point B. A line tangent to the curve at point B is indicated by the dashed line in Fig. 2-4. The intersection of this line with the ordinate gives an improved value y_0 of $(d\theta/d\eta)_{\eta=0}$. The slope of this tangent line is

$$\text{Slope} = \frac{y_B - y_0}{x_B - 0} \tag{2-35}$$

Hence, $y_0 = y_B - x_B(\text{slope})$.

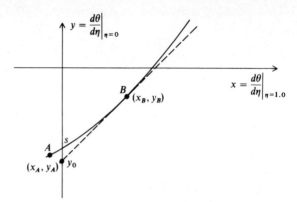

FIGURE 2-4
Newton-Raphson method.

If two values of $(d\theta/d\eta)_{\eta=0}$ are chosen such that points A and B are close together, the slope at point B is approximately equal to the slope of the chord joining A and B.

$$\text{Slope} \approx \frac{y_B - y_A}{x_B - x_A} = \frac{\Delta\theta'(0)}{\Delta\theta'(l)} \qquad (2\text{-}36)$$

This move allows us to calculate an improved value of $(d\theta/d\eta)_{\eta=0}$ from Eq. (2-35). This calculation is the important feedback link that leads to the value of $d\theta/d\eta$ at $\eta = 0$ which satisfies the governing differential equation and both boundary conditions to sufficient accuracy. This iteration technique is referred to as the *Newton-Raphson method*. The decision as to what constitutes "sufficient accuracy" is another choice the analyst must make. It depends upon the accuracy of the mathematical model, the accuracy of the computer, the purpose of the results, and the cost of computation. The incremental step size in $\Delta\eta$ may also be changed to determine the proper balance between the speed and accuracy of computation.

The computer program FIN has been written to integrate the two equivalent first-order differential equations given by Eqs. (2-33) and (2-34). The fourth-order Runge-Kutta technique was used and the Newton-Raphson iteration technique was applied to obtain the correct value of $(d\theta/d\eta)_{\eta=0}$. Since $d\theta/d\eta$ at $\eta = 0$ is unknown, two initial values for this initial gradient, differing by 0.1, are assigned and stored as computer variables G1 and G2. Two integrations between $0 \le \eta \le 1.0$, successively using these two values as boundary conditions, produce two results for $d\theta/d\eta$ at $\eta = 1.0$, identified by I1 and I2 in terms of computer variables. Typical results are represented by Fig. 2-5 where G1 and I1 have been plotted as point A, and G2 and I2 have been plotted as point B. The desired value of $d\theta/d\eta$ at the end of the insulated

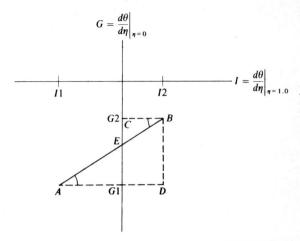

FIGURE 2-5
Iteration plot.

fin is zero. We now use the Newton-Raphson iteration scheme to calculate an improved estimate for $d\theta/d\eta$ at $\eta = 0$. The slope at point B is calculated using the approximate expression given by Eq. (2-36). From Fig. 2-5, we have

$$\left(\frac{d\theta}{d\eta}\right)_{\eta=0} = G2 - I2\left(\frac{G2 - G1}{I2 - I1}\right)$$

Rearranging, the new wall temperature gradient $(d\theta/d\eta)_{\eta=0}$ is given by

$$\left(\frac{d\theta}{d\eta}\right)_{\eta=0} = \frac{(I2)G1 - (I1)G2}{I2 - I1} = (Z)_{\eta=0} \qquad (2\text{-}37)$$

This equation is used to calculate the new value of Z at $\eta = 0$ after the first two integrations. The value of G2 used in the second integration is then defined as G1, and the new value of Z just calculated is defined as G2. If the value of the temperature gradient at the end of the fin is less than 10^{-6}, a solution has been obtained. This degree of accuracy is more than adequate for most solutions. If the value of Z at $\eta = 1.0$ is not less than 10^{-6}, I1 is set equal to I2, and I2 is set equal to the most recent value of Z. Thus, the results of the two most recent integrations are always used in the iteration process to recalculate a new value of Z at $\eta = 0$. The process continues until the value of Z at $\eta = 1.0$ is less than 10^{-6}.

A listing of the program FIN appears below (Listing 2-1). The computer variables are identified in Table 2-2; also see Flow Chart 2-1. The same fourth-order Runge-Kutta algorithm used in Chap. 1 for the program PLATES is employed. As shown in the printout of a typical solution, a solution is obtained by using the Newton-

Raphson iteration technique, and then a printout of the full results can be obtained. In most heat transfer calculations, the wall temperature gradient $Z = (d\theta/d\eta)_{\eta=0}$ is the value of interest since this value is needed to determine the steady-state heat transfer through the fin.

Nondimensional results for a tapered fin are presented in Figs. 2-6 and 2-7. In Fig. 2-6 the value of $N2 = e/l$ is varied between 0.001 and 0.1, while the values of the other nondimensional parameters are maintained constant. The wall temperature gradient is initially increased as N2 increases, and then the gradient begins to decrease. This result is due to the uniform internal generation per unit volume.

In Fig. 2-7 the value of $N4 = \bar{h}l/k$ is varied between 0.001 and 0.1, while the other three parameters are held constant. Values of N4 above 0.1 are not given since the two-dimensional analysis discussed in Chap. 3 is more appropriate for larger Biot numbers. As the Biot number N4 is increased, the change in temperature through the fin increases because of the increase in the internal resistance to heat transfer by conduction, relative to the resistance to heat transfer by convection. Note also that as the Biot number becomes very small, a value is reached where the temperature gradient at the wall changes sign and the nondimensional temperature becomes

Table 2-2   VARIABLES USED IN FIN PROGRAM

Variable	Definition
C1, L, M, N, R	Logic control variables
T8, T9	Print control variables
X9	Maximum value of η
H	Increment $\Delta\eta$
N1	Parameter $q'''l^2/k(T_w - T_\infty)$
N2	Parameter e/l
N3	Parameter b/l
N4	Parameter $\bar{h}l/k$
F1	Tangent of ϕ
A	Arctan ϕ
X	Independent variable η
Y0	Initial value of θ
Z0	Initial value of $d\theta/d\eta$
Y	Dependent variable θ
Z	Derivative $d\theta/d\eta$
F	First derivative $d\theta/d\eta$
G	Second derivative $d^2\theta/d\eta^2$
K1, K2, K3, K4	Runge-Kutta coefficients for θ_{i+1}
L1, L2, L3, L4	Runge-Kutta coefficients for $(d\theta/d\eta)_{i+1}$
G1	Initial gradient $d\theta/d\eta$ at $\eta = 0$
G2	Modified gradient $d\theta/d\eta$ at $\eta = 0$
I1	Results of $d\theta/d\eta$ at $\eta = 1.0$ using G1
I2	Results of $d\theta/d\eta$ at $\eta = 1.0$ using G2
C	Absolute value of $d\theta/d\eta$ at $\eta = 1.0$

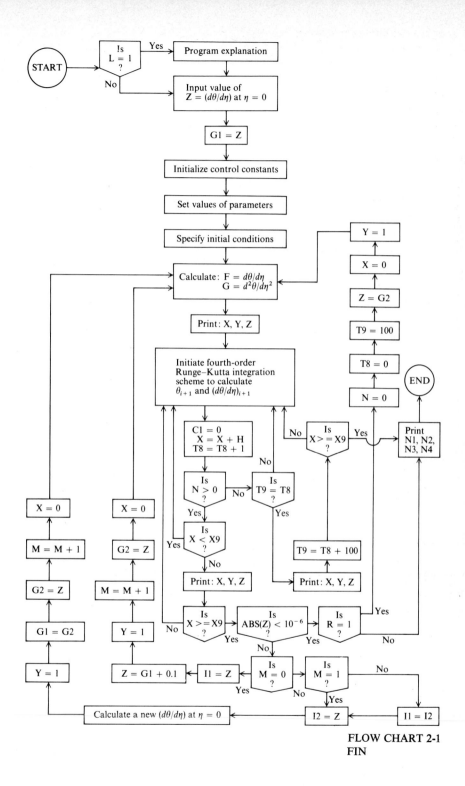

FLOW CHART 2-1
FIN

Listing 2-1 FIN

```
10    PRINT"TYPE 1 IF YOU WANT A PROGRAM EXPLANATION."
12    PRINT
13    PRINT"OTHERWISE, TYPE 2."
14    INPUTL
16    IF L=1 THEN 20
18    IF L=2 THEN 42
20    PRINT
22    PRINT"THIS PROGRAM USES A FOURTH ORDER RUNGE-KUTTA SCHEME"
24    PRINT"TO INTEGRATE TWO EQUIVALENT FIRST ORDER EQUATIONS"
26    PRINT"DESCRIBING THE ONE-DIMENSIONAL, STEADY STATE TEMPERATURE"
27    PRINT"DISTRIBUTION IN A TAPERED FIN, GIVEN BY:"
28    PRINT
30    PRINT"   F=Z"
31    PRINT
32    PRINT"   G=((2*N4*Y/COS(A) +(N3-N2)*Z)/(N2+(N3-N2)*(1-X))-N1"
33    PRINT
34    PRINT" WHERE Y IS A NON-DIMENSIONAL TEMPERATURE,"
35    PRINT" Z IS THE DERIVATIVE OF Y WITH RESPECT TO X,"
36    PRINT" AND X IS A NON-DIMENSIONAL DISTANCE."
37    PRINT
38    PRINT" THE BOUNDARY CONDITIONS ARE:"
39    PRINT
40    PRINT"   AT X=0, Y=1.0; AT X=1.0, Z=0"
42    PRINT
44    PRINT"YOU ARE REQUIRED TO INPUT AN ESTIMATED VALUE OF THE"
45    PRINT"TEMPERATURE GRADIENT Z AT X=0."
50    PRINT
52    PRINT"TYPE YOUR ESTIMATE OF Z AT X=0."
54    INPUTZ
56    LETG1=Z
58    PRINT"     THE RESULTS OF ONE INTEGRATION GIVE:"
59    PRINT
60    REM INITIALIZE CONTROL CONSTANTS.
65    LETN=1
70    LETM=0
80    LETC1=0
85    LETX9=1
90    LETH=0.001
100   REM SET VALUES OF NON-DIMENSIONAL PARAMETERS
110   LETN1=0.1
120   LETN2=0.05
130   LETN3=0.05
140   LETN4=0.01
150   LETF1=ABS(0.5*(N3-N2))
160   LETA=ATN(F1)
1000  REM SET PROGRAM VALUES AND INITIAL CONDITIONS
1030  REM INITIALIZE INDEPENDENT VARIABLE
1040  LETX=0
1050  REM SPECIFY INITIAL CONDITION
1060  LETY0=1.0
1065  LETZ0=Z
1070  LETY=Y0
1075  LETZ=Z0
1080  LETF=Z
1085  LETG=(((2*N4*Y)/COS(A) +(N3-N2)*Z)/(N2+(N3-N2)*(1-X)))-N1
1100  PRINT "DISTANCE","Y","Z"
1110  PRINT
1120  PRINTX,Y,Z
1130  REM
1140  REM FOURTH ORDER RUNGE-KUTTA INTEGRATION SCHEME
1150  REM
1160  LETF=Z
1170  LETG=(((2*N4*Y)/COS(A) +(N3-N2)*Z)/(N2+(N3-N2)*(1-X)))-N1
1190  IF C1=1 THEN 1300
```

Listing 2-1   FIN (*continued*)

```
1200 IF C1=2 THEN 1380
1210 IF C1=3 THEN 1460
1220 LETK1=H*F
1230 LETL1=H*G
1240 LETX=X+0.5*H
1270 LETY=Y+0.5*K1
1275 LETZ=Z+0.5*L1
1280 LETC1=1
1290 GOTO 1160
1300 LETK2=H*F
1310 LETL2=H*G
1350 LETY=Y-0.5*K1+0.5*K2
1355 LETZ=Z-0.5*L1+0.5*L2
1360 LETC1=2
1370 GOTO 1160
1380 LETK3=H*F
1390 LETL3=H*G
1400 LETX=X+0.5*H
1430 LETY=Y-0.5*K2+K3
1435 LETZ=Z-0.5*L2+L3
1440 LETC1=3
1450 GOTO 1160
1460 LETK4=H*F
1470 LETL4=H*G
1510 LETY=Y-K3+(K1+2*K2+2*K3+K4)/6
1515 LETZ=Z-L3+(L1+2*L2+2*L3+L4)/6
1516 REM END OF THE RUNGE-KUTTA INTEGRATION SCHEME
1520 LETC1=0
1540 LETT8=T8+1
1550 REM
1560 REM PRINT SEQUENCE
1570 REM
1580 IF N>0 THEN 1650
1590 IF T9=T8 THEN 1610
1600 GOTO 1160
1610 PRINTX,Y,Z
1620 LETT9=T8+100
1630 IF X>=X9 THEN 3000
1640 GOTO 1160
1650 IF X<X9 THEN 1680
1660 PRINTX,Y,Z
1670 IF X>=X9 THEN 1720
1680 GOTO 1160
1690 REM
1700 REM CONVERGENCE TEST
1710 REM
1720 IF ABS(Z)<1E-6 THEN 2500
2000 IF M=0 THEN 2010
2002 IF M=1 THEN 2070
2004 IF M>=2 THEN 2140
2005 PRINT"THE PARAMETERS USED IN THIS SOLUTION ARE"
2006 PRINT
2008 REM CALCULATE A NEW GRADIENT AT X=0.
2010 PRINT
2012 PRINT"WE NOW CHANGE YOUR INPUT WALL GRADIENT VALUE BY 0.1"
2014 PRINT"AND REPEAT THE INTEGRATION FOR THIS NEW VALUE,OBTAINING:"
2015 PRINT
2016 LETI1=Z
2020 LETZ=G1+0.1
2030 LETY=1
2040 LETM=M+1
2050 LETG2=Z
2060 LETX=0
2065 GOTO 1080
```

Listing 2-1   FIN (*continued*)

```
2070 PRINT
2072 PRINT"WE NOW USE THE NEWTON-RAPHSON EQUATION TO CALCULATE"
2074 PRINT"AN IMPROVED VALUE OF Z AT X=0, AND CONTINUE THE"
2075 PRINT"ITERATION PROCESS UNTIL A SOLUTION IS OBTAINED."
2076 LETI2=Z
2080 LETZ=(I2*G1-I1*G2)/(I2-I1)
2090 LETY=1
2100 LETG1=G2
2105 LETG2=Z
2110 LETM=M+1
2114 PRINT
2118 LETX=0
2121 GOTO 1080
2140 LETI1=I2
2150 GOTO 2076
2500 PRINT
2505 PRINT"DO YOU WISH FULL RESULTS? IF YES TYPE 1, IF NO TYPE 2."
2510 PRINT
2525 INPUTR
2530 IF R=1 THEN 2600
2540 IF R=2 THEN 3000
2600 PRINT
2605 LETN=0
2610 LETT8=0
2620 LETT9=100
2630 LETZ=G2
2640 LETX=0
2650 LETY=1
2660 GOTO 1080
3000 PRINT
3010 PRINT"THE FINAL SOLUTION SATISFIES THE CONVERGENCE REQUIREMENT"
3020 PRINT"THAT Z AT X=1.0 IS LESS THAN 1.0E-6."
3030 PRINT
3040 PRINT"THE PARAMETERS USED IN THIS SOLUTION ARE:"
3050 PRINT
3060 PRINT"N1="N1,"N2="N2,"N3="N3,"N4="N4
4000 END
```

Computer Results 2-1   FIN

```
TYPE 1 IF YOU WANT A PROGRAM EXPLANATION.

OTHERWISE, TYPE 2.
? 1

THIS PROGRAM USES A FOURTH ORDER RUNGE-KUTTA SCHEME
TO INTEGRATE TWO EQUIVALENT FIRST ORDER EQUATIONS
DESCRIBING THE ONE-DIMENSIONAL, STEADY STATE TEMPERATURE
DISTRIBUTION IN A TAPERED FIN, GIVEN BY:

   F=Z

   G=((2*N4*Y/COS(A) +(N3-N2)*Z)/(N2+(N3-N2)*(1-X))-N1

WHERE Y IS A NON-DIMENSIONAL TEMPERATURE,
Z IS THE DERIVATIVE OF Y WITH RESPECT TO X,
AND X IS A NON-DIMENSIONAL DISTANCE.

THE BOUNDARY CONDITIONS ARE:

   AT X=0, Y=1.0; AT X=1.0, Z=0

YOU ARE REQUIRED TO INPUT AN ESTIMATED VALUE OF THE
TEMPERATURE GRADIENT Z AT X=0.
```

Computer Results 2-1 FIN (*continued*)

```
TYPE YOUR ESTIMATE OF Z AT X=0.
? -.5
      THE RESULTS OF ONE INTEGRATION GIVE:

DISTANCE          Y                  Z

0                 I                  -0.5
I.                0.621061           -0.282974

WE NOW CHANGE YOUR INPUT WALL GRADIENT VALUE BY 0.1
AND REPEAT THE INTEGRATION FOR THIS NEW VALUE,OBTAINING:

DISTANCE          Y                  Z

0                 I                  -0.4
I.                0.727862           -0.162299

WE NOW USE THE NEWTON-RAPHSON EQUATION TO CALCULATE
AN IMPROVED VALUE OF Z AT X=0, AND CONTINUE THE
ITERATION PROCESS UNTIL A SOLUTION IS OBTAINED.

DISTANCE          Y                  Z

0                 I                  -0.265508
I.                0.8715             -1.46598 E-7

DO YOU WISH FULL RESULTS? IF YES TYPE 1, IF NO TYPE 2.

? 1

DISTANCE          Y                  Z
0                 I                  -0.265508
0.1               0.974932           -0.23602
0.2               0.952765           -0.207475
0.3               0.933409           -0.179761
0.4               0.916788           -0.152766
0.5               0.902836           -0.126382
0.600001          0.891495           -0.100504
0.700001          0.882722           -7.50285 E-2
0.800001          0.87648            -4.98529 E-2
0.900001          0.872744           -2.48767 E-2
I.                0.8715             -1.46598 E-7

THE FINAL SOLUTION SATISFIES THE CONVERGENCE REQUIREMENT
THAT Z AT X=1.0 IS LESS THAN 1.0E-6.

THE PARAMETERS USED IN THIS SOLUTION ARE:

N1= 0.1        N2= 0.05        N3= 0.05        N4= 0.01
```

greater than unity. This change is due to the internal generation within the fin as specified by the nondimensional parameter N1.

The steady-state heat transfer through the fin is given by Fourier's law. Thus, the heat transfer through the fin in terms of the nondimensional temperature gradient $d\theta/d\eta$ at $\eta = 0$ is

$$q = -\left(kA\frac{dT}{dx}\right)_{x=0} = -\left(kA\frac{T_w - T_\infty}{l}\frac{d\theta}{d\eta}\right)_{\eta=0} \qquad (2\text{-}38)$$

Area A is the base area of the fin per unit depth. Values of $(d\theta/d\eta)_{\eta=0}$ which correspond to the results given in Fig. 2-6 and Fig. 2-7 are given in Table 2-3.

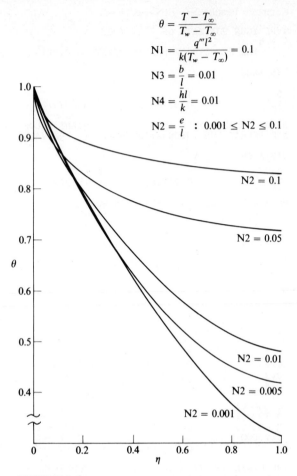

$$\theta = \frac{T - T_\infty}{T_w - T_\infty}$$

$$N1 = \frac{q'''l^2}{k(T_w - T_\infty)} = 0.1$$

$$N3 = \frac{b}{l} = 0.01$$

$$N4 = \frac{hl}{k} = 0.01$$

$$N2 = \frac{e}{l} \ : \ 0.001 \le N2 \le 0.1$$

FIGURE 2-6
Nondimensional profiles for tapered fins with N2 varying.

The mathematical model may be changed to account for a change in the profile of the fin surface. In the next section the profile of the fin is changed from trapezoidal to parabolic.

2-6 Analysis of Parabolic Fins

To show how other longitudinal fins of various profiles can be analyzed, we go back to the general governing differential equation. The general dimensional equation was derived earlier and shown to be (Eq. 2-13)

$$k[2y(x)]\frac{d^2\theta}{dx^2} + 2k\frac{dy(x)}{dx}\frac{d\theta}{dx} + \frac{2y(x)}{T_w - T_\infty}q''' - \frac{2\hbar\theta}{\cos\phi} = 0$$

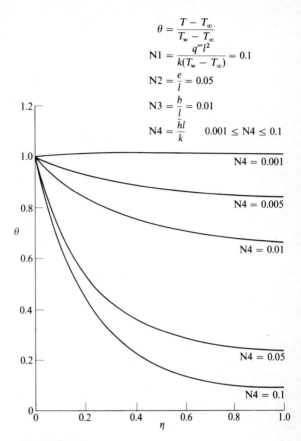

$$\theta = \frac{T - T_\infty}{T_w - T_\infty}$$

$$N1 = \frac{q''' l^2}{k(T_w - T_\infty)} = 0.1$$

$$N2 = \frac{e}{l} = 0.05$$

$$N3 = \frac{b}{l} = 0.01$$

$$N4 = \frac{hl}{k} \qquad 0.001 \le N4 \le 0.1$$

N4 = 0.001

N4 = 0.005

N4 = 0.01

N4 = 0.05

N4 = 0.1

FIGURE 2-7
Nondimensional profiles for tapered fins with N4 varying.

Consider a concave parabolic profile of the form

$$y(x) = \frac{e}{2} + \left(\frac{b}{2} - \frac{e}{2}\right)\left(1 - \frac{x}{l}\right)^2 \qquad (2\text{-}39)$$

We now have

$$\frac{dy(x)}{dx} = (b - e)\frac{x - l}{l^2}$$

The governing differential equation is then written in the form

$$\frac{d^2\theta}{dx^2} = \frac{2\bar{h}\theta/k \cos\phi - 2[(b - e)(x - l)/l^2](d\theta/dx)}{e + (b - e)(1 - x/l)^2} - \frac{q'''}{k(T_w - T_\infty)} \qquad (2\text{-}40)$$

Table 2-3   NONDIMENSIONAL TEMPERATURE GRADIENTS
FOR TRAPEZOIDAL AND CONCAVE PARABOLIC FINS

Fig. 2-6: N1 = 0.1, N3 = 0.01, N4 = 0.01: Trapezoidal fins

N2 =	0.001	0.005	0.01	0.05	0.10
$\left(\dfrac{d\theta}{d\eta}\right)_{\eta=0} =$	-1.1088	-1.1559	-1.1936	-1.2571	-1.1659

Fig. 2-7: N1 = 0.1, N2 = 0.05, N3 = 0.01: Trapezoidal fins

N4 =	0.001	0.005	0.01	0.05	0.10
$\left(\dfrac{d\theta}{d\eta}\right)_{\eta=0} =$	0.0949	-0.5970	-1.2571	-3.7928	-5.2943

Fig. 2-9: N2 = 0.05, N3 = 0.10, N4 = 0.01: Concave fins

N1 =	0.0	0.10	0.50	1.0	5.0
$\left(\dfrac{d\theta}{d\eta}\right)_{\eta=0} =$	-0.1845	-0.1224	$+0.1258$	$+0.4362$	$+2.9187$

Fig. 2-10: N1 = 0.1, N2 = 0.05, N4 = 0.01: Concave fins

N3 =	0.001	0.005	0.01	0.05	0.10
$\left(\dfrac{d\theta}{d\eta}\right)_{\eta=0} =$	-10.5869	-2.4086	-1.2731	-0.2655	-0.1224

The corresponding nondimensional equation for a concave parabolic fin with $y \sim x^2$ is given by

$$\frac{d^2\theta}{d\eta^2} = \frac{2(\bar{h}l/k)\,\theta/\cos\phi + 2[(b/l - e/l)(1 - \eta)](d\theta/d\eta)}{e/l + (b/l - e/l)(1 - \eta)^2} - \frac{q'''l^2}{k(T_w - T_\infty)} \qquad (2.41)$$

Again the same four nondimensional parameters (N1, N2, N3, and N4) can be identified.

Only a few modifications to the computer program FIN, which was used for a tapered trapezoidal fin, are needed to treat this problem. An example of these modifications is as follows:

```
150   (REMOVE)
160   (REMOVE)
1082  LET F2 = ABS((N3 − N2) * (1 − X))
```

1083   LET A = ATN(F2)
1084   LET D = N2 + (N3 − N2) * (1 − X) ↑ 2
1085   LET G = (((2 * N4 * Y)/COS(A) + 2 * (N3 − N2) * (1 − X) * Z)/D) − N1
1162   (same as 1082)
1163   (same as 1083)
1164   (same as 1084)
1070   (same as 1085)

Nondimensional results for a concave parabolic fin without internal generation are given in Fig. 2-8. When these results are compared with the nondimensional results for trapezoidal fins, it is seen that the temperature profile for steady-state heat conduction decreases more rapidly for the concave parabolic fin as the end height is

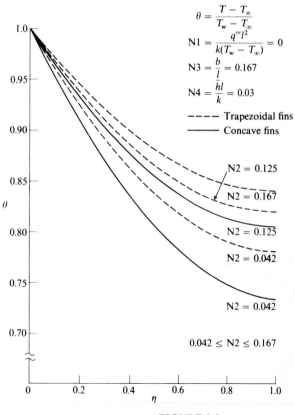

$$\theta = \frac{T - T_\infty}{T_w - T_\infty}$$

$$N1 = \frac{q''' l^2}{k(T_w - T_\infty)} = 0$$

$$N3 = \frac{b}{l} = 0.167$$

$$N4 = \frac{hl}{k} = 0.03$$

- - - - Trapezoidal fins
———— Concave fins

N2 = 0.125

N2 = 0.167

N2 = 0.125

N2 = 0.042

N2 = 0.042

0.042 ≤ N2 ≤ 0.167

FIGURE 2-8
Concave and trapezoidal fins.

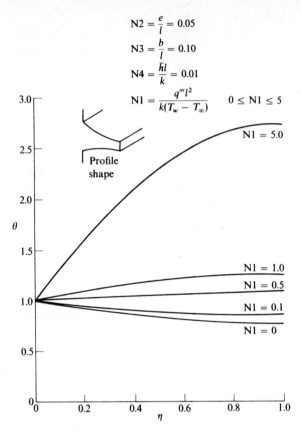

FIGURE 2-9
Nondimensional profiles for parabolic, concave fins with N1 varying.

decreased. This is caused by the combined effect of an increase in resistance to conduction, due to a reduction in cross-sectional area, and a decrease in resistance to
convection, due to an increase in fin surface area.

   Additional results of nondimensional solutions for concave parabolic fins are
shown in Figs. 2-9 and 2-10. In the previous nondimensional results for trapezoidal
fins the parameters N2 and N4 were varied. In the parabolic fin solutions the parameters N1 and N3 are varied to observe the effect of the internal generation and the
base thickness. In Fig. 2-9 the value of N1 is varied between 0 and 5.0, while the
other three parameters are maintained constant. As the value of N1 is increased, the
temperature profile approaches the parabolic form which occurs in a rod with internal
generation and no heat transfer by convection from its surface. This occurs since
the energy removed by convection for the fin surface becomes a smaller percentage of

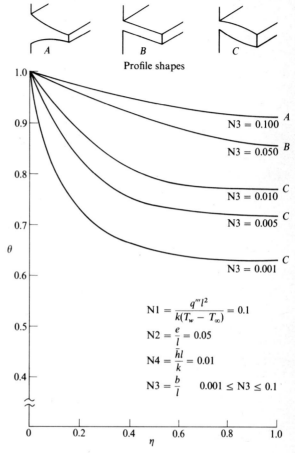

FIGURE 2-10
Nondimensional profiles for parabolic, concave fins with N3 varying.

the total thermal energy conducted through the fin as the internal generation increases. It is also important to observe the sharp change in the temperature distribution within $0.5 \le N1 \le 5.0$ as compared to the gradual change within $0.1 \le N1 \le 1.0$. A similar trend is noted for a tapered trapezoidal fin.

Figure 2-10 shows the effect of varying N3 between 0.001 and 0.1. Increasing the value of N3 reduces the resistance to heat transfer by conduction within the fin. Hence a more uniform temperature distribution is obtained as N3 is increased. As the base height is decreased, thus reducing the value of N3, the temperature gradient at the wall increases. However, the base area of the fin normal to the direction of heat transfer decreases, and the total heat transfer through the fin is proportional to the product of these two quantities. It is important not to make the mistake of looking only at the variation in the temperature gradient at the base of the fin. By using the computer to interact with the mathematical models and to vary the values of the non-dimensional parameters, the effects of fin length, fin thickness, fin profile, internal generation, and environmental conditions may be isolated and analyzed by the investigator. Values of $(d\theta/d\eta)_{\eta=0}$ corresponding to the results given in Fig. 2-9 and Fig. 2-10 appear in Table 2-3.

Reference 2-3 presents analytical solutions to the governing linear equations for fins of concave parabolic and convex parabolic profiles. These solutions are mathematically involved and are limited by assumptions such as steady-state, constant thermal conductivity, constant convection coefficient of heat transfer, and no internal generation. Often, these limiting assumptions can more easily be removed in a numerical analysis than in an analytical analysis.

2-7   Convective Boundary Conditions

A more general formulation of the fin problem considers heat transfer by convection from the end of the fin. For the special case when there is no heat transfer at the end of the fin, the results of these solutions agree with those discussed in the previous section. The convective boundary condition is obtained from the first law of thermodynamics. Since we are interested in the heat transfer across the interface between the end of the fin and the surrounding environment, we define the end face of the fin as a closed thermodynamic system and apply the first law of thermodynamics to this system. The resulting equation for steady-state conditions is a mathematical statement that the heat transfer into the end of the fin equals the heat transfer out of the end of the fin:

$$\delta Q = 0$$

If the environment is at a lower temperature than the end of the fin, the heat transfer into the end of the fin will be by conduction from within the fin. The heat transfer per unit area is given by Fourier's law evaluated at the end of the fin:

$$q'' = -k \left(\frac{dT}{dx}\right)_{x=l}$$

Again we express the convective heat transfer from the fin end by using Newton's law of cooling, Eq. (I-2). The convection heat transfer coefficient at the end surface of the fin is denoted by h_e. Supplementing the first law of thermodynamics with Fourier's and Newton's laws of conduction and convection, we obtain

$$-k \left(\frac{dT}{dx}\right)_{x=l} = h_e(T_{x=l} - T_\infty)$$

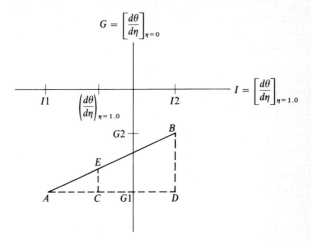

FIGURE 2-11
Iteration plot.

It is desirable to express this equation in nondimensional variables. In terms of our previously defined nondimensional variables, $\eta = x/l$ and $\theta = (T - T_\infty)/(T_w - T_\infty)$, we write

$$\frac{dT}{dx} = \frac{T_w - T_\infty}{l} \frac{d\theta}{d\eta}$$

Substituting into the above equation and solving for the nondimensional temperature gradient yields

$$\left(\frac{d\theta}{d\eta}\right)_{\eta = 1.0} = -\frac{h_e l}{k}(\theta)_{\eta = 1.0} = -N5(\theta)_{\eta = 1.0} \qquad (2\text{-}42)$$

The term N5 is the same Biot number parameter that appeared in the nondimensional governing equation, except that the convection coefficient h_e is the value at the end of the fin rather than along the surface of the fin. If this value is equal to the surface convection coefficient, then N5 = N4. When the end of the fin is adiabatic, the end convection heat transfer coefficient is zero.

An iteration scheme similar to the one used for the adiabatic fin end can be used for the convective end boundary condition. A value of $d\theta/d\eta$ at $\eta = 0$ must be found which allows the temperature profile to satisfy the condition specified in Eq. (2-42) at $\eta = 1.0$. Two initial integrations using assumed values of $d\theta/d\eta$ at $\eta = 0$ are carried out and the values of the gradients involved plotted as points A and B as shown in Fig. 2-11. Since convective heat transfer occurs from the end of the fin, the required value of $(d\theta/d\eta)_{\eta = 1.0}$ is now at some value other than zero, such as point

E. The ordinate value of point *E* specifies the required value of $(d\theta/d\eta)_{\eta=0}$. In terms of the computer variables, by similar triangles we have

$$\frac{G1 - G2}{I2 - I1} = \frac{G1 - (d\theta/d\eta)_{\eta=0}}{(d\theta/d\eta)_{\eta=1.0} - I1} \qquad (2\text{-}43)$$

After two trial integrations, the values of G1, G2, I1, and I2 are known. The true value of $(d\theta/d\eta)_{\eta=0}$ is not known until the final solution is obtained. However, an approximation can be obtained by using the value of $(\theta)_{\eta=1.0}$ obtained from the most recent integration and calculating $(d\theta/d\eta)_{\eta=1.0}$ from Eq. (2-42). This value is given a computer variable name of I0. Then Eq. (2-43) is used to calculate an improved value of $(d\theta/d\eta)_{\eta=0}$. This value is given by

$$\left(\frac{d\theta}{d\eta}\right)_{\eta=0} = G1(I2 - I0) + \frac{G2(I0 - I1)}{I2 - I1} \qquad (2\text{-}44)$$

The process is repeated, and after each integration the value of $(\theta)_{\eta=1.0}$ is used to calculate the new value of $(d\theta/d\eta)_{\eta=0}$ until the calculated value of $(d\theta/d\eta)_{\eta=0}$ differs from the previous value by some arbitrarily small amount, say 10^{-3}. The temperature profile resulting from the final integration represents the solution. For an adiabatic end I0 = 0, and Eq. (2-44) reduces to the form used for the iteration of the adiabatic boundary condition.

Results for a trapezoidal fin with heat transfer by convection at the end are shown in Fig. 2-12. The modifications to the program FIN used to solve the nondimensional tapered fin problem with a convective boundary condition are

```
145   LET N5 = VALUE DESIRED
1720  IF ABS (G2 − G1) < 1.0E − 3 THEN 2500
2078  LET I0 = −N5 ∗ Y
2080  LET Z = (G1 ∗ (I2 − I0) + G2 ∗ (I0 − I1))/(I2 − I1)
3060  PRINT "N1 = "N1, "N2 = "N2, "N3 = "N3, "N4 = "N4, "N5 = "N5
```

For the nondimensional parameters indicated in Fig. 2-12, the results show that the various convective boundary conditions have little effect on the heat transfer and steady-state temperature distribution. This characteristic is due to the small surface area at the end of the fin, specified by $N2 = e/l = 0.001$. Using this observation the investigator may wish to fix $N5 = h_e l/k$ and vary N2 to determine the effect of the ratio e/l on the steady-state temperature distribution. Note that the results in Fig. 2-12 are for a fin with internal generation and that the end height is one-tenth the base height $(N2/N3 = 0.1)$.

The maximum value of N5 as defined in Eq. (2-42) is limited by the formulation of the problem. As the value of N5 increases, the value of θ at $\eta = 1.0$ becomes

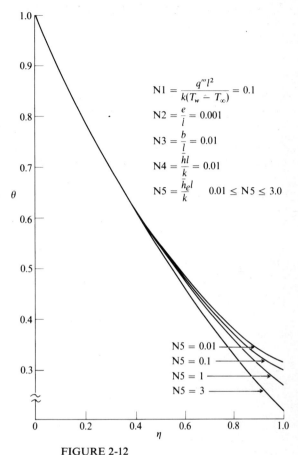

$$N1 = \frac{q'''l^2}{k(T_w - T_\infty)} = 0.1$$

$$N2 = \frac{e}{l} = 0.001$$

$$N3 = \frac{b}{l} = 0.01$$

$$N4 = \frac{\bar{h}l}{k} = 0.01$$

$$N5 = \frac{\bar{h}_e l}{k} \qquad 0.01 \le N5 \le 3.0$$

N5 = 0.01
N5 = 0.1
N5 = 1
N5 = 3

FIGURE 2-12
Tapered fin with convective boundary conditions.

smaller as shown in Fig. 2-12. Physically we can describe a system where N5 is effectively infinite because of an infinite value of h_e. The temperature at the end of the fin then equals the temperature of the environment since the resistance to heat transfer by convection at the end of the fin $1/h_e A$ is zero. Thus the value of $(\theta)_{\eta = 1.0}$ equals zero. However, as N5 is increased, Statement 2078 multiplies a large value N5 times a small value Y, where $Y = (\theta)_{\eta = 1.0}$. Because of the limited accuracy of the computer, acceptable solutions cannot be obtained for N5 > 3.0. This lack of convergence causes the iteration process to continue indefinitely until terminated by an external command. This occurs because the convergence criterion cannot be satisfied. When a time-sharing terminal is used, the investigator can continually monitor the progress of the solution. This procedure allows the investigator to interactively use his judgment in achieving convergence to a solution. As $(\theta)_{\eta=1.0}$ approaches zero, it is necessary that the temperature at the end of the fin approach the isothermal value of T_∞. For larger values of N5 the problem is best solved by formulating the analysis in terms of an isothermal boundary condition at the end of the fin. We will consider this problem in the next section.

For a rectangular fin without internal generation, Eq. (2-17) reduces to

$$\frac{d^2\theta}{dx^2} - \frac{2\bar{h}\theta}{ke} = 0$$

It was previously shown that the solution to this equation can be written in the form

$$\theta = C_1 \exp mx + C_2 \exp (-mx)$$

where $m = (\bar{h}p/kA)^{1/2}$.

The boundary conditions with convection at the end of the fin are

$$\text{at} \quad x = 0 \qquad \qquad \theta = 1.0$$

$$\text{at} \quad x = l \qquad \left(\frac{\partial\theta}{\partial x}\right)_{x=l} = \frac{-\bar{h}}{k}(\theta)_{x=l}$$

These boundary conditions result in the following temperature distribution:

$$\theta = \frac{\cosh m(l-x) + (\bar{h}/mk) \sinh m(l-x)}{\cosh ml + (\bar{h}/mk) \sinh ml} \qquad (2\text{-}45)$$

2-8   Isothermal Boundary Condition

To treat an isothermal boundary condition at the end of the fin, two trial integrations are again made. A plot similar to Figs. 2-5 and 2-11 is also used, but now the abscissa is the value of θ at $\eta = 1.0$, rather than $d\theta/d\eta$ at $\eta = 1.0$. We denote the values of θ at $\eta = 1.0$ after the first two integrations as J1 and J2, respectively. These replace the values of the gradients I1 and I2 used previously. The true value of θ at $\eta = 1.0$ is now a known boundary condition and is read into the computer as N6. In terms of the computer variables, the iteration equation used to calculate the new value of the temperature gradient at the base of the fin is

$$Z = (G1 * (J2 - N6) + G2 * (N6 - J1))/(J2 - J1)$$

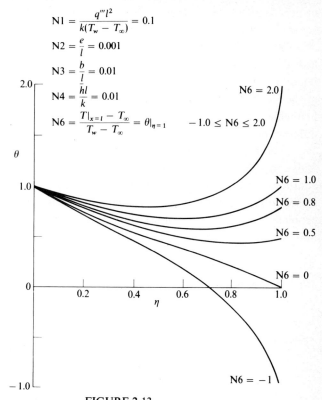

$$N1 = \frac{q''' l^2}{k(T_w - T_\infty)} = 0.1$$

$$N2 = \frac{e}{l} = 0.001$$

$$N3 = \frac{b}{l} = 0.01$$

$$N4 = \frac{hl}{k} = 0.01$$

$$N6 = \frac{T|_{x=l} - T_\infty}{T_w - T_\infty} = \theta|_{\eta=1} \qquad -1.0 \le N6 \le 2.0$$

FIGURE 2-13
Tapered fin with isothermal boundary conditions.

In order to treat the isothermal boundary condition the following modifications to the program FIN are required:

```
 145   LET N6 = VALUE DESIRED
1720   IF ABS(G2 − G1) < 1.0 E − 3 THEN 2500
2016   LET J1 = Y
2076   LET J2 = Y
2080   LET Z = (G1 * (J2 − N6) + G2 * (N6 − J1))/(J2 − J1)
2140   LET J1 = J2
3060   PRINT "N1 = "N1, "N2 = "N2, "N3 = "N3, "N4 = "N4, "N6 = "N6
```

Figure 2-13 gives the results of calculations based on the above modifications. The value of the nondimensional temperature at the end of the fin N6 can be positive or negative and greater or less than unity. The results are again for a fin with internal generation and with a ratio of base height to end height of 10 to 1.

The reader should now be able to calculate the temperature distribution and heat transfer for a variety of fin configurations. Isothermal, adiabatic, or convective boundary conditions can be used, depending on the physics of the problem. Analysis of the results should suggest further investigations which may require a change in the mathematical model, new values of the governing nondimensional parameters, or different boundary conditions.

The governing equations for the preceding fin problems have all been linear equations. Nonlinear equations can be treated in a similar manner without much additional complication. This is not the case for an analytical analysis since very few closed-form solutions to nonlinear equations can be found. In the next section the problem of a radiating fin of arbitrary profile is analyzed. The governing equation is a nonlinear equation for which no closed-form solution is known.

2-9 Radiating Fin

The analysis in this section is concerned with a fin under conditions such that the predominant mode of heat transfer from the surface is radiation. If the temperature of the fin is very high, then convection from the surface can be neglected relative to the radiation from the fin. On the other hand, if the fin is located in a region where the environmental pressure is very low, such as in outer space or in a vacuum chamber, convection will be negligible even at low temperatures. If desired, the overall co-efficient h_T which accounts for both convection and radiation can be determined as discussed in Sec. 1-8. In the analysis that follows, the convection from the fin is assumed to be negligible.

The mathematical model chosen for analysis is a fin of arbitrary profile with one-dimensional, steady-state heat conduction within an isotropic material with constant thermal properties. Uniform internal generation is assumed. The net radiation is the difference between the emitted radiation and the incoming irradiation G. To start the analysis, the first law of thermodynamics is applied to a differential closed system. This system is identical to the one shown in Fig. 2-1. The energy conducted through the left face of the differential system is given by Fourier's law

$$q = -k[2y(x)]\frac{dT}{dx}$$

The energy conducted through the right face is

$$q = -\left(k[2y(x)]\frac{dT}{dx} + \frac{d}{dx}\left\{k[2y(x)]\frac{dT}{dx}\right\}dx\right)$$

The internal generation of thermal energy within the system is

$$q = q'''[2y(x)]\,dx$$

The energy crossing the exposed surfaces of the differential element by thermal radiation must be expressed in terms of the physical laws of radiation. As shown in Chap. 1 the thermal radiation emitted from a surface due to its absolute temperature is proportional to the fourth power of the temperature. When radiation between two surfaces is being considered, the net radiation between the surfaces is a function of the geometry and orientation of the surfaces as well as their absolute temperatures. A shape factor $\mathcal{F}_{1-2}$ is used to account for both the surface resistance to radiation, which is a function of surface emittance, and the space resistance to radiation, which is a function of the amount of radiation emitted by one surface that strikes a second surface.

The equation for the net heat transfer between two gray surfaces is given by Eq. (1-47)

$$q_{net} = A_1 \mathcal{F}_{1-2}\, \sigma(T_1{}^4 - T_2{}^4)$$

For the special case of a small gray body entirely enclosed by black surroundings $\mathcal{F}_{1-2} = \varepsilon_1$ by Eq. (1-50). We will assume that the radiating fin fits this situation. The radiation leaving the top and bottom surfaces is then written

$$q = 2\varepsilon_1 \, \frac{dx}{\cos\phi}\, \sigma T^4 \qquad (2\text{-}46)$$

If we assume that the incoming radiation G is the same for both surfaces and is not a function of x, the radiation absorbed by the fin is

$$q = 2\alpha_1 \, \frac{dx}{\cos\phi}\, G \qquad (2\text{-}47)$$

where α_1 is the absorptance of the fin surface.

The emittance of the fin surface is a function of the surface temperature, and the absorptance is a function of the wavelength of the incoming radiation, which in turn depends upon the temperature of the radiation source. It is useful to define an effective source temperature by equating the absorbed radiation at the fin surface to the emitted radiation from an effective source temperature

$$\alpha_1 G \equiv \varepsilon_1 \sigma T_s{}^4 \qquad (2\text{-}48)$$

where T_s is the effective source temperature.

The first law of thermodynamics states that the net heat transfer must equal zero for steady-state conditions. Thus we have for the differential element

$$\frac{d}{dx}\left\{k[2y(x)]\frac{dT}{dx}\right\}dx + q'''[2y(x)]\,dx + \frac{2\varepsilon_1\sigma\,dx}{\cos\phi}[T_s{}^4 - T^4] = 0$$

Since the thermal conductivity is constant, this equation reduces to

$$2ky(x)\frac{d^2T}{dx^2} + 2k\frac{dy(x)}{dx}\frac{dT}{dx} + q'''[2y(x)] - \frac{2\varepsilon_1\sigma}{\cos\phi}(T^4 - T_s^4) = 0 \qquad (2\text{-}49)$$

Since we no longer have convective heat transfer which depends on the environmental temperature T_∞, we define the nondimensional temperature ratio θ as T/T_w. Again we use the nondimensional independent variable $\eta = x/l$. Using these definitions it can be shown that

$$\frac{dT}{dx} = \frac{T_w}{l}\frac{d\theta}{d\eta} \qquad \frac{d^2T}{dx^2} = \frac{T_w}{l^2}\frac{d^2\theta}{d\eta^2}$$

If one considers a tapered fin with the profile given by

$$y(x) = \frac{e}{2} + (l-x)\left(\frac{b}{2}-\frac{e}{2}\right)\bigg/l$$

the governing equation becomes

$$\left[e + \frac{(l-x)(b-e)}{l}\right]\frac{T_w}{l^2}\frac{d^2\theta}{d\eta^2} - \frac{b-e}{l}\frac{T_w}{l}\frac{d\theta}{d\eta}$$

$$+\frac{q'''}{k}\left[e + \frac{(l-x)(b-e)}{l}\right] - \frac{2\varepsilon_1\sigma}{k\cos\phi}(T^4 - T_s^4) = 0$$

Rearranging we have

$$\left[\frac{e}{l} + \left(\frac{b}{l}-\frac{e}{l}\right)(1-\eta)\right]\frac{d^2\theta}{d\eta^2} - \left(\frac{b}{l}-\frac{e}{l}\right)\frac{d\theta}{d\eta}$$

$$+\frac{q'''l^2}{kT_w}\left[\frac{e}{l} + \left(\frac{b}{l}-\frac{e}{l}\right)(1-\eta)\right] - \frac{2\varepsilon_1\sigma l T_w^3}{k\cos\phi}(\theta^4 - \theta_s^4) = 0$$

Solving for the highest derivative gives

$$\frac{d^2\theta}{d\eta^2} = \frac{(2\varepsilon_1\sigma l T_w^3/k\cos\phi)(\theta^4 - \theta_s^4) + (b/l - e/l)(d\theta/d\eta)}{e/l + (b/l - e/l)(1-\eta)} - \frac{q'''l^2}{kT_w} \qquad (2\text{-}50)$$

The nondimensional parameters that govern the solution are now obvious. For the purpose of this analysis we define

$$N1 = \frac{q'''l^2}{kT_w}$$

$$N2 = \frac{e}{l}$$

$$N3 = \frac{b}{l}$$

$$\qquad\qquad (2\text{-}51)$$

$$N7 = \frac{\varepsilon_1\sigma l}{k}T_w^3$$

The first three nondimensional parameters have the same form as those for a fin with only convection from the surface. The fourth parameter arises due to the radiation from the fin surface. A fifth parameter can also be defined as θ_s. This is a measure of the incoming irradiation. Based on the definition of the effective source temperature, $\theta_s{}^4$ is given as

$$\theta_s{}^4 = \frac{T_s{}^4}{T_w{}^4} = \frac{\alpha_1 G}{\varepsilon_1 \sigma T_w{}^4} = N8 \qquad (2\text{-}52)$$

Equation (2-50) can then be written

$$\frac{d^2\theta}{d\eta^2} = \frac{2(N7)(\theta^4 - N8)/\cos\phi + (N3 - N2)(d\theta/d\eta)}{N2 + (N3 - N2)(1 - \eta)} - N1 \qquad (2\text{-}53)$$

It should be observed that the resulting differential equation (2-53) is nonlinear due to the θ^4 term. This distinction is the major one between this equation and the governing equation for fins with only convection from the surface. The equation is still second-order, and two boundary conditions are required to specify the problem. If an adiabatic end boundary condition is appropriate, the two boundary conditions are

$$\begin{aligned} &\text{at} \quad \eta = 0 \quad &&\theta = 1.0 \\ &\text{at} \quad \eta = 1.0 \quad &&\frac{d\theta}{d\eta} = 0 \end{aligned} \qquad (2\text{-}54)$$

Results for the radiating fin obtained from computer solutions to Eq. (2-50) with an adiabatic end boundary condition, Eq. (2-54), are shown in Figs. 2-14 and 2-15. Modifications to the FIN program used to obtain these results are listed below:

```
140   LET N7 = (VALUE DESIRED)
145   LET N8 = 0
1082  LET D = N2 + (N3 − N2) * (1 − X)
1085  LET G = (((2 * N7 * (Y ↑ 4 − N8))/COS(A) + (N3 − N2) * Z)/D) − N1
1165  Same as 1082
1170  Same as 1085
3060  PRINT "N1 = "N1, "N2 = "N2, "N3 = "N3
3070  PRINT
3080  PRINT "N7 = "N7, "N8 = "N8
```

The results in Fig. 2-14 are for a fin of rectangular profile with $N2 = N3 = 0.02$, no internal generation ($N1 = 0$), and no incoming radiation ($N8 = 0$). The effect of varying the radiation parameter N7 over two orders of magnitude is shown. Physically

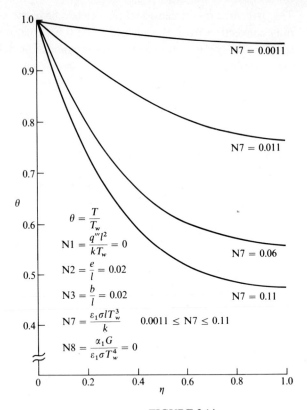

FIGURE 2.14
Radiating fin with N7 varying.

this parameter can be interpreted as the ratio of radiated energy to conducted energy. The results given agree with other published solutions for nontapered radiating fins without internal generation. (One such solution is given on p. 393 of Ref. 2-7.) The reader can now interact with the mathematical model to study the effect of internal generation by varying N1, the effect of incoming radiation by varying N8, and the effect of geometry by varying N2 or N3. The mathematical model may be changed to study effects such as fin configuration or variable properties.

One can define an ideal heat transfer using the base temperature of a fin of unit depth as

$$q_{ideal} = 2l\varepsilon_1\sigma T_w^4 \qquad (2\text{-}55)$$

For example, if $l = 0.25$ ft (0.0763 m), $\varepsilon_1 = 0.16$, $k = 50$ Btu/h-ft-°F (86.5 W/m-°C), $b = 0.005$ ft (0.00152 m), and $T_w = 2,000°R$ (1,110 K), we calculate

$$q_{ideal} = 2.20 \times 10^3 \text{ Btu/h} = 6.45 \times 10^2 \text{ W}$$

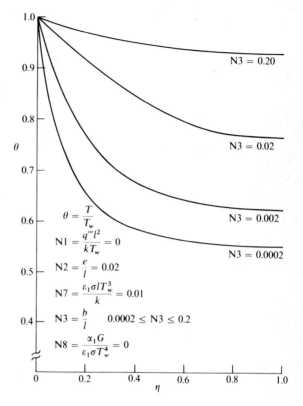

The figure contains the following labels:

$$\theta = \frac{T}{T_w}$$

$$N1 = \frac{q'''l^2}{kT_w} = 0$$

$$N2 = \frac{e}{l} = 0.02$$

$$N7 = \frac{\varepsilon_1 \sigma l T_w^3}{k} = 0.01$$

$$N3 = \frac{b}{l} \qquad 0.0002 \le N3 \le 0.2$$

$$N8 = \frac{\alpha_1 G}{\varepsilon_1 \sigma T_w^4} = 0$$

$$N3 = 0.20$$
$$N3 = 0.02$$
$$N3 = 0.002$$
$$N3 = 0.0002$$

FIGURE 2-15
Radiating tapered fin with N3 varying.

and
$$N7 = 0.011$$

The nondimensional temperature gradient at the base of the fin obtained from the solution given in Fig. 2-14 for $N7 = 0.011$ was -0.5673. The actual heat transfer through the fin is calculated from Fourier's law:

$$q = -k(b)(1)\left(\frac{dT}{dx}\right)_{x=0} = -kb\,\frac{T_w}{l}\left(\frac{d\theta}{dx}\right)_{\eta=0}$$

Thus,
$$q = 1.13 \times 10^3 \text{ Btu/h}$$
$$= 3.31 \times 10^2 \text{ W}$$

The fin efficiency is defined as the ratio of q to q_{ideal}

$$\eta = \frac{q}{q_{ideal}} = 0.514 \qquad (2\text{-}56)$$

Similar profiles which show the effects of taper and internal generation can be obtained by the reader. Figure 2-15 shows the results for a tapered fin obtained by varying N3 between 0.20 and 0.0002. Since N2 was fixed at 0.02, the results for N3 < 0.02 are for fins of a reverse taper where the end height is larger than the base height. The nondimensional temperature gradients corresponding to the results given in Figs. 2-14 and 2-15 are given in Table 2-4.

2-10   Radiating Fin-tube Assembly

Another modification to a radiating fin analysis includes the effects of radiation from surrounding structures. Consider the fin-tube radiator assembly shown in Fig. 2-16. Provided that the additional assumption of black surfaces ($\alpha_1 = \varepsilon_1 = 1$) is made, the only change required in the analysis is the addition of appropriate shape factors between the fin and surrounding structures. For a black surface, no radiation will be reflected between the surfaces. The surface areas of adjacent tubes and fins seen by the differential area dA_x indicated in Fig. 2-16 are denoted by A_1 and A_2, respectively. The radiation falling on the top and bottom surface of the differential area is

$$q = 2A_1 F_{A_1 - dA_x}\, \sigma T_w{}^4 + 2A_2\, F_{A_2 - dA_x}\, \sigma T_w{}^4 \qquad (2\text{-}57)$$

The term $F_{A_1 - dA_x}$ is the shape factor for blackbody radiation and is discussed in Chap. 1. When the energy contribution expressed by Eq. (2-57) is added to the energy

Table 2-4   NONDIMENSIONAL TEMPERATURE GRADIENTS FOR RADIATING FINS

Fig. 2-14: N1 = 0, N2 = 0.02, N3 = 0.02, N8 = 0

N7 =	0.0011	0.011	0.06	0.110
$\left(\dfrac{d\theta}{d\eta}\right)_{\eta=0} =$	−0.0969	−0.5673	−1.5089	−2.0716

Fig. 2-15: N1 = 0, N2 = 0.02, N7 = 0.01, N8 = 0

N3 =	0.0002	0.002	0.02	0.2
$\left(\dfrac{d\theta}{d\eta}\right)_{\eta=0} =$	−14.9771	−2.6742	−0.5340	−0.0861

Fig. 2-18: N1 = 0, N2 = 0.02, N3 = 0.02, N7 = 0.011, N8 = 0

N9 =	0.1	0.5	1.0	5.0
$\left(\dfrac{d\theta}{d\eta}\right)_{\eta=0} =$	−0.5427	−0.4788	−0.4269	−0.2646

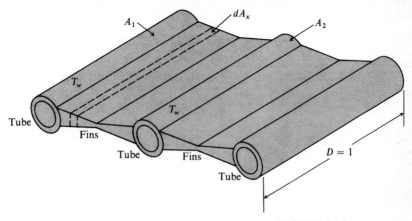

FIGURE 2-16
Fin-tube radiator assembly.

terms which occur in the derivation of the governing differential equation for the temperature distribution in the fin, we have

$$\frac{d}{dx}\left\{ k[2y(x)]\frac{dT}{dx}\right\} dx + q'''[2y(x)]\,dx$$

$$+ \frac{2\sigma\,dx}{\cos\phi}(T_s^4 - T^4) + 2\sigma T_w^4(A_1 F_{A_1-dA_x} + A_2 F_{A_2-dA_x}) = 0 \qquad (2\text{-}58)$$

The term $A_1 F_{A_1-dA_x}\sigma T_w^4$ is the total energy radiated by the tube surface area A_1 seen by dA_x which falls directly on dA_x. In a similar manner, the total energy radiated by dA_x which falls directly on A_1 is $dA_x F_{dA_x-A_1}\sigma T^4$. The net radiation between A_1 and dA_x may be written as either

$$q = A_1 F_{A_1-dA_x}\sigma(T_w^4 - T^4) \qquad (2\text{-}59)$$

or

$$q = dA_x F_{dA_x-A_1}\sigma(T_w^4 - T^4)$$

since $A_1 F_{A_1-d_x} = dA_x F_{dA_x-A_1}$ by use of the reciprocity relation. For the fin-tube assembly of unit depth we can write

$$A_1 F_{A_1-dA_x} = dA_x F_{dA_x-A_1} = dx(1)F_{dA_x-A_1} \qquad (2\text{-}60)$$

and

$$A_2 F_{A_2-dA_x} = dx F_{dA_x-A_2} \qquad (2\text{-}61)$$

For constant thermal conductivity, the equation reduces to

$$2ky(x)\frac{d^2T}{dx^2} + 2k\frac{dy(x)}{dx}\frac{dT}{dx} + q'''[2y(x)]$$

$$- \frac{2\sigma}{\cos\phi}(T^4 - T_s^4) + 2\sigma T_w^4(F_{dA_x-A_1} + F_{dA_x-A_2}) = 0 \qquad (2\text{-}62)$$

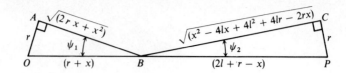

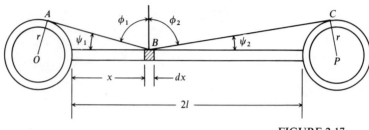

FIGURE 2-17
Assembly end view.

In terms of the nondimensional variables, the governing equation is

$$\frac{d^2\theta}{d\eta^2} =$$

$$\frac{[2(N7)/\cos\phi][\theta^4 - N8 - \cos\phi(F_{dA_x-A_1} + F_{dA_x-A_2})] + (N3 - N2)(d\theta/d\eta)}{N2 + (N3 - N2)(1 - \eta)} - N1 \quad (2\text{-}63)$$

This equation is the governing one for the tapered, radiating fin-tube assembly shown in Fig. 2-16, provided that the taper angle is small. If the taper angle is not small, the radiation emitted from the surface of the adjacent fin will also affect the steady-state temperature distribution. We consider rectangular fins in the subsequent analysis.

The shape factors for a fin-tube radiator assembly consisting of rectangular fins are given in Table 1-2.

$$F_{dA_x-A_1} = 0.5(1 - \sin\phi_1) \quad (2\text{-}64)$$

$$F_{dA_x-A_2} = 0.5(1 - \sin\phi_2) \quad (2\text{-}65)$$

The angles ϕ_1 and ϕ_2 are defined in Fig. 2-17. Two right triangles (AOB and CPB) can be used to express $\sin\phi_1 = \cos\psi_1$ and $\sin\phi_2 = \cos\psi_2$. From Fig. 2-17 it can be seen that

$$\sin\phi_1 = \cos\psi_1 = \frac{\sqrt{2rx + x^2}}{r + x} \quad (2\text{-}66)$$

$$\sin\phi_2 = \cos\psi_2 = \frac{\sqrt{x^2 - 4lx + 4l^2 + 4lr - 2rx}}{2l + r - x} \quad (2\text{-}67)$$

The shape factors vary with x as the integration is carried out over the fin length.

Table 2-5   ADDITIONAL
VARIABLES FOR
RADIATING FIN-
TUBE ASSEMBLY

Variable	Definition
D	$N2 + (N3 - N2)(1 - \eta)$
F2	Shape factor $F_{dA_x - A_1}$
F3	Shape factor $F_{dA_x - A_2}$
N7	Parameter $\sigma l T_w^3 / k$
N8	Parameter $\alpha_1 G / \sigma T_w^4$
N9	Parameter r/l
S1	$\sin \phi_1$ (Eq. (2-66))
S2	$\sin \phi_2$ (Eq. (2-67))

It is now convenient to introduce a nondimensional parameter $N9 = r/l = \xi$ along with $\eta = x/l$. One can then write

$$\sin \phi = \frac{\sqrt{2\xi\eta + \eta^2}}{\xi + \eta} \qquad (2\text{-}68)$$

$$\sin \phi_2 = \frac{\sqrt{\eta^2 - 4\eta + 4 + 4\xi - 2\xi\eta}}{2 + \xi - \eta} \qquad (2\text{-}69)$$

The necessary computer program to obtain the nondimensional solution to the nonlinear equations for a radiating fin-tube assembly can now be written. Suggested modifications to program FIN are listed below. In addition to the variables identified in Table 2-2, the computer variables in Table 2-5 are used.

Modifications used to obtain the results shown in Fig. 2-18 are

```
140   LET N7 = 0.011
145   LET N8 = 0
148   LET N9 = 5
150   LET F1 = ABS(0.5 * (N3 − N2))
160   LET A  = ATN (F1)
1080  LET S1 = (2 * N9 * X + X * X) ↑ 0.5/(N9 + X)
1082  LET S2 = (X * X − 4 * X + 4 + 4 * N9 − 2 * N9 * X) ↑ 0.5/(2 + N9 − X)
1085  LET F2 = 0.5 * (1 − S1)
1086  LET F3 = 0.5 * (1 − S2)
1088  LET D  = N2 + (N3 − N2) * (1 − X)
1090  LET F  = Z
```

(Continued on next page)

```
1095   LET G  = (((2 * N7 * (Y ↑ 4 − N8 − COS(A) * (F2 + F3)))/COS(A)) +
                 (N3 − N2) * 2)/D − N1
1160   = 1080
1162   = 1082
1164   = 1084
1166   = 1086
1168   = 1088
1170   = 1090
1180   = 1095
3060   PRINT "N1 = "N1, "N2 = "N2, "N3 = "N3
3070   PRINT
3080   PRINT "N7 = "N7, "N8 = "N8, "N9 = "N9
```

Results for fins of rectangular profile are given in Fig. 2-18, where the parameter N9 has been varied. These results indicate the effect of increasing the ratio of tube radius to fin length. As the ratio $r/l = \xi$ increases, the temperature gradient at the base of the fin decreases, and this decreases the heat transfer from the fin. Thus a small value of the tube radius improves the efficiency of the fins. However, a smaller flow area for the working fluid in the tubes increases the pressure drop and pumping power required

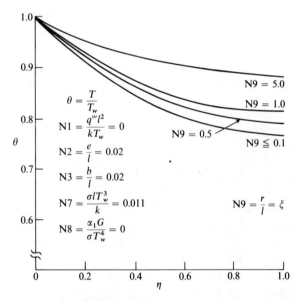

FIGURE 2-18
Radiating assembly profiles with N9 varying.

by the system. The nondimensional wall temperature gradients corresponding to the results given in Fig. 2-18 appear in Table 2-4. The results in Fig. 2-14 can also be obtained by setting $\xi = r/l = \frac{1}{2}b/l = \frac{1}{2}N3 = 0.01$. For this value of ξ the diameter of the tube would equal the thickness of the fin, and no radiation would fall on the fin from the tube surface.

Many other modifications may be necessary for practical problems. For example, the wall temperature of the tubes has been assumed constant. If the working fluid inside the tubes is changing phase, then this assumption is good. On the other hand, the heat transfer from the fluid often causes a significant temperature gradient along the tube axis. This problem is discussed in Chap. 7. The fin-tube assembly may be studied for various thermodynamic cycles rejecting thermal energy through the fins. It is usually necessary to design fins of minimum weight for space applications, and this requirement may dictate the working fluid, material, geometry, and boundary conditions for the fin. The reader is referred to Refs. 2-9 through 2-12 for further discussion on the subject.

2-11   Fin Design

It is important to be aware of the existing analytical techniques for finding optimum fin designs. In Ref. 2-4, it is shown that high fin efficiencies are possible by designing a fin of inverse parabolic profile since a larger portion of the material is concentrated near the base of the fin where the temperature is highest. It is also shown that to produce the same heat transfer rate, a triangular fin requires only 4.0 percent more material than the optimum parabolic fin of least profile area. On the other hand, a rectangular fin requires over 50 percent more material to produce the same effect. Costs of manufacturing and maintenance usually dictate straight fin surfaces such as rectangular, triangular, or trapezoidal fins.

The total weight of a fin is proportional to the ratio of the material density to thermal conductivity. It is common to see aluminum fins on heat exchangers and engine cylinder heads. The reason for this choice becomes obvious when one compares the weight requirements of aluminum to another material such as copper:

$$\frac{(\rho/k)_{Al}}{(\rho/k)_{Cu}} = \frac{1.43}{2.50} = 0.57$$

This equation indicates that approximately 43 percent of the fin weight can be saved by using aluminum rather than copper even though the thermal conductivity of copper is much higher than aluminum. Savings of up to 90 percent of fin weight are realized when aluminum is used to replace such metals as nickel and iron.

Equations for the optimum profile area of fins of various profiles are given in Ref. 2-4. They take the form

$$A_p = \frac{C}{h^2 k} \left(\frac{q_w}{T_w} \right)^3 \qquad (2\text{-}70)$$

where C is a constant which depends upon the fin profile. The volume of a fin per unit depth is equal to A_p, the profile area. If the heat transfer to be removed by the fin is doubled, the designer must either increase the profile area by a factor of 8, as indicated in Eq. (2-70), or provide two fins instead of one. This latter choice is more attractive, and it explains why fin arrays usually consist of thin, closely spaced fins. There is an obvious limitation on the fin spacing which is governed by the requirement of a convective flow field around the fins. This is a function of the boundary-layer thickness that develops between the fin surfaces. Boundary-layer flows are discussed in Chaps. 5 and 6.

The thickness of a fin varies inversely with the thermal conductivity. Thus, copper may be more advantageous than aluminum if fin weight is no problem. Structural requirements also become important in thin fins. An increase in the convection heat transfer coefficient reduces the thickness and material requirements of a fin. Since the convection coefficient depends on the velocity of the flow over the fin, different fin thicknesses should be used for large and small velocities. The optimum performance of a fin can occur only at one velocity.

The numerical approach discussed above can be applied to many similar fin problems. An analysis of heat transfer from cylindrical surfaces by circular fins is a good example. Cylindrical coordinates can be used to specify the temperature distribution as a function of radius. The reader is referred to Refs. 2-4 and 2-5 for the derivation of the governing equations for circular fins of various profiles. Sometimes it is more effective to extend a surface by means of pins or spines rather than by fins. This method creates more exposed surface area for heat transfer by convection. However, the value of the convection coefficient will also change. Reference 2-5 provides further discussion on pins and spines.

Although the discussion has been based on the classical definition of a fin, it should be realized that the formulation of many heat transfer problems results in the same boundary-value problem even though the system being analyzed would not normally be considered a fin. Heat transfer problems illustrating this fact arise in the following physical situations:

1 a turbine blade fixed at one end to the rotor and surrounded by hot gases
2 a stirring device with one end submerged in hot liquid and the upper surface exposed to the environment

3 a steel rod being heated in a furnace in such a manner that an axial gradient exists within the rod

4 a fever thermometer with one end in contact with body tissues and the surface area of the glass exposed to the environment

5 a structural member attached at both ends to hot surfaces and its surface area exposed to the environment

This chapter has been concerned with problems which have been governed by ordinary differential equations, both linear and nonlinear. These equations were obtained because of the ubiquitous assumption of one-dimensional, steady-state heat conduction within the fins. In the next chapter we deal with problems which must be analyzed to allow for two-dimensional conduction. Because of the existence of more than one independent variable, the resulting governing differential equations will be partial differential equations rather than ordinary differential equations. The objective will continue to be the presentation of numerical methods which will aid in the formulation, analysis, and understanding of thermal systems, and complement the classical analytical methods of solution.

References

2-1 KREITH, F.: "Principles of Heat Transfer," 2d ed., International Textbook, Scranton, Pa., 1966.

2-2 REYNOLDS, W. C.: "Thermodynamics," McGraw-Hill, New York, 1965.

2-3 KRAUS, A. D.: "Extended Surfaces," Spartan Books, New York, 1964.

2-4 SCHNEIDER, P. J.: "Conduction Heat Transfer," Addison-Wesley, Reading, Mass., 1957.

2-5 JAKOB, M.: "Heat Transfer," vol. 1, Wiley, New York, 1959.

2-6 HOLMAN, J. P.: "Heat Transfer," 3d ed., McGraw-Hill, New York, 1971.

2-7 JAMES, M. L., G. M. SMITH, and J. C. WOLFORD: "Analog and Digital Computer Methods in Engineering Analysis," International Textbook, Scranton, Pa., 1965.

2-8 SPARROW, E. M., and R. D. CESS: "Radiation Heat Transfer," Brooks/Cole, Belmont, Calif., 1966.

2-9 SPARROW, E. M., V. K. JONSSON, and W. J. MINKOWYCZ: Heat Transfer from Fin-tube Radiators Including Longitudinal Heat Conduction and Radiant Interchange between Longitudinally Nonisothermal Finite Surfaces, NASA TN D-2077, December 1963.

2-10 KREBS, R. P., H. C. HALLER, and B. M. AUER: Analysis and Design Procedures for a Flat, Direct-condensing Central Finned-tube Radiator, NASA TN D-2474, September 1964.

2-11 SAULE, A. V., R. P. KREBS, and B. M. AUER: Design Analysis and General Characteristics of Flat-plate Central-fin-tube Sensible Heat Space Radiators, NASA TN D-2839, 1965.

2-12 AUER, B. M., and A. V. SAULE: Computer Program Details for Design of Sensible Heat Space Radiators, NASA TN D-2840, June 1965.

Problems[1]

2-1 The governing equation for a rectangular fin with no internal generation is given by Eq. (2-19).

(a) Define $\eta = x/l$, and express the governing equation in nondimensional form. Identify the resulting nondimensional parameters.

(b) Show that the solution to Eq. (2-19) is given by Eq. (2-21) for an adiabatic boundary condition at the end of the fin. What is the corresponding solution in terms of the nondimensional independent variable η?

(c) Obtain two additional solutions to the nondimensional equation obtained in part (a) for the following two sets of isothermal boundary conditions:

1	at	$\eta = 0$	$\theta = 1$	2	at	$\eta = 0$	$\theta = 1$
	at	$\eta = 1$	$\theta = 2$		at	$\eta = 1$	$d\theta/d\eta = 0.1\theta$

(d) What is the governing equation for a rectangular fin with uniform internal generation? Express the solution to this equation in terms of the sum of a homogeneous and a particular solution. Evaluate the two integration constants for an adiabatic end boundary condition.

2-2 A load-bearing fin with uniform internal generation has been designed such that $b/l = e/l = 0.01$. Because of structural problems it is necessary to increase the end height e by a factor of 10. How much will this change increase the temperature at the end of the fin? The values of N1, N3, and N4 are equal to those in Fig. 2-6, so this figure may be used to obtain a solution. Let $T_w = 500°C$ and $T_\infty = 100°C$. Obtain the nondimensional gradients $(d\theta/d\eta)_{\eta=0}$ from Table 2-3. What is the change in heat transfer at the fin base due to the ratio of e/l? The length of the fin remains fixed at 2.0 cm.

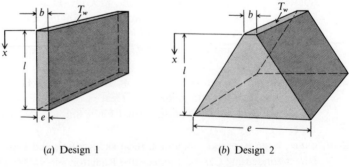

(a) Design 1   (b) Design 2

FIGURE 2-2P
Design of load-bearing fins.

[1] Problems whose numbers are followed by a superscript italic c should be analyzed by obtaining solutions on a digital computer.

2-3 A reverse-tapered fin has the value of $e/l = 0.05$, and $b/l = 0.01$. Because of nuclear fission within the material, a value of $N1 = q'''l^2/k(T_w - T_\infty) = 0.1$ exists. The length l is 0.5 ft, and $k = 100$ Btu/h-ft-°R. During normal operation the fin is being cooled by forced convection of an inert gas which produces a convection coefficient of $\bar{h} = 20$ Btu/h-ft²-°R, with $T_w = 800°F$ and $T_\infty = 200°F$. What is the temperature at $x = l$ under steady-state conditions? (Use Fig. 2-7.) Without forced convection the convection coefficient is reduced by a factor of 10. What is the temperature at $x = l$ for this situation? How small can $\bar{h}$ be before heat transfer occurs from the fin to the wall?

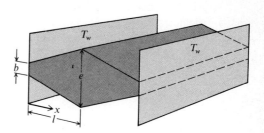

FIGURE 2-3P
Reverse-tapered fin.

2-4 The solution for a parabolic, concave fin is given by Fig. 2-9. For the fixed values of N2, N3, and N4 indicated in the figure, analyze the following situation. (Notice that the effect of internal generation is small for small values of N1, but that it increases rapidly as N1 increases).

(a) Give three possible fin designs for a fixed geometry which will give an approximate uniform temperature distribution throughout the fin.

(b) What is the percent change in the temperature at the end of the fin when N1 changes by an order of magnitude from 0.1 to 1.0?

(c) What is the percent change in the temperature at the end of the fin when N1 changes by an order of magnitude from 0.5 to 5.0?

(d) How much heat transfer per unit area must be removed by the wall to maintain steady-state conditions when $N1 = 5.0$? The conductivity of the fin is $k = 50.0$ W/m-K, and $l = 0.05$ m. The temperatures are $T_w = 200°C$ and $T_\infty = 0.0°C$.

2-5 The spacing of parabolic fins on a flat wall is being considered. Consider the effect of varying the ratio b/l for fins of unit depth. Use the results presented in Fig. 2-10. For the value of $0.001 \leq b/l \leq 0.1$, plot the results of the total heat transfer at the wall (Btu/h) vs. the ratio b/l. Obtain the nondimensional gradients from Table 2-3. Use $k = 100$ Btu/h-ft-°R, $T_w = 500°F$, $T_\infty = 1,000°F$, and $l = 0.10$ ft. Discuss the results and explain any maximum or minimum values that occur. Can this figure be used to study the effect of varying l and holding b constant? If so, explain how.

2-6 A double-wall container encloses a flowing, biological liquid at 200 K. Supporting fins which can supply energy to the liquid from electrical dissipation within the tapered fins are located between the walls. The ends of the fins are in contact with the inner liquid container and are isothermal at a temperature equal to the liquid temperature. Assume that steady-state conditions exist, and use Fig. 2-13 to analyze the problem.

(a) At what location in the fin will $T = T_\infty$ if $T_w = 600$ K and $T_\infty = 400$ K?

(b) If T_∞ is reduced to 200 K, what will be the temperature within the fin at $\eta = 0.5$?

(c) To what must T_∞ be reduced for zero heat transfer across the end of the fin? Is this physically possible?

(d) If T_w is cooled to the liquid temperature and $T_\infty = 600$ K, what will be the highest temperature in the fin, and where will it occur?

(e) Under the conditions given in part (b), in which direction is the heat transfer at $\eta = 1.0$? Under the conditions given in part (d), in which direction is the heat transfer at $\eta = 1.0$? Comparing the conditions of parts (b) and (d), in which case will the heat transfer across the end of the fin be the larger?

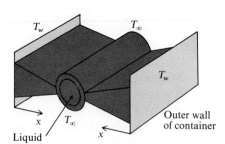

FIGURE 2-6P
Double-wall container.

2-7   A fin to dissipate the energy removed from a working fluid during condensation is being designed for an outer space vehicle. Assume that the operation conditions are specified by the nondimensional parameters given in Fig. 2-15. Use this figure to investigate the thermal effect of changing from a rectangular fin with N2 = N3 to a tapered fin of the same material with N3/N2 = 10.0. Massive fins are necessary for this type of application to provide the necessary surface area. The length is fixed at 10.0 ft, and $T_w = 700°$R. Assume a gray fin surface ($\alpha_1 = \varepsilon_1 = 0.4$). Calculate the heat transfer removed per unit area for each fin design, and compare the end fin temperatures. What is the ratio of the total heat transfer rates? What is the weight ratio of the two fins?   (See Table 2-4.)

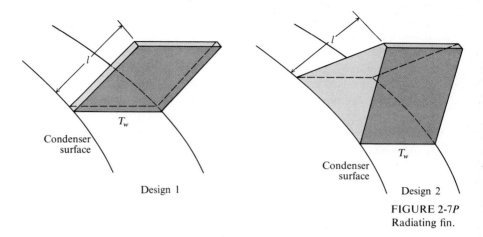

Design 1

Design 2

FIGURE 2-7P
Radiating fin.

2-8   Two radiating fin designs are shown in Fig. 2-8P. The only difference between the
two is the ratio r/l. It is necessary to build several designs and vary the value of r to
meet the different flow rates and the Reynolds number requirements for the flow
through the central tube. In design 1 the ratio $r/l = 0.1$, and in design 2 $r/l = 1.0$.
Neglect radiation from adjacent fins and the environment, but include radiation from
the tube to the fin. All radiating surfaces are black ($\alpha_1 = \varepsilon_1 = 1.0$). Use the results
given in Fig. 2-18 to compare the heat transfer removed by the four fins for designs
1 and 2. The thermal conductivity of the fin material is 50 Btu/h-ft-°R, and the fin
length is 1.0 ft. What is the radiation emitted by the tube surface? For each design,
compare the total heat transfer per unit length from a tube with four fins to the radiation from a bare tube. What is the percent increase in the total heat transfer due to the
addition of fins for each design?   (See Table 2-4.)

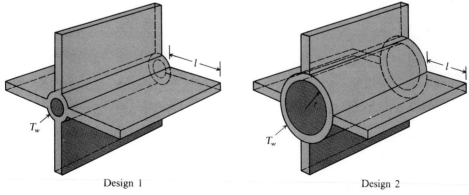

Design 1

Design 2

FIGURE 2-8P
Radiating fin-tube assembly.

2-9 Because of economic reasons, the following question has been raised: Can the rectangular aluminum cooling fins in a piece of electronic equipment be replaced by plastic fins which are molded in the shape of tapered fins with 10^{-2} ft $\leq e \leq 10^{-4}$ ft? The existing fin design is shown in Fig. 2-9P. A fan circulates air over the fins and creates an average value of 20.0 Btu/h-ft²-°R for the convection coefficient; and $T_w = 190°F$, $T_\infty = 70°F$, and $k = 80$ Btu/h-ft-°R.

(a) Calculate the total heat transfer removed by the existing fin assembly by using the analytical solution for an insulated end boundary condition.

(b) Assume that the existing rectangular fins are to be replaced by plastic tapered fins ($k = 15.0$ Btu/h-ft-°R) with the same base width and length as the rectangular fins. Calculate the total heat transfer for a proposed fin assembly, by either an analytical or numerical method.

(c) What temperature difference $T_w - T_\infty$ must be maintained for the proposed fin assembly to give the same heat transfer obtained with the aluminum fins?

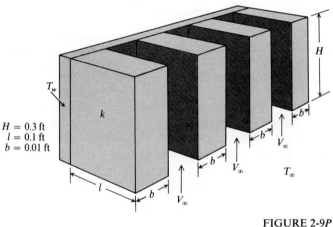

$H = 0.3$ ft
$l = 0.1$ ft
$b = 0.01$ ft

FIGURE 2-9P
Cooling fins.

2-10 Structural members support two isothermal surfaces as shown in Fig. 2-10P. The members lose thermal energy by convection to the air gaps with $\bar{h} = 5.0$ W/m²-K. Treat the structural members as trapezoidal fins with isothermal boundary conditions at each end. The thermal conductivity of the steel members is 50 W/m-K.

(a) Show that the steady-state temperature distribution in the fins can be assumed to be one-dimensional since Bi < 0.1.

(b) Calculate the total heat transfer per unit assembly depth when $T_w = 300°C$, $T_\infty = 150°C$; $\bar{h} = 5.0$ W/m²-K, 10.0 W/m²-K, and 15.0 W/m²-K.

(c)ᶜ Discuss what effects on q_w and T_e would be expected if the cross-sectional shape were changed to a concave shape, with the same values of b and e. Obtain numerical solutions to support your discussion.

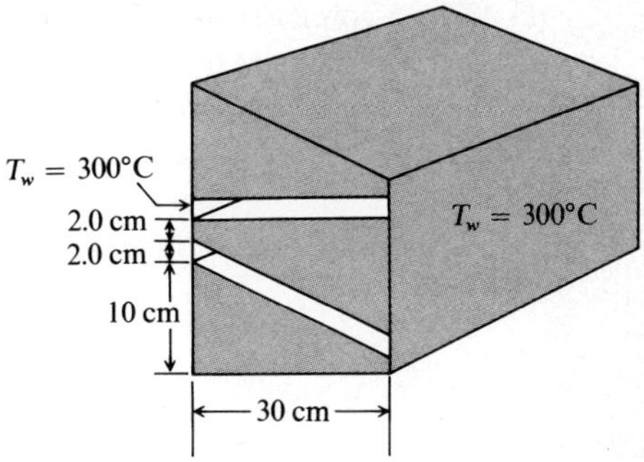

FIGURE 2-10P
Structural members.

$T_w = 300°C$

2.0 cm
2.0 cm

10 cm

$T_w = 300°C$

30 cm

2-11ᶜ It has been determined that the local convection heat transfer coefficient for a vertical fin surface, such as shown in Fig. 2-2P, is given by

$$h_x = 0.21(T - T_\infty)^{1/3}$$

where $T(°F)$ is the local fin surface temperature.

(a) Derive the governing equation for a vertical fin of rectangular profile using this local value of the convection heat transfer coefficient, rather than a constant average value. (The determination of local convection heat transfer coefficients is discussed in detail in Chaps. 5 and 6.)

(b) Obtain a numerical solution for the temperature distribution in the fin. Compare the results with the analytical solution obtained by assuming that the average convection heat transfer coefficient is given by

$$\bar{h} = 0.21(T_w - T_\infty)^{1/3}$$

2-12ᶜ The governing equation for a constant-property, circular fin of rectangular profile is given by

$$\frac{d^2T}{dr^2} + \frac{1}{r}\frac{dT}{dr} - \frac{2\bar{h}}{kb}T = 0 \qquad r_i \leq r \leq r_o$$

for steady-state, one-dimensional, radial conduction with no internal generation. Assume that the following two boundary conditions apply:

$$\text{at} \qquad r = r_i \qquad T = T_w$$

$$\text{at} \qquad r = r_o \qquad \frac{dT}{dr} = 0$$

(Note: As shown in Ref. 2-5, the analytical solution to this problem is expressed in terms of Bessel and Hankel functions.)

(a) Define $\bar{r} = r/r_o$ and $\theta = (T - T_\infty)/(T_w - T_\infty)$. Write the governing equation in nondimensional form.

(b) Write a computer program, and obtain nondimensional solutions for this problem for various values of the important nondimensional parameters.

(c) Compare the performance of the circular fin to a straight, rectangular fin with the same length and same surface area.

(d) Describe some design applications which would require circular fins and some which would require straight fins.

FIGURE 2-12P
Circular fin.

2-13   Consider a one-dimensional, rectangular fin with an adiabatic end boundary condition. The fin operates at high temperature, and heat transfer by both radiation and convection from the fin is important. Assume that the gray body shape factor is equal to the emittance of the fin surface ($\mathscr{F}_{1-2} = \varepsilon_1$). Starting from basic conservation principles, derive the governing differential equation which specifies the steady-state temperature distribution within the fin. List all assumptions which are inherent in the mathematical model. Define appropriate nondimensional parameters, and put the governing equation in nondimensional form.

2-14^c Consider the thermal design of a rear supporting strut for a small gas turbine hydrofoil boat. The base of the strut is attached to the exhaust section of the turbine. Under steady-state operating conditions the base of the strut will be at a temperature of 300°C. The tip of the strut will always be submerged and may be considered isothermal at a temperature equal to the water temperature. The thermal conductivity of the steel alloys to be used in construction is expected to be between 60.0 and 90.0 W/m-K. The length of the strut is 0.3 m, and the end thickness of the strut must be at least 6.0 cm to attach to the hydrofoil surface. The base thickness of the strut can vary between 6.0 cm and 24.0 cm. Preliminary design sketches indicate that the strut will be shielded from the air stream. Therefore an average convection heat transfer coefficient can be assumed to equal 15.0 W/m²-K. Structural considerations require

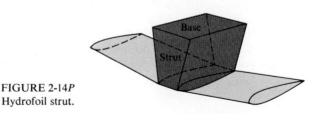

FIGURE 2-14P
Hydrofoil strut.

the determination of the temperature gradients that will occur within the strut for various designs. Make this analysis for several possible designs.

2-15ᶜ Because of the variation of electrical resistance with temperature, the effect of internal generation due to electrical dissipation is sometimes more properly analyzed by using a mathematical model which includes a variable internal generation term. For most conductors, the generation can be assumed to vary linearly with temperature, and one can write

$$\frac{q'''}{k} = \frac{q_o'''}{k}(1 + \beta T)$$

where β is a temperature coefficient of the electrical resistivity of the metal and q_o''' is a constant generation rate.

The design of an electrical power system has a part which is shaped like a tapered fin and carries a high current. The temperature distribution within the part has been calculated by using the one-dimensional fin equation with constant internal generation given by the equation

$$q''' = \frac{i^2 R_e}{V}$$

where $i =$ current flow, $R_e =$ electrical resistance, and $V =$ total volume of the part. It has been determined that the part material actually has a variable internal generation rate with $\beta = 6.1 \times 10^{-3} R_e^{-1}$.

You are asked to determine how much error is present in the previous calculations which were based upon a value of $\beta = 0$. If the error depends upon the magnitude of certain important nondimensional parameters, present your answer with this in mind.

2-16ᶜ A vertical surface transfers energy to the environment by free convection. It is suggested that a corrugated surface be used to increase the heat transfer and the strength

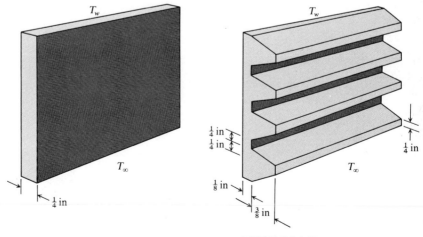

FIGURE 2-16*P*
Vertical surfaces with free convection.

of the surface. The corrugated surface may be treated as a surface with tapered fins. Because of the short length of the fins, a convective end boundary condition should be used in the analysis. The average convection coefficient over all exposed surfaces is 1.5 Btu/h-ft²-°R. Calculate the ratio of the total steady-state heat transfer by convection from the corrugated surface to that for the flat surface. How could you change the design to increase the heat transfer from the corrugated surface?

2-17ᶜ A radiating fin assembly in outer space similar to the one shown in Fig. 2-16 operates as follows. The tube attached to the base of the fin contains a liquid metal at 400°F. The tube attached to the tip of the fin contains a liquid gas at −230°F. The fin is rectangular in cross section. The fin-tube assembly is in orbit around Venus, and convective heat transfer is negligible. The fin material is aluminum ($k = 110$ Btu/h-ft-°R). Write a computer program, and obtain solutions for the one-dimensional, steady-state temperature distribution as a function of various values of irradiation from the sun. The range of interest for the incoming radiation varies from 500 Btu/h-ft² to 2,000 Btu/h-ft². Plot your results in terms of the important nondimensional parameters. Assume black surfaces.

TWO-DIMENSIONAL CONDUCTION

3-1 Introduction to Multidimensional Analysis

The problem of steady-state, two-dimensional conduction is a special case of the general three-dimensional, transient conduction problem. The conservation of energy principle is applied to a properly chosen differential element, and Fourier's law of conduction is used to derive the governing equation for the temperature distribution. Since the dependent variable, temperature, is a function of more than one independent variable, we expect the temperature distribution to be specified by a partial differential equation.

The resulting partial differential equation, along with associated boundary conditions, forms the mathematical model which will predict the temperature field and heat transfer rates. The solution to a partial differential equation can be obtained by analytical or numerical means. Numerical solutions are especially useful when irregular geometry or irregular boundary conditions are involved. Both types of solutions will be discussed in this chapter.

3-2   Mathematical Model

We now derive the governing partial differential equation for three-dimensional, transient heat conduction in rectangular coordinates. Since the temperature is allowed to vary in all three directions, it is necessary to apply the conservation-of-energy principle to a differential element of dimensions dx, dy, and dz. Such an element is shown in Fig. 3-1. This element, which is a closed thermodynamic system, is located entirely within the solid material. Thus, the energy crossing the boundaries of this system occurs only by conduction due to temperature gradients and internal generation.

We assume a stationary, homogeneous, isotropic material. The conduction of energy across the differential element can be accounted for by using Fourier's law. For the net heat conduction across each of the orthogonal surfaces normal to the x, y, and z directions shown in Fig. 3-1, one obtains

$$q_{x+dx/2} - q_{x-dx/2} = \left(q_x + \frac{\partial q_x}{\partial x} \frac{dx}{2} \right) - \left(q_x + \frac{\partial q_x}{\partial x} \frac{-dx}{2} \right)$$

$$= \frac{\partial q_x}{\partial x} dx \tag{3-1}$$

and in a similar manner

$$q_{y+dy/2} - q_{y-dy/2} = \frac{\partial q_y}{\partial y} dy \tag{3-2}$$

and

$$q_{z+dz/2} - q_{z-dz/2} = \frac{\partial q_z}{\partial z} dz \tag{3-3}$$

A first-order Taylor series expansion is used in the above three equations.

To determine the generation rate within the differential system, the internal generation per unit volume q''' is multiplied by the differential volume $dx\, dy\, dz$. The internal generation is treated as a source from which energy enters the system. The same mathematical model for internal generation was used in Chap. 2 in the analysis of one-dimensional conduction.

The conservation of energy principle requires that the internal generation rate equal the net heat transfer rate crossing the surfaces by conduction plus the rate of change of internal energy within the system. The rate of change of internal energy can be expressed as $dU/dt = mc(dT/dt) = \rho\, dx\, dy\, dz\, c(dT/dt)$, as shown in Sec. 1-9. Thus one can write

$$q''' \, dx\, dy\, dz = \frac{\partial q_x}{\partial x} dx + \frac{\partial q_y}{\partial y} dy + \frac{\partial q_z}{\partial z} dz$$

$$+ \rho\, dx\, dy\, dz\, c\, \frac{dT}{dt}$$

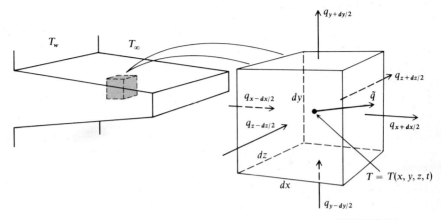

FIGURE 3-1
Differential element.

Using Fourier's law of conduction, $q_x = -kA_x(\partial T/\partial x)$ and $A_x = dy\,dz$. Also, $q_y = -k\,dz\,dx\,(\partial T/\partial y)$, and $q_z = -k\,dx\,dy\,(\partial T/\partial z)$. These equalities lead to the following equation, which is called the *energy equation for conduction:*

$$\frac{\partial}{\partial x}\left(k\frac{\partial T}{\partial x}\right) + \frac{\partial}{\partial y}\left(k\frac{\partial T}{\partial y}\right) + \frac{\partial}{\partial z}\left(k\frac{\partial T}{\partial z}\right) + q''' = \rho c\,\frac{dT}{dt} \qquad (3\text{-}4)$$

In general, we can write $T = T(x, y, z, t)$, and since T is a point function, the differential dT can be written

$$dT = \frac{\partial T}{\partial x}\,dx + \frac{\partial T}{\partial y}\,dy + \frac{\partial T}{\partial z}\,dz + \frac{\partial T}{\partial t}\,dt$$

Dividing by dt and rearranging we obtain[1]

$$\frac{dT}{dt} = \frac{\partial T}{\partial t} + u\frac{\partial T}{\partial x} + v\frac{\partial T}{\partial y} + w\frac{\partial T}{\partial z} \qquad (3\text{-}5)$$

where $u = dx/dt$, $v = dy/dt$, and $w = dz/dt$.

For a system which is stationary relative to the defined coordinate system, $u = v = w = 0$, and it follows that $dT/dt = \partial T/\partial t$. The governing partial differential equation is then written

$$\frac{\partial}{\partial x}\left(k\frac{\partial T}{\partial x}\right) + \frac{\partial}{\partial y}\left(k\frac{\partial T}{\partial y}\right) + \frac{\partial}{\partial z}\left(k\frac{\partial T}{\partial z}\right) + q''' = \rho c\,\frac{\partial T}{\partial t} \qquad (3\text{-}6)$$

---

[1] The derivative in Eq. (3-5) is called a *substantial* derivative and is often written as DT/Dt to distinguish it from an ordinary derivative.

It is instructive to compare the preceding results with the mathematical model obtained for one-dimensional conduction in Chap. 2. In the one-dimensional analysis the temperature was assumed to vary only in the x direction. This assumption allowed the differential element to be defined such that part of its surface was exposed to the environment. (See Fig. 2-1.) Since convective heat transfer occurred across this exposed surface, the convection heat transfer coefficient appeared in the governing differential equation. Since no convection occurs across the differential element shown in Fig. 3-1, the convection heat transfer coefficient appears in the mathematical statement of the boundary conditions rather than in the governing differential equation.

3-3   Governing Equations

Equation (3-6) forms the basis for many special situations which allow some terms in this general equation to be neglected. If we assume constant thermal conductivity and two-dimensional conduction in the x and y directions, the energy equation (3-6) reduces to

$$\alpha\left(\frac{\partial^2 T}{\partial x^2} + \frac{\partial^2 T}{\partial y^2}\right) + \frac{q'''}{\rho c} = \frac{\partial T}{\partial t} \qquad (3\text{-}7)$$

The term α in Eq. (3-7) is defined as the *thermal diffusivity* and is the ratio $k/\rho c$. Physically, it can be considered as a ratio of the rate of energy conducted to the energy stored per unit area. This interpretation becomes clearer by writing $k/\rho c = kA(\Delta T/l)/(\rho c V \, \Delta T/l^2)$, where $l = V/A$.

For steady-state, two-dimensional conduction the temperature distribution is not a function of time, and the governing equation can be written as

$$\frac{\partial^2 T}{\partial x^2} + \frac{\partial^2 T}{\partial y^2} + \frac{q'''}{k} = 0 \qquad (3\text{-}8)$$

Equation (3-8) is called the *Poisson equation*.

If we define nondimensional variables, $\theta = (T - T_1)/(T_0 - T_1)$, $\bar{y} = y/l$, $\bar{x} = x/l$, where T_0, T_1, and l are chosen reference temperatures and a reference length, respectively, then the Poisson equation can be written as

$$\frac{\partial^2 \theta}{\partial \bar{x}^2} + \frac{\partial^2 \theta}{\partial \bar{y}^2} + \frac{q'''l^2}{k(T_0 - T_1)} = 0 \qquad (3\text{-}9)$$

The term $q'''l^2/k(T_0 - T_1)$ is the familiar nondimensional heat generation term. When the internal generation is zero, Eq. (3-9) reduces to the *Laplace equation*.

The Poisson equation arises in many physical applications other than heat transfer. Examples are found in the theory of incompressible flow at low velocities, theories of electricity and magnetism which include the effects of charge density and pole strength, and the St. Venant theory of torsion. The Laplace equation is associated with irrotational flow of an incompressible flow and with determining electrical and magnetic potential fields. Thus many different physical problems can be formulated in terms of the same nondimensional equation. Often the nondimensional boundary conditions are also the same. When this is true, the solutions which give the nondimensional dependent variable as a function of the nondimensional independent variables are identical. Analytical solutions to the Poisson and Laplace equations have been reported in the literature for many different boundary conditions. For a complete treatment of analytical solutions in two-dimensional heat conduction, the reader is referred to Refs. 3-2, 3-3, and 3-4. An analytical solution using separation of variables is presented in the following section.

3-4 Analytical Analysis

The purpose of the formulation and analysis of heat conduction problems is to obtain the unique temperature distribution which satisfies the particular boundary conditions and the governing differential equation. Once the temperature distribution is known, the heat transfer may be calculated using Fourier's law. In a two-dimensional field, the local heat transfer rate is the vector sum of the two orthogonal components which are in turn determined from the temperature gradients. The equation for the heat transfer may be written as

$$q = -kA\,\nabla T = -kA_x \frac{\partial T}{\partial x} i - kA_y \frac{\partial T}{\partial y} j \qquad (3\text{-}10)$$

For steady-state, constant property heat conduction with no internal generation, the governing equation is given by the Laplace equation

$$\frac{\partial^2 \theta}{\partial x^2} + \frac{\partial^2 \theta}{\partial y^2} = 0 \qquad (3\text{-}11)$$

We desire a solution for the temperature distribution for the rectangular region shown in Fig. 3-2. Isothermal boundary conditions are specified in the figure. Using these boundary conditions, we define θ as follows:

$$\theta = \frac{T - T_1}{T_2 - T_1} \qquad (3\text{-}12)$$

The boundary conditions are then

$$
\begin{array}{lll}
\text{at} & x = 0 & \theta = 0 \\
\text{at} & y = 0 & \theta = 0 \\
\text{at} & x = w & \theta = 0 \\
\text{at} & y = l & \theta = 1.0
\end{array}
\qquad (3\text{-}13)
$$

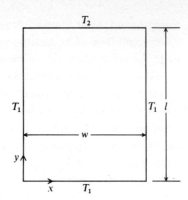

FIGURE 3-2
Analytical model.

Note that the definition of θ has produced three homogeneous boundary conditions. This result simplifies the analytical solution to the Laplace equation.

The isothermal boundary conditions are defined in a manner which allows an analytical solution to be obtained in a straightforward manner. The reason is that the entire surface at $x = 0$ is isothermal. In a similar manner the entire surfaces at $y = 0$ and $x = w$ are isothermal. Analytical complications arise when irregular boundary conditions are encountered. An example of an irregular boundary condition is when part of a surface is isothermal and part is adiabatic.

An analytical solution for the region specified in Fig. 3-2 can be obtained by applying the classical separation of variables technique. Separation of variables is accomplished by assuming that the final solution to the Laplace equation can be expressed as the product of two functions, each a function of a single independent variable. Mathematically, this is written as

$$\theta(x, y) = X(x)Y(y) \qquad (3\text{-}14)$$

Differentiating and substituting Eq. (3-14) into Eq. (3-11) yields

$$-\frac{1}{X}\frac{d^2X}{dx^2} = \frac{1}{Y}\frac{d^2Y}{dy^2} \qquad (3\text{-}15)$$

Since each side of Eq. (3-15) can be varied independently of the other, the equality will be valid for all x and y only if both sides are equal to the same constant λ^2. If the value of λ^2 is equal to zero, the solutions to the two equations resulting from Eq. (3-15) give

$$X = C_1 x + C_2 \qquad (3\text{-}16)$$
$$Y = C_3 y + C_4 \qquad (3\text{-}17)$$

From Eq. (3-14) we then have

$$\theta = (C_1 x + C_2)(C_3 y + C_4) \qquad (3\text{-}18)$$

However, an attempt to satisfy the boundary conditions given in Eqs. (3-13) leads to a trivial solution. Thus, λ^2 must not equal zero. We can still choose a plus or minus sign for the constant λ^2, but only one choice will allow all boundary conditions to be satisfied. To make the proper choice, note that both boundary conditions on x are homogeneous. It seems reasonable to choose the sign of λ^2 so that the solution of X as a function of x is expressed in terms of $\sin \lambda x$ and $\cos \lambda x$. We know that the value of these trigonometric functions may be made equal to zero by a proper choice of λ. Thus we set both sides of Eq. (3-15) equal to $+\lambda^2$. This results in

$$\frac{d^2 X}{dx^2} + \lambda^2 X = 0 \qquad (3\text{-}19)$$

$$\frac{d^2 Y}{dy^2} - \lambda^2 Y = 0 \qquad (3\text{-}20)$$

The corresponding solutions to these two equations are

$$X = C_1 \cos \lambda x + C_2 \sin \lambda x \qquad (3\text{-}21)$$
$$Y = C_3 e^{-\lambda y} + C_4 e^{\lambda y} \qquad (3\text{-}22)$$

It can now be seen that a choice of $-\lambda^2$ would lead to exponentials in the solution for $X(x)$ which could not satisfy both homogeneous boundary conditions on x.

Substituting Eqs. (3-21) and (3-22) into the assumed product form, Eq. (3-14) yields

$$\theta(x, y) = (C_1 \cos \lambda x + C_2 \sin \lambda x)(C_3 e^{-\lambda y} + C_4 e^{\lambda y}) \qquad (3\text{-}23)$$

The four constants of integration are determined by applying the four boundary conditions listed in Eqs. (3-13). Since the entire vertical surface at $x = 0$ is isothermal for $0 < y < l$, the first boundary condition requires that $\theta(0, y) = 0$. From Eq. (3-23) we write

$$0 = C_1(C_3 e^{-\lambda y} + C_4 e^{\lambda y}) \qquad (3\text{-}24)$$

Since choosing $C_3 = C_4 = 0$ would result in a trivial solution, we conclude that $C_1 = 0$. The second boundary condition at $y = 0$ is also valid for all $0 \le x \le w$ and requires that $\theta(x, 0) = 0$. This condition leads to

$$0 = C_2 \sin \lambda x \, (C_3 + C_4) \qquad (3\text{-}25)$$

This equation can be satisfied for $C_3 = -C_4$. The third boundary condition at $x = w$, $0 < y < l$, requires that $\theta(w, y) = 0$, and we obtain

$$0 = C_2 \sin \lambda w \, (-C_4 e^{-\lambda y} + C_4 e^{\lambda y})$$
$$= C_2 C_4 (e^{\lambda y} - e^{-\lambda y}) \sin \lambda w \qquad (3\text{-}26)$$

To satisfy this equality, the values of λ are chosen such that $\sin \lambda w = 0$. This requirement means that $\lambda = n\pi/w$, $n = 1, 2, \ldots$.

The expression for $\theta(x, y)$ now has the form:

$$\theta(x, y) = C_2 C_4 \sin \frac{n\pi x}{w} (e^{n\pi y/w} - e^{-n\pi y/w}) \qquad n = 1, 2, \ldots$$

Introducing $\sinh \lambda y = (e^{\lambda y} - e^{-\lambda y})/2$, the above equation can be written

$$\theta(x, y) = 2C_2 C_4 \sin \frac{n\pi x}{w} \sinh \frac{n\pi y}{w} \qquad n = 1, 2, \ldots$$

In the Laplace equation neither the dependent variable θ nor its derivatives occur to a degree higher than unity. Thus, the equation is linear, and the principle of superposition allows the solution to be written as a sum of the solutions for each value of n. It is also customary to combine the constants and write the solution as

$$\theta(x, y) = \sum_{n=1}^{\infty} C_n \sin \frac{n\pi x}{w} \sinh \frac{n\pi y}{w} \qquad (3\text{-}27)$$

The remaining constants in the series are determined by applying the fourth boundary condition at $y = l$ which specifies $\theta(x, l) = 1.0$. This results in

$$1.0 = \sum_{n=1}^{\infty} C_n \sin \frac{n\pi x}{w} \sinh \frac{n\pi l}{w} \qquad (3\text{-}28)$$

The values of C_n are determined by comparing Eq. (3-28) with the expression for unity in terms of a Fourier series over the interval $0 \le x \le w$. This Fourier series is

$$1.0 = \frac{2}{\pi} \sum_{n=1}^{\infty} \frac{(-1)^{n+1} + 1}{n} \sin \frac{n\pi x}{w} \qquad (3\text{-}29)$$

Comparison of Eqs. (3-28) and (3-29) indicates how to choose C_n so that Eq. (3-28) will be a Fourier series representation of unity. The proper expression for C_n is

$$C_n = \frac{2}{\pi \sinh(n\pi l/w)} \frac{(-1)^{n+1} + 1}{n} \qquad n = 1, 2 \ldots \qquad (3\text{-}30)$$

The general equation which gives the values for C_n is based on the infinite series expansion of an arbitrary function $f(x)$ over a specified interval using orthogonal functions. When a function $g(x)$ satisfies the expression

$$\int_a^b g_m(x) g_n(x) \, dx = 0 \qquad m \ne n \qquad a \le x \le b$$

then the infinite set of functions $g_1(x), g_2(x), \ldots, g_n(x), \ldots, g_m(x), \ldots$ is called orthogonal. An infinite series expansion of $f(x)$ in terms of orthogonal functions is written

$$f(x) = \sum_{n=1}^{\infty} C_n g_n(x)$$

Expanding the series, multiplying each side of this equation by $g_n(x)$, integrating each term between a and b, and using the definition of orthogonal functions leads to

$$\int_a^b f(x)g_n(x)\, dx = C_n \int_a^b g_n^2(x)\, dx$$

or

$$C_n = \frac{\int_a^b f(x)g_n(x)\, dx}{\int_a^b g_n^2(x)\, dx}$$

The sine function is orthogonal. We can view Eq. (3-28) as a Fourier sine series expansion, $g(x) = \sin(n\pi x/w)$, of the function $f(x) = 1.0$ over the interval $0 \le x \le w$ along the surface $y = l$. The two integrations required to determine C_n give

$$\int_a^b g_n^2(x)\, dx = \int_0^w \sin^2 \frac{n\pi x}{w}\, dx$$

$$= \frac{w}{n\pi}\left(\frac{n\pi x}{2w} - \frac{1}{4}\sin\frac{2n\pi x}{w}\right)_0^w = \frac{w}{2}$$

and

$$\int_a^b f(x)g_n(x)\, dx = \int_0^w 1.0 \sin\frac{n\pi x}{w}\, dx$$

$$= \frac{-w}{n\pi}\left(\cos\frac{n\pi x}{w}\right)_0^w = \frac{-w}{n\pi}\left(\cos n\pi - 1\right)$$

It follows that

$$C_n = \frac{2}{n\pi}(-\cos n\pi + 1) = \frac{2}{\pi}\frac{(-1)^{n+1} + 1}{n}\qquad n = 1, 2, \ldots$$

Equation (3-28) takes the form of an infinite series expansion of $f(x) = 1.0$ over $0 \le x \le w$ if we define the coefficients in Eq. (3-28) as given by Eq. (3-30).

The final solution for the steady-state, two-dimensional temperature distribution is given by

$$\theta(x, y) = \frac{2}{\pi}\sum_{n=1}^{\infty}\frac{(-1)^{n+1} + 1}{n}\sin\frac{n\pi x}{w}\frac{\sinh(n\pi y/w)}{\sinh(n\pi l/w)}\qquad (3\text{-}31)$$

The mathematical advantage of having uniform boundary conditions over a surface is now apparent. If irregular boundary conditions occur on a surface, the analytical analysis is more involved, and a numerical analysis offers advantages.

3-5 Finite Difference Equations

As in the differential formulation, the *finite difference analysis* begins with the definition of a model. However, the region to be analyzed is now pictured as being made up of small, but finite, volume elements. The difference in the resulting equations for the

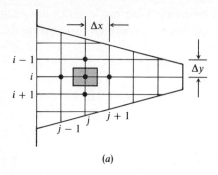

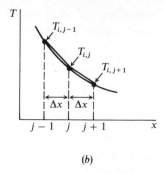

(a)

(b)

FIGURE 3-3
Grid network.

finite difference analysis as compared to the differential analysis is due to this important distinction in the definition of the model. The finite element used to obtain the mathematical model is an approximation of a differential element. As the size of the finite elements decreases, the difference between the results produced by a finite difference model and a differential model decreases.

In the finite difference analysis of two-dimensional conduction, an element of unit depth and dimensions Δx and Δy on the sides is chosen as the system to which the conservation of energy principle is applied. The sum of all finite elements forms a grid network as shown in Fig. 3-3a. The center of each finite volume is called a *node*. A temperature is assigned to each node, and it is assumed that this one value gives the temperature of the entire element. For the moment, only the interior nodes will be considered. Nodes which are located on a boundary are considered later. Note that the rectangular grid shown is made up of rows and columns. The rows i and columns j are measured from the upper left corner of the grid. This method corresponds to the way matrices are normally stored in a computer.

An interior, finite element of unit depth with width Δx and height Δy is shown shaded in Fig. 3-3a. This geometry corresponds to the grid spacing. The surrounding nodes locate the center of adjacent elements. A typical temperature profile in the x direction across the elements is indicated in Fig. 3-3b. The curved line represents the temperature variation through the solid material. In a finite difference formulation the temperature gradient is calculated as though a linear profile exists between nodes as shown in Fig. 3-3b. At the same time, the nodal representation of the temperature field inplies a step change in temperature across the interface between two adjacent elements. This formulation is obviously more accurate as the distance between nodes decreases.

The principle of conservation of energy for steady-state heat transfer is now applied to an internal node. The temperature gradient between nodes is constant. We can account for the heat transfer crossing the boundaries of this system using Fourier's law. To establish a consistent sign convection, the heat transfer between nodes i, j and $i, j - 1$ is written as follows:

$$q = k \, \Delta y \, (1) \, \frac{T_{i, j-1} - T_{i, j}}{\Delta x} \qquad (3\text{-}32)$$

If the temperature of the central node $T_{i, j}$ is larger than $T_{i, j-1}$, the value for q is negative, and the heat transfer is out of the system; and conversely a positive value of q is obtained for the reverse situation. If one uses similar equations for the other three surrounding nodes and accounts for the internal generation, application of the conservation of energy principle will lead to the following equation:

$$k \, \Delta y \, (1) \, \frac{T_{i, j-1} - T_{i, j}}{\Delta x} + k \, \Delta x \, (1) \, \frac{T_{i-1, j} - T_{i, j}}{\Delta y}$$

$$+ k \, \Delta y \, (1) \, \frac{T_{i, j+1} - T_{i, j}}{\Delta x} + k \, \Delta x \, (1) \, \frac{T_{i+1, j} - T_{i, j}}{\Delta y}$$

$$+ q''' \, \Delta x \, \Delta y \, (1) = 0$$

For a rectangular grid it is convenient to choose $\Delta x = \Delta y$. Then the above equation reduces to

$$T_{i, j-1} + T_{i-1, j} + T_{i, j+1} + T_{i+1, j} - 4T_{i, j} + \frac{q'''(\Delta x)^2}{k} = 0 \qquad (3\text{-}33)$$

This is the finite difference equation for all interior nodes for steady-state, two-dimensional conduction.

It is desirable to use the same nondimensional temperature that was defined in the analytical solution in Sec. 3-3. Equation (3-33) then becomes

$$\theta_{i, j-1} + \theta_{i-1, j} + \theta_{i, j+1} + \theta_{i+1, j} - 4\theta_{i, j} + \frac{q'''(\Delta x)^2}{k(T_2 - T_1)} = 0 \qquad (3\text{-}34)$$

where $\theta = (T - T_1)/(T_2 - T_1)$. The last term in Eq. (3-34) is the finite difference form of the nondimensional internal generation parameter. If we introduce the nondimensional parameters $\overline{\Delta x} = \Delta x/l$ and $\text{N1} = q''' l^2/k(T_2 - T_1)$, the internal generation parameter can be written as the product $(\overline{\Delta x})^2 \text{N1}$.

The finite difference form of an equation can also be obtained directly from the governing equation. Referring again to Fig. 3-3b, a forward finite difference approximation to the derivative dT/dx at $x_{i, j}$ is based on an assumed constant linear temperature gradient between nodes and is given by

$$\left(\frac{dT}{dx}\right)_{i, j} \approx \frac{T_{i, j+1} - T_{i, j}}{\Delta x} \qquad (3\text{-}35)$$

In a similar manner a backward finite difference approximation to the derivative at the node i, j is given by

$$\left(\frac{dT}{dx}\right)_{i,j} \approx \frac{T_{i,j} - T_{i,j-1}}{\Delta x} \qquad (3\text{-}36)$$

A more accurate approximation to the temperature derivative at a node is given by a central finite difference which is based on the temperatures of both the forward and backward nodes. This approximation is expressed as

$$\left(\frac{dT}{dx}\right)_{i,j} \approx \frac{T_{i,j+1} - T_{i,j-1}}{2\Delta x} \approx \frac{T_{i,j+1/2} - T_{i,j-1/2}}{\Delta x} \qquad (3\text{-}37)$$

The second derivative based upon a central difference approximation is then

$$\left(\frac{d^2T}{dx^2}\right)_{i,j} \approx \frac{1}{(\Delta x)}\left[\left(\frac{dT}{dx}\right)_{i,j+1/2} - \left(\frac{dT}{dx}\right)_{i,j-1/2}\right] \qquad (3\text{-}38)$$

We can now express each of the first derivatives in terms of a central difference approximation using equations similar to Eq. (3-37). When we approximate the derivative at location $i, j + \frac{1}{2}$, the forward node is $T_{i,j+1}$, and the backward node is $T_{i,j}$. Equation (3-38) then takes the form

$$\left(\frac{d^2T}{dx^2}\right)_{i,j} \approx \frac{1}{\Delta x}\left(\frac{T_{i,j+1} - T_{i,j}}{\Delta x} - \frac{T_{i,j} - T_{i,j-1}}{\Delta x}\right)$$

which becomes

$$\left(\frac{d^2T}{dx^2}\right)_{i,j} \approx \frac{1}{(\Delta x)^2}(T_{i,j+1} - 2T_{i,j} + T_{i,j-1}) \qquad (3\text{-}39)$$

In a similar manner

$$\left(\frac{d^2T}{dy^2}\right)_{i,j} \approx \frac{1}{(\Delta y)^2}(T_{i+1,j} - 2T_{i,j} + T_{i-1,j}) \qquad (3\text{-}40)$$

If one uses the finite difference expressions developed above, the Poisson equation (3-8) can be written

$$\frac{1}{(\Delta x)^2}(T_{i,j+1} - 2T_{i,j} + T_{i,j-1}) + \frac{1}{(\Delta y)^2}(T_{i+1,j} - 2T_{i,j} + T_{i-1,j}) + \frac{q'''}{k} = 0$$

$$(3\text{-}41)$$

Choosing $\Delta x = \Delta y$ we again obtain Eq. (3-33):

$$T_{i,j+1} + T_{i,j-1} + T_{i+1,j} + T_{i-1,j} - 4T_{i,j} + \frac{q'''(\Delta x)^2}{k} = 0$$

3-6   Errors Associated with a Finite Difference Approximation

The size of the error introduced by using a finite difference approximation to specify the temperature field depends on the size of the finite volume elements. Since the results obtained from a finite difference model approach the results obtained from a differential model in the limit as $\Delta x \to 0$ and $\Delta y \to 0$, a Taylor series expansion can be used to determine the error introduced by the finite difference approximation. For a finite interval $\Delta x = x - x_{i,j}$, additional terms in a Taylor series expansion of the temperature about the point $x_{i,j}$ must be retained to improve the accuracy. The Taylor series expansion of T about $x_{i,j}$ in the x direction is

$$T = T_{i,j} + \left(\frac{\partial T}{\partial x}\right)_{i,j}(x - x_{i,j}) + \frac{1}{2!}\left(\frac{\partial^2 T}{\partial x^2}\right)_{i,j}(x - x_{i,j})^2$$

$$+ \frac{1}{3!}\left(\frac{\partial^3 T}{\partial x^3}\right)_{i,j}(x - x_{i,j})^3 + \frac{1}{4!}\left(\frac{\partial^4 T}{\partial x^4}\right)_{i,j}(x - x_{i,j})^4 + \cdots$$

For $x = x_{i,j+1}$, the value of $x - x_{i,j} = \Delta x$. Thus

$$T_{i,j+1} = T_{i,j} + \left(\frac{\partial T}{\partial x}\right)_{i,j}\Delta x + \frac{1}{2!}\left(\frac{\partial^2 T}{\partial x^2}\right)_{i,j}(\Delta x)^2$$

$$+ \frac{1}{3!}\left(\frac{\partial^3 T}{\partial x^3}\right)_{i,j}(\Delta x)^3 + \frac{1}{4!}\left(\frac{\partial^4 T}{\partial x^4}\right)_{i,j}(\Delta x)^4 + \cdots \qquad (3\text{-}42)$$

For $x = x_{i,j-1}$, the value of $x - x_{i,j} = -\Delta x$. Thus

$$T_{i,j-1} = T_{i,j} - \left(\frac{\partial T}{\partial x}\right)_{i,j}\Delta x + \frac{1}{2!}\left(\frac{\partial^2 T}{\partial x^2}\right)_{i,j}(\Delta x)^2$$

$$- \frac{1}{3!}\left(\frac{\partial^3 T}{\partial x^3}\right)_{i,j}(\Delta x)^3 + \frac{1}{4!}\left(\frac{\partial^4 T}{\partial x^4}\right)_{i,j}(\Delta x)^4 - \cdots \qquad (3\text{-}43)$$

Adding Eqs. (3-42) and (3-43) yields

$$T_{i,j+1} + T_{i,j-1} = 2T_{i,j} + \left(\frac{\partial^2 T}{\partial x^2}\right)_{i,j}(\Delta x)^2 + \frac{1}{12}\left(\frac{\partial^4 T}{\partial x^4}\right)_{i,j}(\Delta x)^4 + \cdots \qquad (3\text{-}44)$$

If we neglect the last term, the central finite difference approximation to the second derivative is obtained from Eq. (3-44):

$$\left(\frac{\partial^2 T}{\partial x^2}\right)_{i,j} = \frac{1}{(\Delta x)^2}(T_{i,j+1} - 2T_{i,j} + T_{i,j-1})$$

The truncation error involved by omitting the last term is of the order $(\Delta x)^4$.

Subtracting Eq. (3-43) from Eq. (3-42) results in

$$T_{i,j+1} - T_{i,j-1} = 2\left(\frac{\partial T}{\partial x}\right)_{i,j}\Delta x + \frac{1}{3}\left(\frac{\partial^3 T}{\partial x^3}\right)_{i,j}(\Delta x)^3 + \cdots \qquad (3\text{-}45)$$

Again if we neglect the last term and solve for the temperature derivative, the central finite difference approximation for the first derivative is obtained:

$$\left(\frac{\partial T}{\partial x}\right)_{i,j} = \frac{T_{i,j+1} - T_{i,j-1}}{2\,\Delta x}$$

The truncation error introduced by this finite difference approximation is of the order $(\Delta x)^3$.

As the size of the volume elements is decreased, the accuracy of the computations increases, but so does the required computer time. A compromise on element size must be made based upon the accuracy required and the cost of computation. In some problems it may be useful to choose one size of grid network for one region and another smaller size for regions of high temperature gradients. It may also be desirable to use grid networks made up of shapes other than squares or rectangles in order to match the shape of a boundary.

3-7 Finite Difference Boundary Conditions

When nodes are located on the surface, the nodal equations must account for energy crossing the surface as well as the conduction to the surface from within the solid. Boundary conditions at the surface may specify constant temperature, insulation, convection, radiation, or a combination of these items. The nodal equations for surface nodes can be determined by applying the conservation of energy principle at the surface. The dimensions of the finite volumes near a surface are usually different from interior volumes, and the area across which heat conduction occurs must be recalculated. Consider the surface node shown in Fig. 3-4. Energy is conducted across three surfaces of the element, and convection occurs across the exposed surface.

The energy terms which account for conduction, convection, and internal generation can be set equal to zero for steady-state heat transfer.

$$k\,\Delta y\,\frac{T_{i,j-1} - T_{i,j}}{\Delta x} + k\,\frac{\Delta x}{2}\,\frac{T_{i-1,j} - T_{i,j}}{\Delta y}$$

$$+ \bar{h}\,\Delta y\,(T_\infty - T_{i,j}) + k\,\frac{\Delta x}{2}\,\frac{T_{i+1,j} - T_{i,j}}{\Delta y}$$

$$+ q'''\,\frac{\Delta x}{2}\,\Delta y = 0 \qquad (3\text{-}46)$$

When $\Delta x = \Delta y$, the above equation reduces to

$$T_{i,j} = \frac{T_{i+1,j} + T_{i-1,j} + 2T_{i,j-1} + 2\bar{h}\,\Delta x\,T_\infty/k + q'''(\Delta x)^2/k}{4 + 2\bar{h}\,\Delta x/k} \qquad (3\text{-}47)$$

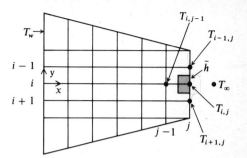

FIGURE 3-4
Surface node.

When there is no convective heat transfer from the surface and no internal generation within the material, Eq. (3-47) reduces to

$$T_{i,j} = \frac{T_{i+1,j} + T_{i-1,j} + 2T_{i,j-1}}{4} \qquad (3\text{-}48)$$

To express the surface nodal equations in terms of a nondimensional temperature θ, reference temperatures must be chosen. The form of the convective heat transfer term in Eq. (3-46) leads to a choice of a reference temperature T_∞. Introducing $\theta = (T - T_\infty)/(T_w - T_\infty)$ results in the following nondimensional surface nodal equation:

$$\theta_{i,j} = \frac{\theta_{i+1,j} + \theta_{i-1,j} + 2\theta_{i,j-1} + q'''(\Delta x)^2/k(T_w - T_\infty)}{4 + 2\bar{h}\,\Delta x/k} \qquad (3\text{-}49)$$

The temperature T_w is the wall temperature at $x = 0$ as shown in Fig. 3-4. The reference temperature difference $T_w - T_\infty$ is chosen to give a value $\theta = 1$ at $x = 0$. Using the definition $\overline{\Delta x} = \Delta x/l$, we finally write

$$\theta_{i,j} = \frac{\theta_{i+1,j} + \theta_{i-1,j} + 2\theta_{i,j-1} + (\overline{\Delta x})^2 \text{N1}}{4 + 2\,\overline{\Delta x}\,\text{Bi}} \qquad (3\text{-}50)$$

where the Biot number is $\text{Bi} = \bar{h}l/k$.

Consider the node shown in Fig. 3-5 at the corner formed by a perfectly insulated (adiabatic) surface and a surface with convection. Conduction occurs across the two interior surfaces, and convection occurs across the lower exposed surface. The nodal equation for $\Delta x = \Delta y$ becomes

$$k\,\frac{\Delta x}{2}\,\frac{T_{i-1,j} - T_{i,j}}{\Delta y} + k\,\frac{\Delta y}{2}\,\frac{T_{i,j+1} - T_{i,j}}{\Delta x}$$

$$+\,\bar{h}\,\frac{\Delta x}{2}\,(T_\infty - T_{i,j}) + \frac{q'''(\Delta x)^2}{4} = 0$$

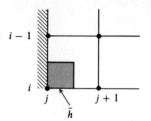

FIGURE 3-5
Corner node.

Solving for $T_{i,j}$, one obtains

$$T_{i,j} = \frac{T_{i-1,j} + T_{i,j+1} + \bar{h}\,\Delta x\, T_\infty/k + q'''(\Delta x)^2/2k}{2 + \bar{h}\,\Delta x/k} \qquad (3\text{-}51)$$

Defining $\theta = (T - T_\infty)/(T_w - T_\infty)$ this can be written in nondimensional form as

$$\theta_{i,j} = \frac{\theta_{i-1,j} + \theta_{i,j+1} + \frac{1}{2}(\overline{\Delta x})^2 \mathrm{N1}}{2 + \overline{\Delta x}\,\mathrm{Bi}} \qquad (3\text{-}52)$$

Another situation requiring a special nodal equation is an interior node near a curved boundary. Such a node might occur in the analysis of two-dimensional conduction in a parabolic fin. Consider the node i, j shown in Fig. 3-6. The two nearby surface temperatures are denoted by T_A and T_B. The conservation of energy principle requires that

$$k\left(\frac{\Delta l}{2} + b\,\frac{\Delta l}{2}\right)\frac{T_{i,j-1} - T_{i,j}}{\Delta l} + k\left(\frac{\Delta l}{2} + a\,\frac{\Delta l}{2}\right)\frac{T_{i+1,j} - T_{i,j}}{\Delta l}$$

$$+ k\left(\frac{\Delta l}{2} + b\,\frac{\Delta l}{2}\right)\frac{T_A - T_{i,j}}{a\,\Delta l} + k\left(\frac{\Delta l}{2} + a\,\frac{\Delta l}{2}\right)\frac{T_B - T_{i,j}}{b\,\Delta l}$$

$$+ q'''\left(\frac{\Delta l}{2} + a\,\frac{\Delta l}{2}\right)\left(\frac{\Delta l}{2} + b\,\frac{\Delta l}{2}\right) = 0$$

This equation can be simplified to

$$\frac{2}{1+a}T_{i,j-1} + \frac{2}{1+b}T_{i+1,j} + \frac{2}{a(1+a)}T_A$$

$$+ \frac{2}{b(1+b)}T_B - 2\left(\frac{1}{a} + \frac{1}{b}\right)T_{i,j} + \frac{q'''(\Delta l)^2}{k} = 0 \qquad (3\text{-}53)$$

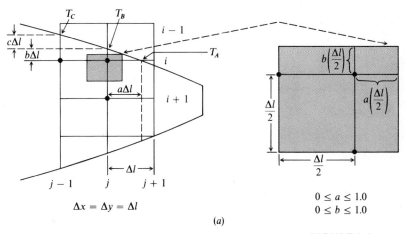

$$\Delta x = \Delta y = \Delta l$$

$$(a)$$

$$0 \le a \le 1.0$$
$$0 \le b \le 1.0$$

FIGURE 3-6a
Near-surface node.

Defining $\theta = (T - T_\infty)/(T_w - T_\infty)$, $\overline{\Delta l} = \Delta l/l$ and $N1 = q'''l^2/k(T_w - T_\infty)$, the nondimensional form of this equation is

$$\theta_{i,j} =$$

$$\frac{[2/(1 + a)]\theta_{i,j-1} + [2/(1 + b)]\theta_{i+1,j} + [2/a(1 + a)]\theta_A + [2/b(1 + b)]\theta_B + (\overline{\Delta l})^2 N1}{2(1/a + 1/b)}$$

$$(3\text{-}54)$$

If the surface temperature is known, the near-surface nodal equations together with the interior nodal equations specify the temperature distribution. When convective heat transfer occurs from the surface, additional equations for the temperatures of the surface nodes must be used. The enlarged grid shown in Fig. 3-6b identifies near-surface nodes and interior nodes. Note that the location of the surface temperatures is fixed by the extension of rows and columns in the grid network. We now focus our attention on the derivation of the nodal equation for the surface temperature node T_B. We must account for conduction between T_B and the two adjacent surface nodes (T_A and T_C) and the near-surface node $T_{i,j}$. Also, heat transfer by convection occurs between T_B and T_∞. We use the linear distance between the surface nodes to specify the temperature gradients. If one uses the nomenclature given in Fig. 3-6b, the distance between T_A and T_B is $\Delta l \sqrt{a^2 + b^2}$, and the distance between T_B and T_C is $\Delta l \sqrt{1 + c^2}$. The surface area per unit depth for convection across the temperature difference $T_B - T_\infty$ is equal to $\frac{1}{2} \Delta l \sqrt{a^2 + b^2} + \frac{1}{2} \Delta l \sqrt{1 + c^2}$. The values of a and b

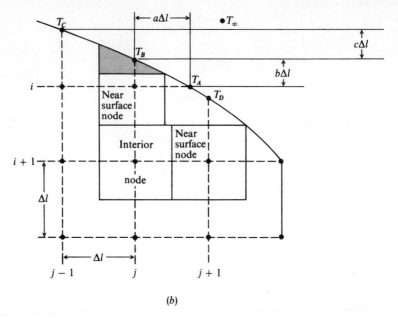

(b)

FIGURE 3-6b
Grid network for surface nodes.

depend on the geometry of the cross section, and they vary from surface node to surface node.

The term representing conduction between $T_{i,j}$ and T_B is written as

$$k\left(\frac{\Delta l}{2} + a\frac{\Delta l}{2}\right)\frac{T_{i,j} - T_B}{b\,\Delta l} = k\frac{a+1}{2b}(T_{i,j} - T_B)$$

The term for convection between T_B and T_∞ is

$$\bar{h}(\tfrac{1}{2}\Delta l\sqrt{a^2 + b^2} + \tfrac{1}{2}\Delta l\sqrt{1 + c^2})(T_\infty - T_B)$$

$$= \frac{\bar{h}\,\Delta l}{2}(\sqrt{a^2 + b^2} + \sqrt{1 + c^2})(T_\infty - T_B)$$

In using Fourier's law to express the heat conduction terms between the surface nodes, we must know the area normal to the conduction. If we limit ourselves to rectangular "building blocks" to approximate the finite elements, we represent the triangular shaded area shown in Fig. 3-6b with a rectangular area as shown in Fig. 3-6c. The height or width of the rectangle is determined by the value of $b\,\Delta l/2$ or $a\,\Delta l/2$ for

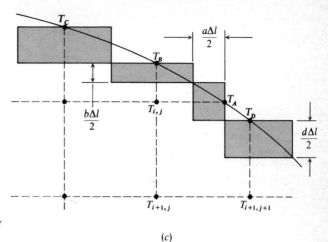

FIGURE 3-6c
Approximate grid network for
surface nodes.

(c)

each surface node. The area normal to the direction of heat conduction between T_B and T_A is then approximated by $(b \, \Delta l/2)(1)$. The finite difference expression for the heat transfer by conduction between T_A and T_B is then written

$$k \, \frac{b \, \Delta l}{2} \frac{T_A - T_B}{\Delta l \sqrt{a^2 + b^2}} = \frac{kb}{2} \frac{T_A - T_B}{\sqrt{a^2 + b^2}}$$

In a similar manner the heat transfer by conduction between T_C and T_B is specified by

$$k \, \frac{b \, \Delta l}{2} \frac{T_C - T_B}{\Delta l \sqrt{1 + c^2}} = \frac{kb}{2} \frac{T_C - T_B}{\sqrt{1 + c^2}}$$

Assuming no internal generation and setting the sum of the energy terms equal to zero as required by the conservation of energy principle yields

$$\frac{\bar{h} \, \Delta l}{k} \left(\sqrt{a^2 + b^2} + \sqrt{1 + c^2} \right) T_\infty + \frac{a + 1}{b} T_{i,j}$$

$$+ \frac{b}{\sqrt{a^2 + b^2}} T_A + \frac{b}{\sqrt{1 + c^2}} T_C$$

$$= \left[\frac{\bar{h} \, \Delta l}{k} \left(\sqrt{a^2 + b^2} + \sqrt{1 + c^2} \right) + \frac{a + 1}{b} \right.$$

$$\left. + \frac{b}{\sqrt{a^2 + b^2}} + \frac{b}{\sqrt{1 + c^2}} \right] T_B \qquad (3\text{-}55)$$

If we use the previous definitions of θ, Bi, and $\overline{\Delta l}$, we can write

$$\theta_{i,j} = \left\{ \left[\overline{\Delta l}\, \text{Bi}(\sqrt{a^2 + b^2} + \sqrt{1 + c^2}) + \frac{a+1}{b} + \frac{b}{\sqrt{a^2 + b^2}} + \frac{b}{\sqrt{1 + c^2}} \right] \theta_B \right.$$

$$\left. - \frac{b}{\sqrt{a^2 + b^2}}\, \theta_A - \frac{b}{\sqrt{1 + c^2}}\, \theta_C \right\} \bigg/ \frac{a+1}{b} \qquad (3\text{-}56)$$

We can now classify three general types of nodal equations. They are the interior equations, the surface equations, and the near-surface equations. The near-surface equations are needed for a boundary which does not match the grid network used in the finite difference formulation. These finite difference equations can be used in a numerical iteration process to obtain the nodal temperature distribution throughout the solid region. The method of carrying out the numerical iteration on a computer is the topic of the following section.

3-8   Numerical Iteration

The solution to steady-state, two-dimensional heat conduction problems requires the determination of the nodal temperatures in the finite difference grid. The temperature of each node must satisfy the appropriate nodal equation. For example, the nondimensional equation for the interior nodes is given by Eq. (3-34). Solving for the central node one obtains

$$\theta_{i,j} = \frac{\theta_{i,j-1} + \theta_{i-1,j} + \theta_{i,j+1} + \theta_{i+1,j} + (\overline{\Delta x})^2 \text{N1}}{4} \qquad (3\text{-}57)$$

Here $\theta = (T - T_1)/(T_2 - T_1)$. At the right vertical surface the nodal equation for a surface with an adiabatic boundary condition is Eq. (3-50) with Bi $= 0$. The Biot number is zero because the convective heat transfer is zero for a surface with an adiabatic boundary condition. We can define $\theta = (T - T_1)/(T_2 - T_1)$ when there is no convection at the surface because T_∞ does not enter into the model. The resulting equation is written

$$\theta_{i,j} = \frac{\theta_{i+1,j} + \theta_{i-1,j} + 2\theta_{i,j-1} + (\overline{\Delta x})^2 \text{N1}}{4} \qquad (3\text{-}58)$$

The equation for an adiabatic node on the left vertical surface is the same as Eq. (3-58) except that the term $\theta_{i,j-1}$ is replaced by $\theta_{i,j+1}$.

In order to calculate the temperature at each node in the grid, a scheme must be devised so that calculations at each node are made in a sequential manner by means of the proper nodal equation until all nodal temperatures have been determined. All

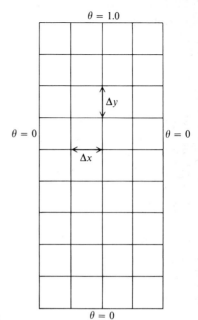

$\theta = 1.0$

$\theta = 0$ $\theta = 0$

Δy

Δx

$\theta = 0$

FIGURE 3-7
Rectangular grid.

nodes must be assigned an initial value to begin the calculations. These initial values should be based on known boundary conditions and estimated temperatures.

Consider a two-dimensional grid as shown in Fig. 3-7. The temperature along the upper surface is constant at T_2. It is assumed that the other exposed surfaces are maintained constant at T_1. The values of the nondimensional, isothermal boundary conditions are shown in Fig. 3-7 for $\theta = (T - T_1)/(T_2 - T_1)$.

The ten by five grid network shown is made up of finite volume elements of unit depth with $\Delta x = \Delta y$. In a ten by five grid the total length l is $9\,\Delta y$, and the total width is $4\,\Delta x$. If $\bar{y} = y/l$ and $\bar{x} = x/l$, then $0 \le \bar{y} \le 1$ and $0 \le \bar{x} \le \frac{4}{9}$. In this example all surface temperatures are known, and the chosen grid network exactly fits the contour of the solid. Therefore, only the nodal equation for the interior nodes is needed to determine the temperature distribution.

The discussion that follows relates to program SSHC on pages 134–135. This program is written to solve the temperature distribution in the rectangular region with the isothermal boundary conditions shown in Fig. 3-7. It also incorporates provisions for specifying an adiabatic boundary condition over all or part of the vertical surfaces. A list of the variables used in this program is given in Table 3-1. The rows and columns in the computer program are denoted by i's and j's, respectively. It is necessary to assign nondimensional temperatures to each of the 50 nodes before the nodal equations are used to recalculate the proper temperatures. The temperatures of the surface nodes are known from the boundary conditions, but the interior nodes are unknown. The temperatures of the interior nodes may be greater or less than the exposed surface temperatures, depending on the magnitude of the internal generation. In order to compare the numerical solution with the analytical solution in Sec. 3-4, we take the internal generation to be zero. If we consider a limiting case of no internal generation and adiabatic boundary conditions for the entire vertical surfaces at $\bar{x} = 0$ and $\bar{x} = \frac{4}{9}$, the nondimensional temperature distribution is a one-dimensional linear profile between θ_2 and θ_1. This linear distribution is initially assigned to the nodes as indicated in Statements 300 to 350.

The logic statements which apply the finite difference equations to the proper nodes begin in Statement 420. The variables N and U appearing in Statements 405 and 410 are counting variables. The number of iterations is indicated by the value of N, and the number of nodes which are unsatisfactory after each iteration, based on the chosen convergence criterion, is indicated by U. The convergence criterion is that the most recently calculated nodal temperature does not differ from the previously calculated value by more than 0.001. The lower and upper horizontal surfaces (I = 10 and I = 1) are held fixed at T1 = 0 and T2 = 1.0, respectively. The vertical surfaces are assumed to be either adiabatic or isothermal at T1 = 0. The number of nodes on the left face measured downward from row 1 to which an adiabatic boundary condition is applied is denoted by I1. The corresponding number of nodes on the right face is denoted by I2. For the problem posed in Fig. 3-7, I1 and I2 are both zero.

Table 3-1   VARIABLES FOR PROGRAM SSHC

Variable	Definition
T(I, J)	Current nodal temperatures
X(I, J)	Newly calculated nodal temperatures
Y(I, J)	Difference in nodal temperatures
Z(I, J)	Truncated nodal temperatures
I1, I2	Number of insulated nodes on left and right surfaces
N1	$q''l^2/k(T_w - T_\infty)$
D	$\overline{\Delta x = \Delta y = \frac{1}{9}}$
T1	$\theta_1 = (T_1 - T_1)/(T_2 - T_1) = 0$
T2	$\theta_2 = (T_2 - T_1)/(T_2 - T_1) = 1.0$
Y1	Number of grid points along $\bar{x} = 0$
X1	Number of grid points along $\bar{y} = 0$
M, P, N	Counting variables
E1	Convergence criterion
U	Number of nodes not satisfying convergence criterion

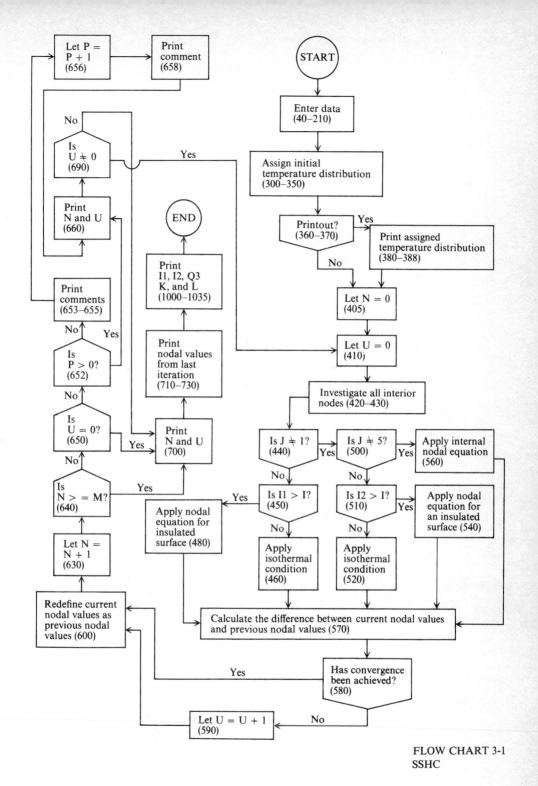

FLOW CHART 3-1
SSHC

Listing 3-1   SSHC

```
10    PRINT"TYPE 1 IF YOU WANT A PROGRAM EXPLANATION."
12    PRINT
13    PRINT"OTHERWISE, TYPE 2."
14    INPUTL
15    PRINT
16    IF L=1 THEN 20
18    IF L=2 THEN 40
20    PRINT
21    PRINT"THIS PROGRAM SOLVES THE TWO-DIMENSIONAL LAPLACE OR"
22    PRINT"POISSON EQUATION  FOR STEADY STATE CONDUCTION IN A SOLID"
23    PRINT"IN A RECTANGULAR GEOMETRY SPECIFIED BY A 10X5 GRID."
24    PRINT
25    PRINT"THE GOVERNING EQUATION IS EXPRESSED BY USE OF"
26    PRINT"A FINITE DIFFERENCE APPROXIMATION."
27    PRINT
28    PRINT"THE TOP AND BOTTOM SURFACES ARE ISOTHERMAL."
29    PRINT
30    PRINT"THE TWO VERTICAL SIDES ARE EITHER ADIABATIC OR ISOTHERMAL."
31    PRINT
32    PRINT"THE NUMBER OF ADIABATIC NODES, MEASURED DOWN FROM THE TOP,"
33    PRINT"ARE SPECIFIED BY THE INPUT VALUES OF I1 AND I2"
34    PRINT" FOR THE LEFT AND RIGHT SURFACES RESPECTIVELY."
35    PRINT
36    PRINT"A LINEAR TEMPERATURE PROFILE IN THE SOLID IS INITIALLY ASSIGNED."
37    PRINT
38    PRINT"THIS PROFILE IS MODIFIED UNTIL ALL BOUNDARY CONDITIONS"
39    PRINT"AND THE DIFFERENTIAL EQUATION ARE SATISFIED."
40    DIM Y(10,5),T(10,5),X(10,5),Z(10,5)
50    READ I1,I2,N1,D
60    READ T1,T2
95    REM ASSIGN GRID SIZE
100   LETY1=10
120   LETX1=5
190   REM INITIALIZE CONTROL CONSTANTS
200   LETM=50
202   LETP=0
210   LETE1=0.001
290   REM ASSIGN INITIAL TEMPERATURE DISTRIBUTION
300   FOR I=1 TO Y1
310   FOR J=1 TO X1
320   LETA=I
330   LETT(I,J)=T2+(T1-T2)*(A-1)/(Y1-1)
340   NEXT J
350   NEXT I
355   PRINT
360   PRINT"DO YOU WANT A PRINT OUT OF THE INITIALLY ASSIGNED"
362   PRINT"LINEAR TEMPERATURE DISTRIBUTION?"
364   PRINT
366   PRINT"IF YES TYPE 1.  IF NO TYPE 2"
370   INPUTB6
372   IF B6=1 THEN 380
374   IF B6=2 THEN 400
380   PRINT"THE ASSIGNED TEMPERATURE DISTRIBUTION IS:"
382   FOR I=1 TO Y1
383   FOR J=1 TO X1
384   LETY(I,J)=INT(100*T(I,J)+0.5)/100
385   NEXT J
387   NEXT I
388   MAT PRINT Y;
400   PRINT
402   PRINT"THE ITERATION OF THE FINITE DIFFERENCE EQNS BEGINS NOW."
404   REM INITIALIZE COUNTING VARIABLES
405   LETN=0
410   LETU=0
420   FOR J=1 TO X1
430   FOR I=2 TO Y1-1
```

Listing 3–1   SSHC (*continued*)

```
440   IF J<>1 THEN 500
450   IF I1>=I THEN 480
460   LETX(I,1)=T1
470   IF I1<I THEN 570
480   LETX(I,J)=(T(I+1,J)+2*T(I,J+1)+T(I-1,J)+N1*(D^2))/4
490   IF I1>=I THEN 570
500   IF J<>X1 THEN 560
510   IF I2>=I THEN 540
520   LETX(I,X1)=T1
530   IF I2<I THEN 570
540   LETX(I,J)=(T(I+1,J)+2*T(I,J-1)+T(I-1,J)+N1*(D^2))/4
550   IF I2>=I THEN 570
560   LETX(I,J)=(T(I,J+1)+T(I,J-1)+T(I+1,J)+T(I-1,J)+N1*(D^2))/4
570   LETY(I,J)=X(I,J)-T(I,J)
580   IF ABS(Y(I,J))<E1 THEN 600
590   LETU=U+1
600   LETT(I,J)=X(I,J)
610   NEXT I
620   NEXT J
630   LETN=N+1
640   IF N>=M THEN 700
650   IF U=0 THEN 700
652   IF P>0.0 THEN 660
653   PRINT"AFTER EACH ITERATION THE U VALUE INDICATES THE NUMBER"
654   PRINT"OF GRID POINTS WHERE A SOLUTION HAS NOT YET BEEN"
655   PRINT"OBTAINED TO SUFFICIENT ACCURACY."
656   LETP=P+1
657   PRINT
658   PRINT"IT. NO.        U VALUE"
659   PRINT
660   PRINTN,U
670   PRINT
690   IF U<>0.0 THEN 410
698   PRINT
700   PRINTN,U
710   PRINT
715   PRINT"THE FINAL SOLUTION IS"
720   FOR I=1 TO Y1
722   FOR J=1 TO X1
724   LETZ(I,J)=INT(1000*T(I,J)+0.5)/1000
726   NEXT J
728   NEXT I
729   PRINT
730   MAT PRINT Z;
1000  PRINT
1032  PRINT"I1="I1,"I2="I2,"N1="N1
1033  PRINT
1034  PRINT"DELTA X/L="D
1035  PRINT
1040  DATA 0,0,0,.111
1045  DATA 0,1.0
4000  END
```

The computer logic first determines if a node is a surface node or an interior node. If J is not equal to one, left-surface nodal equations are not required and the computer logic jumps from Statement 440 to 500. Then at 500 if J is not equal to five, right-surface nodal equations are not required, and the logic jumps to Statement 560 where the nondimensional interior nodal equation (3-34) is applied to the node. Throughout the calculations, $X(I, J)$ represents the current calculated values in the nodal matrix, and $T(I, J)$ represents the previously calculated values in the nodal matrix. Statement

570 defines Y(I, J), which is the difference between the current nodal value and the previously calculated nodal value. With the exception of the isothermal surface nodes, the values of all nodes are recalculated during each successive iteration.

The value of Y(I, J) indicates the convergence of the calculations. As the calculations proceed, the difference between the newly calculated nodal values and the previously calculated values becomes smaller, and the correct temperature distribution is approached. The allowable difference is a function of the accuracy required in the calculations. This allowable difference is determined by the programmer and is read into the computer as variable E1 in Statement 210. Statement 580 determines whether the convergence criterion has been met for each node. If not, the value of U is increased in increments of one for each node that does not satisfactorily meet the convergence criterion. Then the calculated nodal values of X(I, J) are defined as the previously calculated nodal values T(I, J) in Statement 600 for use in the next iteration.

When a surface node is encountered, a different logical path is followed. If J = 1 at Statement 440, then Statement 450 determines if the node falls on an adiabatic surface or on an isothermal surface. If the number of adiabatic nodes on the left surface denoted by I1 is greater than or equal to the row value I, then the nondimensional adiabatic nodal equation is used in Statement 480. If I1 is less than I, the node is assigned the isothermal value T1 in Statement 460. A similar determination is made for the nodes on the right face (J = 5) in Statements 500 to 550. The nondimensional nodal equation for the right vertical surface appears in Statement 540.

After one complete iteration of all 40 nodes, the value of N is increased by an increment in Statement 630 to account for the number of iterations completed. If convergence has not been achieved after a maximum number of iterations (M = 50), provision is made to terminate the iteration procedure in Statement 640. The user defines the desired value in Statement 200. If a solution has been obtained, the iteration is terminated by Statement 640 which is satisfied when all nodes fall within the convergence criterion set in Statement 580. Statement 652 is controlled by the print variable P. The value of P is initially zero as defined in Statement 202, and this value allows the print Statements 654 to 656 to be accomplished after the first iteration only. The program concludes by printing out the final temperature matrix along with parameters I1, I2, and N1.

3-9   Numerical Solutions

The computer program SSHC (see Flow Chart 3-1 and Listing 3-1) was used to obtain a solution to the problem specified in Fig. 3-7. A printout of the solution follows.

The values of the nondimensional nodal temperatures obtained numerically are within 2.0 percent of the values obtained from the analytical solution given by Eq. (3-31). This agreement was obtained with a ten by five grid network. More accurate results, at the expense of more computer time, may be obtained by using a finer grid. The numerical results are shown in graphical form in Fig. 3-8.

The resulting temperature profiles shown in Fig. 3-8 are symmetrical about $x/l = 0.222$. This value results from the choice of a ten by five grid which results in $0 \leq \bar{x} \leq \frac{4}{9}$. The center of the grid is located at $\bar{y} = 0.5$ and $\bar{x} = 0.222$. To show the effect of adiabatic boundary conditions on the vertical sides, the same program was run with $I1 = 6$ and $I2 = 4$. This choice of values corresponds to the top six nodes on the left surface and the top four nodes on the right surface being adiabatic. The results are given in Fig. 3-9. The temperature profiles at values of y/l near unity resemble the linear profiles that exist throughout the material if both vertical surfaces are entirely adiabatic. The temperature profiles at small values of y/l resemble the profiles obtained in Fig. 3-8.

Many other combinations of adiabatic and isothermal boundary conditions along the vertical surfaces can be analyzed. The effect of internal generation can be included by specifying a nonzero value for N1. If there is no internal generation and if the boundary conditions are either isothermal or adiabatic, the temperature profile is independent of the material thermal conductivity. This is true in the results given in Figs. 3-8 and 3-9.

Computer Results 3-1   SSHC
```
TYPE I IF YOU WANT A PROGRAM EXPLANATION.

OTHERWISE, TYPE 2.
? I


THIS PROGRAM SOLVES THE TWO-DIMENSIONAL LAPLACE OR
POISSON EQUATION  FOR STEADY STATE CONDUCTION IN A SOLID
IN A RECTANGULAR GEOMETRY SPECIFIED BY A IOX5 GRID.

THE GOVERNING EQUATION IS EXPRESSED BY USE OF
A FINITE DIFFERENCE APPROXIMATION.

THE TOP AND BOTTOM SURFACES ARE ISOTHERMAL.

THE TWO VERTICAL SIDES ARE EITHER ADIABATIC OR ISOTHERMAL.

THE NUMBER OF ADIABATIC NODES, MEASURED DOWN FROM THE TOP,
ARE SPECIFIED BY THE INPUT VALUES OF II AND I2
FOR THE LEFT AND RIGHT SURFACES RESPECTIVELY.

A LINEAR TEMPERATURE PROFILE IN THE SOLID IS INITIALLY ASSIGNED.

THIS PROFILE IS MODIFIED UNTIL ALL BOUNDARY CONDITIONS
AND THE DIFFERENTIAL EQUATION ARE SATISFIED.

DO YOU WANT A PRINT OUT OF THE INITIALLY ASSIGNED
LINEAR TEMPERATURE DISTRIBUTION?

IF YES TYPE I.  IF NO TYPE 2
? 2
```

Computer Results 3-1 SSHC (*continued*)

THE ITERATION OF THE FINITE DIFFERENCE EQNS BEGINS NOW.
AFTER EACH ITERATION THE U VALUE INDICATES THE NUMBER
OF GRID POINTS WHERE A SOLUTION HAS NOT YET BEEN
OBTAINED TO SUFFICIENT ACCURACY.

IT. NO.	U VALUE
1	40
2	24
3	24
4	24
5	24
6	24
7	24
8	24
9	24
10	23
11	21
12	20
13	16
14	13
15	7
16	0

THE FINAL SOLUTION IS

```
1   1    1    1    1
0   0.433  0.533  0.433  0
0   0.198  0.265  0.198  0
0   0.093  0.128  0.093  0
0   0.045  0.062  0.044  0
0   0.022  0.03   0.021  0
0   0.01   0.014  0.01   0
0   0.005  0.007  0.005  0
0   0.002  0.003  0.002  0
0   0    0    0    0
```

I1= 0   I2= 0   NI= 0

DELTA X/L= 0.111

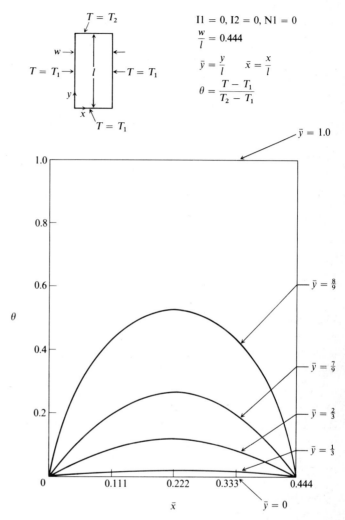

FIGURE 3-8
Two-dimensional conduction in nine by four rectangle with isothermal boundaries.

3-10   Convective Boundary Conditions

To further extend the analysis using a rectangular geometry, the isothermal boundary conditions on the vertical surfaces are replaced by convective boundary conditions. If one uses an energy balance for surface nodes undergoing convection, the following two nodal equations can be derived for the left vertical surface and the right vertical surface (cf. Sec. 3-5); they apply to a square grid network with $\overline{\Delta x} = \overline{\Delta y}$:

$$\theta_{i,j} = \frac{\theta_{i+1,j} + 2\theta_{i,j+1} + \theta_{i-1,j} + (\overline{\Delta x})^2 N1}{4 + 2\overline{\Delta x}\ Bi} \qquad (3\text{-}59)$$

$$\theta_{i,j} = \frac{\theta_{i+1,j} + 2\theta_{i,j-1} + \theta_{i-1,j} + (\overline{\Delta x})^2 N1}{4 + 2\overline{\Delta x}\ Bi} \qquad (3\text{-}60)$$

where

$$\theta = \frac{T - T_\infty}{T_w - T_\infty}$$

$$N1 = \frac{q''' l^2}{k(T_w - T_\infty)}$$

$$Bi = \frac{\bar{h} l}{k}$$

$$\overline{\Delta x} = \frac{\Delta x}{l}$$

The nondimensional ratio $\bar{h}l/k$ was defined earlier as a Biot number and was shown to be proportional to the ratio of internal resistance to external resistance. When the value of this parameter is very small, Eqs. (3-59) and (3-60) reduce to the equations which were used to obtain the solution given in Fig. 3-9.

In general the convection heat transfer coefficient and environmental temperature at the left face are not equal to the value of these parameters at the right face. Suppose that the nodal equations for the left surface and the interior nodes are derived in terms of $\theta = (T - T_{\infty,L})/(T_w - T_{\infty,L})$ where $T_{\infty,L}$ is the environmental temperature adjacent to the left surface. When the energy balance for a node on the right surface is written, the convection is expressed in terms of $T_{\infty,R} - T_{i,j}$. The resulting form of Eq. (3-60) for a right surface node will contain an additional nondimensional term given by $2\overline{\Delta x}\ Bi_R\, \theta_{\infty,R}$ as shown in Eq. (3-61):

$$\theta_{i,j} = \frac{\theta_{i+1,j} + 2\theta_{i,j-1} + \theta_{i-1,j} + (\overline{\Delta x})^2 N1 + 2\overline{\Delta x}\ Bi_R\, \theta_{\infty,R}}{4 + 2\overline{\Delta x}\ Bi} \qquad (3\text{-}61)$$

where

$$\theta_{\infty,R} = \frac{T_{\infty,R} - T_{\infty,L}}{T_w - T_{\infty,L}}$$

When $T_{\infty,R} = T_{\infty,L}$, Eq. (3-61) reduces to the form of Eq. (3-60).

A grid network with $\Delta x = \Delta y$ is not the best choice for all types of boundary conditions. For the rectangular region in Fig. 3-9, the boundary conditions do not

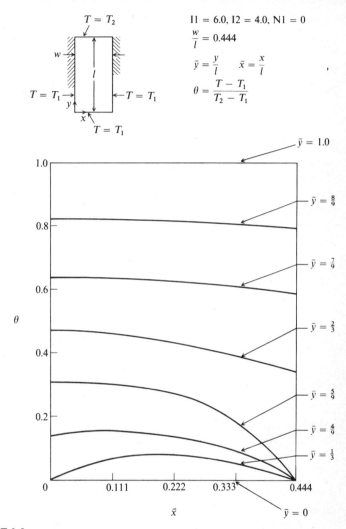

$T = T_2$

w

$T = T_1 \rightarrow$ $\leftarrow T = T_1$

$T = T_1$

$T = T_1$

$I1 = 6.0, I2 = 4.0, N1 = 0$

$\dfrac{w}{l} = 0.444$

$\bar{y} = \dfrac{y}{l}$ $\bar{x} = \dfrac{x}{l}$

$\theta = \dfrac{T - T_1}{T_2 - T_1}$

$\bar{y} = 1.0$

$\bar{y} = \frac{8}{9}$

$\bar{y} = \frac{7}{9}$

$\bar{y} = \frac{2}{3}$

$\bar{y} = \frac{5}{9}$

$\bar{y} = \frac{4}{9}$

$\bar{y} = \frac{1}{3}$

$\bar{x}$

$\bar{y} = 0$

FIGURE 3-9
Two-dimensional conduction in nine by four rectangular grid with isothermal
and adiabatic boundaries.

produce a sharp gradient across the width of the region. Thus a grid network with
$\Delta x = \Delta y$ gives good agreement with the analytical solution even though there are only
half as many nodes across the width of the surface as there are along the length of the
surface. When large temperature gradients occur, a finer grid is necessary to obtain
accurate results. For this reason, when convective boundary conditions are specified

at a surface, the choice of the grid network becomes more critical. The explanation of the need for a finer grid can be found in the physical interpretation of the Biot number. As the Biot number increases, the magnitude of the internal resistance to conduction relative to the external resistance to convection increases. Thus, more nodes are required to represent the interior of the solid in the finite difference approximation. The necessary grid size can be determined by solving the problem several times for successively smaller grids until no appreciable change occurs in the resulting temperature distribution.

One way to increase the ratio of the number of internal nodes to the number of external nodes in the geometry being considered is to choose a rectangular grid with $\Delta y = 2\Delta x$ rather than a square grid. This grid is useful when convective boundary conditions are used along $\bar{x} = 0$ and $\bar{x} = 0.444 = w/l$. It increases the number of internal nodes normal to the surfaces undergoing convection. Large temperature gradients occur in this direction as the value of the Biot number increases. This change in nodal geometry requires that the governing nodal equations be rederived. The resulting equations are given below for $T_{\infty, R} = T_{\infty, L}$, $h_R = h_L$, and $\theta = (T - T_{\infty})/(T_w - T_{\infty})$.

For convection at the left vertical surface and $\Delta y = 2\Delta x$

$$\theta_{i, j} = \frac{\theta_{i+1, j} + 8\theta_{i, j+1} + \theta_{i-1, j} + 4(\overline{\Delta x})^2 \text{N1}}{10 + 8\,\overline{\Delta x}\,\text{Bi}} \tag{3-62}$$

For convection at the right vertical surface and $\Delta y = 2\Delta x$

$$\theta_{i, j} = \frac{\theta_{i+1, j} + 8\theta_{i, j-1} + \theta_{i-1, j} + 4(\overline{\Delta x})^2 \text{N1}}{10 + 8\,\overline{\Delta x}\,\text{Bi}} \tag{3-63}$$

For adiabatic surfaces the equations are identical except that the Biot number Bi is zero.

For interior nodes the equation to be satisfied for $\Delta y = 2\,\Delta x$ is

$$\theta_{i, j} = \frac{\theta_{i-1, j} + \theta_{i+1, j} + 4\theta_{i, j+1} + 4\theta_{i, j-1} + 4(\overline{\Delta x})^2 \text{N1}}{10} \tag{3-64}$$

The solution for a two-dimensional fin with an isothermal boundary condition at both ends is shown in Fig. 3-10. Figure 3-10 shows that the temperature profile fo Bi $= 0.9$ is not one-dimensional, especially for values of $\bar{y}$ near unity. These results were obtained for $T_1 = T_{\infty}$. When T_1 and T_{∞} are not equal, we define $\theta = (T - T_{\infty})/(T_2 - T_{\infty})$, and at the lower isothermal surface we must specify the proper value of $\theta = (T_1 - T_{\infty})/(T_2 - T_{\infty})$.

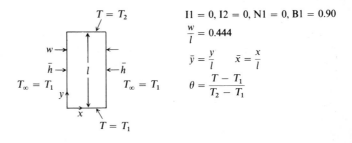

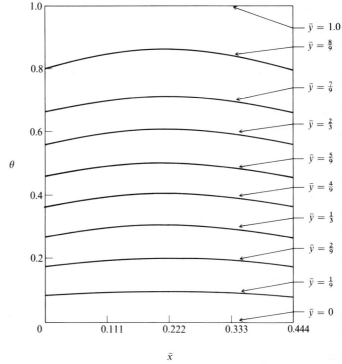

FIGURE 3-10
Two-dimensional conduction with convective and isothermal boundary conditions.

The modifications to the program SSHC used to obtain the results shown in Fig. 3-10 are listed below:

```
20  PRINT"A 19 × 17 GRID, Y = 2X"
40  DIM Y(19, 17), T(19, 17), X(19, 17), Z(19, 17)
70  READ B1
```

(Continued on next page)

```
100   LET Y1 = 19
120   LET X1 = 17
460   LET X(I, J) = (T(I + 1, J) + 8*T(I, J + 1) + T(I − 1, J)
              + 4*(D↑2)*N1)/(10 + 8*D*B1)
480   LET X(I, J) = (T(I + 1, J) + 8*T(I, J + 1) + T(I − 1, J)
              + 4*(D↑2)*N1)/10
520   LET X(I, J) = (T(I + 1, J) + 8*T(I, J − 1) + T(I − 1, J)
              + 4*(D↑2)*N1)/(10 + 8*D*B1)
540   LET X(I, J) = (T(I + 1, J) + 8*T(I, J − 1) + T(I − 1, J)
              + 4*(D↑2)*N1)/10
560   LET X(I, J) = (T(I − 1, J) + T(I + 1, J) + 4*T(I, J + 1)
              + 4*T(I, J − 1) + 4*(D↑2)*N1)/10
1040  DATA 0, 0, 0, 0.0278
1050  DATA 0.90
```

A solution for $\theta = 2.0$ at the lower isothermal surface is given in Fig. 3-11 for a Biot number of 10. The abscissa is $\bar{y}$ rather than $\bar{x}$. For this value of Bi = 10, the temperature field is clearly two-dimensional.

Boundary conditions or nondimensional parameters can be varied to solve a number of different problems. By observing the results and making changes based on the results of these modifications an improved design can be obtained. The results in Fig. 3-11 show that there is a negative temperature gradient causing heat transfer from the wall to the fin base. Note that y is now measured from the top surface where $\theta = 1.0$. A practical design problem might be to determine what modifications to the fin can be made to prevent heat transfer from the fin base at $\bar{y} = 0$. The addition of insulation on the surfaces near the base where $\theta = 1.0$ would reverse the temperature gradient. By varying the value of I1 and I2 and observing the resulting temperature field, the minimum number of adiabatic nodes can be determined. Another possibility is to leave the surfaces exposed to the environment and add significant internal generation to obtain the desired effect. The numerical analysis can be used to determine the required value of the internal generation by varying N1. It is also possible to consider anisotropic materials where the value of the thermal conductivity in one direction may not equal the thermal conductivity in another direction, for example $k_y = 2k_x$. The governing equation is then written

$$\frac{\partial}{\partial x}\left(k_x \frac{\partial T}{\partial x}\right) + \frac{\partial}{\partial y}\left(k_y \frac{\partial T}{\partial y}\right) = 0 \qquad (3\text{-}65)$$

If one treats k_x and k_y as constants and uses the previous definition of θ, the finite difference approximation to the Laplace equation is written

$$\frac{\theta_{i+1, j} - 2\theta_{i, j} + \theta_{i-1, j}}{(\Delta x)^2} + 2\frac{\theta_{i, j+1} - 2\theta_{i, j} + \theta_{i, j-1}}{(\Delta y)^2} = 0 \qquad (3\text{-}66)$$

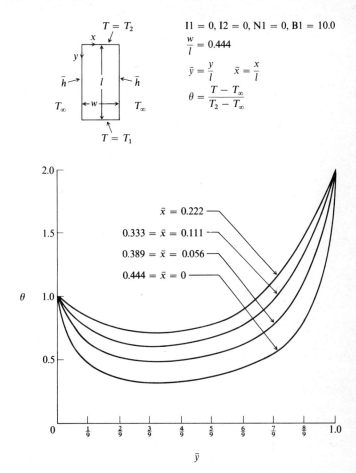

FIGURE 3-11
Two-dimensional conduction in a rectangular fin with isothermal ends.

3-11   Comparison of One-dimensional and Two-dimensional Results

It has been pointed out that when adiabatic boundary conditions apply entirely over the two exposed surfaces of a fin, the resulting temperature is one-dimensional, varying linearly along the length of the body. We now show how the magnitude of the Biot number determines whether a one-dimensional analysis can be used to approximate the temperature field in a fin with isothermal boundary conditions at $\bar{y} = 0$ and $\bar{y} = 1$.

Four solutions which compare one-dimensional results with two-dimensional results for the temperature distribution in a fin are given in Fig. 3-12. The one-dimen-

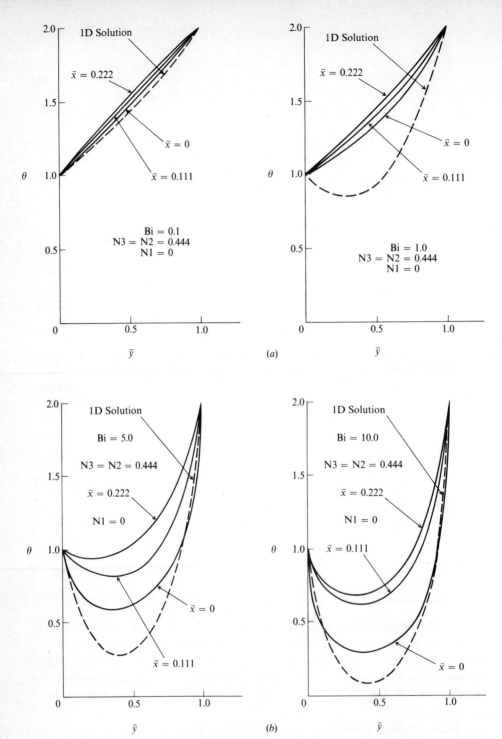

FIGURE 3-12
Comparison of one-dimensional and two-dimensional solutions.

sional results are given by the solution to Eq. (2-27). For a rectangular fin with no internal generation and one-dimensional conduction in the y direction, the governing equation is

$$\frac{d^2\theta}{d\bar{y}^2} - \frac{2\bar{h}l^2}{ke}\theta = 0 \qquad (3\text{-}67)$$

where $2hl^2/ke = 2\,\text{Bi}/\text{N2}$. We have defined $\text{Bi} = \bar{h}l/k$ and $\text{N2} = e/l$. The solution to Eq. (3-67) is

$$\theta = C_1 \exp\left(\sqrt{\frac{2\,\text{Bi}}{\text{N2}}}\,\bar{y}\right) + C_2 \exp\left(-\sqrt{\frac{2\,\text{Bi}}{\text{N2}}}\,\bar{y}\right)$$

with boundary conditions

$$\text{at} \qquad \bar{y} = 0 \qquad \theta = 1$$
$$\text{at} \qquad \bar{y} = 1 \qquad \theta = 2$$

Application of these boundary conditions gives

$$C_1 = \frac{2 - \exp\left(-\sqrt{2\,\text{Bi}/\text{N2}}\right)}{\exp\left(\sqrt{2\,\text{Bi}/\text{N2}}\right) - \exp\left(-\sqrt{2\,\text{Bi}/\text{N2}}\right)}$$

and

$$C_2 = 1 - C_1$$

and the nondimensional temperature profile is specified. For the ten by five grid under consideration, the value of N2 is $\frac{4}{9} = 0.444$.

The results in Fig. 3-12 show that the one-dimensional results agree closely with the two-dimensional results for a Biot number equal to 0.1. For larger Biot numbers there is a distinct difference between the one-dimensional and two-dimensional solutions. The differences between the results produced by a one-dimensional model and a two-dimensional model can also be shown for other boundary conditions at $\bar{y} = 1$. We now see the reason for limiting the value of the Biot number to less than 0.1 when a one-dimensional analysis is used. This procedure was followed in Chap. 2.

3-12   Two-dimensional Conduction in Fins

In Sec. 3-11 the analysis of conduction in a two-dimensional fin was made with isothermal boundary conditions at each end. The next logical extension is to analyze fins with either adiabatic or convective boundary conditions at the end. When the isothermal boundary condition is replaced by an adiabatic or convective boundary condition, special corner nodal equations are needed. The equation for the corner node formed by the intersection of one surface with convection and a second adiabatic surface is given by Eq. (3-52). The equation for a corner node with convection from both of the intersecting surfaces is derived in a similar manner.

When there is convection from both of the intersecting surfaces, separate finite difference equations for each corner node ($J = 1$ and $J = 17$) and a third equation for the remaining nodes on the end surface are required to specify the temperature distribution. For the general situation we consider the environment temperature at the end of the fin T_∞ not equal to $T_{\infty, L}$ or $T_{\infty, R}$. We also consider the Biot number at the end of the fin $(\bar{h}l/k)_E$ not equal to $(\bar{h}l/k)_L$ or $(\bar{h}l/k)_R$. We define $\theta = (T - T_\infty)/(T_w - T_\infty)$ for all nodal equations. The conservation of energy principle applied to the end surface nodes results in the following three equations:

For the left corner

$$\theta_{i,j} = \frac{\theta_{i-1,j} + 4\theta_{i,j+1} + 2(\overline{\Delta x})^2 N1 + 4\overline{\Delta x}\ Bi_L(T_{\infty,L} - T_\infty)/(T_w - T_\infty)}{2\overline{\Delta x}\ Bi_E + 4\overline{\Delta x}\ Bi_L + 5} \tag{3-69}$$

For the right corner

$$\theta_{i,j} = \frac{\theta_{i-1,j} + 4\theta_{i,j-1} + 2(\overline{\Delta x})^2 N1 + 4\overline{\Delta x}\ Bi_R(T_{\infty,R} - T_\infty)/(T_w - T_\infty)}{2\overline{\Delta x}\ Bi_E + 4\overline{\Delta x}\ Bi_R + 5} \tag{3-70}$$

For the remaining end surface nodes

$$\theta_{i,j} = \frac{2\theta_{i,j-1} + \theta_{i-1,j} + 2\theta_{i,j+1} + 2(\overline{\Delta x})^2 N1}{2\overline{\Delta x}\ Bi_E + 5} \tag{3-71}$$

The program modifications to SSHC required to treat convection at the fin end are made by using the statements shown below in addition to the previous changes made to the SSHC computer program. We consider the special situation when $T_\infty = T_{\infty, L} = T_{\infty, R}$ and $(Bi)_L = (Bi)_R$.

```
 30   PRINT"2D FIN, CONVECTION END B.C."
 70   READ B1, B2
430   FOR I = 2 TO Y1
431   IF I<>Y1 THEN 440
432   IF J = 1.0 THEN 436
433   IF J = X1 THEN 438
434   LET X(I, J) = (2*T(I, J − 1) + T(I − 1, J) + 2*T(I, J + 1) + 2*(D↑2)
                 *N1)/(2*D*B2 + 5)
435   GO TO 570
436   LET X(I, J) = (T(I − 1, J) + 4*T(I, J + 1) + 2*(D↑2)*N1)/(2*B2*D
                 + 4*D*B1 + 5)
437   GO TO 570
```

438   LET X(I, J) = (T(I − 1, J) + 4∗T(I, J − 1) + 2∗(D↑2)∗N1)/(2∗B2∗D
          + 4∗D∗B1 + 5)
439   GO TO 570
1050   DATA 1.0, 1.0

When a solution is obtained with the modifications given, the results show the behavior due to replacing an isothermal boundary condition with a convective boundary condition at the end of the fin ($\bar{y} = 1.0$). The results for the two-dimensional temperature distribution in fins with either a convective or adiabatic boundary condition at the end are given in Fig. 3-13 for a Biot number of 1.0. The difference between the solutions for the two different boundary conditions increases as $(Bi)_E$ increases, as can be shown by obtaining additional solutions.

Another extension to the analysis is to consider two-dimensional conduction in a fin of varying cross section, such as a tapered or parabolic fin. The finite difference approach can be used to analyze the two-dimensional problem for Biot numbers greater than 0.1 by including logic statements to identify the nodes on the end of each row in the matrix which represents the fin cross section. The proper nodal equations for the end nodes on each row would be surface nodal equations. Equations for near-surface nodes may also be required. These types of equations were discussed in Sec. 3-7.

3-13   Cylindrical and Spherical Coordinates

The general conduction equation in cylindrical coordinates for constant thermal conductivity is given by Eq. (3-72) (cf. Ref. 3-3). This equation is derived by applying the conservation principle to a differential element defined in cylindrical coordinates:

$$\frac{\partial^2 T}{\partial r^2} + \frac{1}{r}\frac{\partial T}{\partial r} + \frac{1}{r^2}\frac{\partial^2 T}{\partial \phi^2} + \frac{\partial^2 T}{\partial z^2} + \frac{q'''}{k} = \frac{1}{\alpha}\frac{\partial T}{\partial t} \qquad (3\text{-}72)$$

The radial variable is denoted by r, the axial variable by z, and the circumferential variable by ϕ. When there is no axial temperature variation and steady-state conduction exists, the governing equation reduces to

$$\frac{\partial^2 T}{\partial r^2} + \frac{1}{r}\frac{\partial T}{\partial r} + \frac{1}{r^2}\frac{\partial^2 T}{\partial \phi^2} = -\frac{q'''}{k} \qquad (3\text{-}73)$$

This mathematical model assumes conduction in the radial direction and around the circumference of the cylinder.

Consider the hollow cylinder shown in Fig. 3-14. The transformation of the governing partial differential equation to a finite difference equation has been discussed in Sec. 3-5. If one uses the nodal representation shown in Fig. 3-14, and denotes the

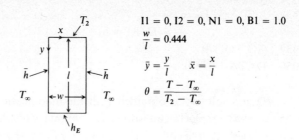

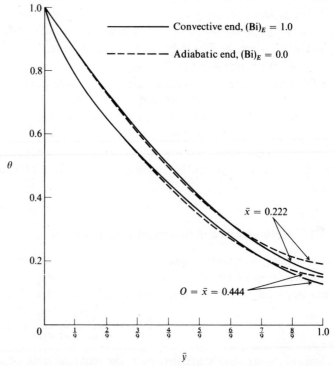

FIGURE 3-13
Two-dimensional conduction in fins with convective or adiabatic ends.

radial distance between nodes as δ, a central finite difference approximation can be used to express Eq. (3-73) as a finite difference equation:

$$\frac{1}{\delta^2}\left(T_{i,\,j+1} - 2T_{i,\,j} + T_{i,\,j-1}\right) + \frac{1}{r_j}\frac{T_{i,\,j+1} - T_{i,\,j-1}}{2\delta}$$

$$+ \frac{1}{r_j{}^2}\frac{T_{i+1,\,j} - 2T_{i,\,j} + T_{i-1,\,j}}{(\Delta\phi)^2} + \frac{q'''}{k} = 0$$

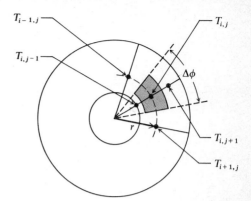

FIGURE 3-14
Cylindrical element $T = f(r, \phi)$.

This equation can be arranged to give

$$T_{i,j} = \left[2r_j{}^2(\Delta\phi)^2(T_{i,j+1} + T_{t,j-1}) + r_j(\Delta\phi)^2\delta(T_{i,j+1} - T_{i,j-1}) \right.$$

$$\left. + 2\delta^2(T_{i+1,j} + T_{i-1,j}) + \frac{2q'''}{k} r_j{}^2(\Delta\phi)^2\delta^2 \right] \Bigg/$$

$$[4r_j{}^2(\Delta\phi)^2 + 4\delta^2] \qquad (3\text{-}74)$$

This equation applies for all interior nodes. Surface nodes require special equations. We define $\theta = (T - T_1)/(T_0 - T_1)$ by use of suitable reference temperatures. We also define nondimensional variables $\bar{r}_j = r_j/R$ and $\bar{\delta} = \delta/R$, where R is a reference radius. The nondimensional form of Eq. (3-74) is then written

$$\theta_{i,j} = [2\bar{r}_j{}^2(\Delta\phi)^2(\theta_{i,j+1} + \theta_{i,j-1}) + \bar{r}_j\bar{\delta}(\Delta\phi)^2(\theta_{i,j+1} - \theta_{i,j-1})$$

$$+ 2\bar{\delta}^2(\theta_{i+1,j} + \theta_{i-1,j}) + 2\bar{r}_j{}^2(\Delta\phi)^2\bar{\delta}^2 \text{N1}]/[4\bar{r}_j{}^2(\Delta\phi)^2 + 4\bar{\delta}^2] \qquad (3\text{-}75)$$

where $\text{N1} = \dfrac{q'''R^2}{k(T_0 - T_1)}$

When convection occurs at a surface, such as the inner surface of the hollow cylinder, a surface nodal equation is required. Consider the inner surface node shown in Fig. 3-15. For steady-state heat conduction, the heat transfer into the volume element around the surface node must equal the heat transfer out of the element.

The temperature of the fluid inside the hollow cylinder is T_B, the bulk temperature. The bulk temperature is an average temperature across the tube. To derive an expression for the bulk temperature, consider the convected energy given by $\dot{m}h = \dot{m}c_p T_B$, where $\dot{m} = \rho AV$. The average enthalpy across the tube cross section is h, the

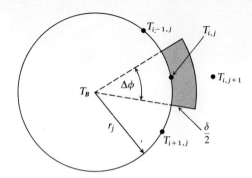

FIGURE 3-15
Inner surface node.

average velocity is V, the average density is ρ, and the average specific heat is c_p. In terms of local values we can write

$$\dot{m} = \int_0^R \rho(2\pi r)u \, dr \qquad (3\text{-}76)$$

and

$$\dot{m}c_p T_B = \int_0^R \rho(2\pi r)uc_p T \, dr \qquad (3\text{-}77)$$

In general, one writes $\rho = \rho(r)$, $u = u(r)$, $c_p = c_p(r)$, and $T = T(r)$. For incompressible flow with constant specific heat, the equation for T_B becomes

$$T_B = \frac{\int_0^R uTr \, dr}{\int_0^R ur \, dr} \qquad (3\text{-}78)$$

Methods for determining the velocity and temperature profiles for internal convection are discussed in Chap. 6. In this chapter we treat T_B as a known reference temperature.

Assuming unit depth in the axial direction and equating the net heat transfer equal to zero gives

$$k \frac{\delta}{2} \frac{T_{i-1,j} - T_{i,j}}{r_j \, \Delta\phi} + k\left(r_j + \frac{\delta}{2}\right) \Delta\phi \, \frac{T_{i,j+1} - T_{i,j}}{\delta}$$

$$+ k \frac{\delta}{2} \frac{T_{i+1,j} - T_{i,j}}{r_j \, \Delta\phi} + \bar{h}r_j \, \Delta\phi \, (T_B - T_{i,j})$$

$$+ q''' r_j \, \Delta\phi \, \frac{\delta}{2} = 0$$

Multiplying each term by $2r\delta \,\Delta\phi$ and solving for $T_{i,j}$ gives

$$T_{i,j} = \left[\delta^2(T_{i-1,j} + T_{i+1,j}) + 2\left(r_j^2 + \frac{r_j\delta}{2}\right)(\Delta\phi)^2 T_{i,j+1}\right.$$
$$\left. + \frac{2\bar{h}\delta}{k}\, r_j^2(\Delta\phi)^2 T_B + \frac{q'''r_j^2(\Delta\phi)^2\delta^2}{k}\right] \Bigg/$$
$$\left[2\delta^2 + 2\left(r_j^2 + \frac{r_j\delta}{2}\right)(\Delta\phi)^2 + \frac{2\bar{h}\delta}{k}\, r_j^2(\Delta\phi)^2\right] \qquad (3\text{-}79)$$

Because of the convection term we choose $\theta = (T - T_B)/(T_0 - T_B)$. We are still free to choose T_0. It might be an inlet fluid temperature or an isothermal surface temperature. The nondimensional form of Eq. (3-79) is

$$\theta_{i,j} = \left[\delta^2(\theta_{i-1,j} + \theta_{i+1,j}) + 2\left(\bar{r}_j^2 + \frac{\bar{r}_j\delta}{2}\right)(\Delta\phi)^2\theta_{i,j+1}\right.$$
$$\left. + \bar{r}_j^2(\Delta\phi)^2\delta^2 \mathrm{N}1\right]$$
$$\left[2\delta^2 + 2\left(\bar{r}_j^2 + \frac{\bar{r}_j\delta}{2}\right)(\Delta\phi)^2 + 2\delta\bar{r}_j^2(\Delta\phi)^2\mathrm{Bi}\right] \qquad (3\text{-}80)$$

where $\mathrm{Bi} = \bar{h}R/k$ and $\mathrm{N}1 = q'''R^2/k(T_0 - T_B)$

When the cylinder is solid rather than hollow, special consideration must be given to the central node at $r = 0$. At $r = 0$ the second term in Eq. (3-72) $(1/r)(\partial T/\partial r)$ is indeterminate. This term can be evaluated by using L'Hopital's rule:

$$\lim_{r\to 0}\frac{1}{r}\frac{\partial T}{\partial r} = \lim_{r\to 0}\frac{(\partial/\partial r)(\partial T/\partial r)}{(\partial/\partial r)(r)} = \frac{\partial^2 T}{\partial r^2}$$

In a like manner the third term in Eq. (3-72) becomes

$$\lim_{r\to 0}\frac{1}{r^2}\frac{\partial^2 T}{\partial\phi^2} = \lim_{r\to 0}\frac{(\partial/\partial r)(\partial^2 T/\partial\phi^2)}{2r} = \lim_{r\to 0}\frac{(\partial^2/\partial r^2)(\partial^2 T/\partial\phi^2)}{2} = 0$$

If one uses these results, the governing partial differential equation at $r = 0$ can be written

$$\frac{\partial^2 T}{\partial r^2} + \frac{\partial^2 T}{\partial r^2} + \frac{q'''}{k} = 0 \qquad (3\text{-}81)$$

or in finite difference form

$$\frac{2T_{i,j+1} - 4T_{i,j} + 2T_{i,j-1}}{\delta^2} + \frac{q'''}{k} = 0 \qquad (3\text{-}82)$$

Because of symmetry at $r = 0$, $T_{i,j+1} = T_{i,j-1}$. Since all radial vectors emanate from the central node, it is appropriate to write

$$T_{i,j+1}|_{\text{avg}} = T_{i,j-1}|_{\text{avg}}$$

where $T|_{\text{avg}}$ is the arithmetic mean for the T values on a circle of radius δ around $r = 0$. The finite difference equation for the central node at $r = 0$ may now be written

$$\frac{4T_{i,j+1}|_{\text{avg}} - 4T_{i,j}}{\delta^2} + \frac{q'''}{k} = 0$$

upon rearranging

$$T_{i,j} = \frac{q'''\delta^2}{4k} + T_{i,j+1}|_{\text{avg}} \qquad r = 0 \qquad (3\text{-}83)$$

In nondimensional form we write

$$\theta_{i,j} = \delta^2 N1 + \theta_{i,j+1}|_{\text{avg}} \qquad (3\text{-}84)$$

Another two-dimensional model of practical importance is two-dimensional steady-state conduction in the radial and axial directions. For this model $T = T(r, z)$, and Eq. (3-72) reduces to

$$\frac{\partial^2 T}{\partial r^2} + \frac{1}{r}\frac{\partial T}{\partial r} + \frac{\partial^2 T}{\partial z^2} + \frac{q'''}{k} = 0 \qquad (3\text{-}85)$$

In finite difference form

$$\frac{1}{\delta^2}(T_{i,j+1} - 2T_{i,j} + T_{i,j-1}) + \frac{1}{r_j}\left(\frac{T_{i,j+1} - T_{i,j-1}}{2\delta}\right)$$

$$+ \frac{1}{\delta^2}(T_{i+1,j} - 2T_{i,j} + T_{i-1,j}) + \frac{q'''}{k} = 0$$

Solving for $T_{i,j}$ gives

$$T_{i,j} = \frac{1}{4}\left[\left(1 + \frac{\delta}{2r_j}\right)T_{i,j+1} + \left(1 - \frac{\delta}{2r_j}\right)T_{i,j-1}\right.$$

$$\left. + T_{i+1,j} + T_{i-1,j} + \frac{q'''\delta^2}{k}\right] \qquad (3\text{-}86)$$

In nondimensional form

$$\theta_{i,j} = \frac{1}{4}\left[\left(1 + \frac{\delta}{2\bar{r}_j}\right)\theta_{i,j+1} + \left(1 - \frac{\delta}{2\bar{r}_j}\right)\theta_{i,j-1}\right.$$

$$\left. + \theta_{i+1,j} + \theta_{i-1,j} + \delta^2 N1\right] \qquad (3\text{-}87)$$

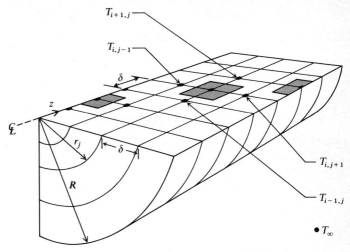

FIGURE 3-16
Cylindrical element $T = f(r, z)$.

Equation (3-87) applies to the interior nodes in Fig. 3-16. The network has been chosen such that $\Delta r = \Delta z = \delta$. Figure 3-16 shows a cross section of an interior nodal element. It is actually a concentric cylindrical shell of thickness Δr and width Δz.

The central nodes at $r = 0$ again require special analysis. As before, the second term in Eq. (3-85) can be written

$$\lim_{r \to 0} \frac{1}{r} \frac{\partial T}{\partial r} = \frac{\partial^2 T}{\partial r^2}$$

The finite difference form of the governing equation is then written

$$\frac{2}{\delta^2}(T_{i,\,j+1} - 2T_{i,\,j} + T_{i,\,j-1}) + \frac{1}{\delta^2}(T_{i+1,\,j} - 2T_{i,\,j} + T_{i-1,\,j})$$

$$+ \frac{q'''}{k} = 0 \qquad r = 0$$

Combining terms and introducing θ and $\bar{\delta}$ gives

$$2(\theta_{i,\,j+1} + \theta_{i,\,j-1}) + \theta_{i+1,\,j} + \theta_{i-1,\,j} - 6\theta_{i,\,j} + \delta^2 \text{N1} = 0$$

By symmetry, $\theta_{i,\,j-1} = \theta_{i,\,j+1}$ at $\bar{r} = 0$, and one can write

$$\theta_{i,\,j+1} - \tfrac{3}{2}\theta_{i,\,j} + \tfrac{1}{4}(\theta_{i+1,\,j} + \theta_{i-1,\,j}) + \tfrac{1}{4}\delta^2 \text{N1} = 0 \qquad (3\text{-}88)$$

This equation gives the expression for $\theta_{i,\,j}$ when $\bar{r} = 0$.

A surface node is also shown in Fig. 3-16. By using an energy balance on this node which is undergoing convection with the environment, one obtains

$$\theta_{i,j} = \frac{(1 - \tfrac{1}{2}\delta)\theta_{i,\,j-1} + \tfrac{1}{2}(\theta_{i-1,\,j} + \theta_{i+1,\,j}) + \left(\dfrac{\bar{\delta}^2}{2}\right)\text{N1}}{2 + \delta\text{Bi} - \tfrac{1}{2}\delta} \tag{3-89}$$

The extension to spherical coordinates follows the same reasoning used in cylindrical coordinates. The general conduction equation in spherical coordinates for constant thermal conductivity is (cf. Ref. 3-3)

$$\frac{1}{r^2}\frac{\partial}{\partial r^2}(rT) + \frac{1}{r^2 \sin \phi}\frac{\partial}{\partial \phi}\left(\sin \phi \frac{\partial T}{\partial \phi}\right)$$

$$+ \frac{1}{r^2 \sin^2 \phi}\frac{\partial^2 T}{\partial \psi^2} + \frac{q'''}{k} + \frac{1}{\alpha}\frac{\partial T}{\partial t} \tag{3-90}$$

For two-dimensional, steady-state conduction in a sphere with $T = f(r, \phi)$

$$\frac{\partial^2 T}{\partial r^2} + \frac{2}{r}\frac{\partial T}{\partial r} + \frac{1}{r^2 \sin \phi}\frac{\partial}{\partial \phi}\left(\sin \phi \frac{\partial T}{\partial \phi}\right) = -\frac{q'''}{k} \tag{3-91}$$

The finite difference approximation to this equation may be written directly with the precaution that special equations are again needed when $r = 0$. At $r = 0$

$$\frac{\partial^2 T}{\partial r^2} + \frac{2}{r}\frac{\partial T}{\partial r} = 3\left(\frac{\partial^2 T}{\partial r^2}\right)_{r=0} = \frac{3T_{i+1,\,j} - 6T_{i,\,j} + 3T_{i-1,\,j}}{(\Delta r)^2}$$

By symmetry, $T_{i+1,\,j} = T_{i-1,\,j}$, and the final form of the finite difference equation at $r = 0$ is

$$\frac{6T_{i+1,\,j} - 6T_{i,\,j}}{(\Delta r)^2} = -\frac{q'''}{k}$$

or

$$T_{i,\,j} = \frac{q'''(\Delta r)^2}{6k} + T_{i+1,\,j} \tag{3-92}$$

and in terms of nondimensional values

$$\theta_{i,\,j} = \tfrac{1}{6}(\overline{\Delta r})^2\text{N1} + \theta_{i+1,\,j} \tag{3-93}$$

where $\text{N1} = q'''R^2/k(T_0 - T_1)$.

The computer program SSHC can be modified to analyze conduction in cylindrical and spherical coordinates. The necessary program changes will be discussed in the next section, along with typical results for steady-state, two-dimensional conduction in cylindrical coordinates.

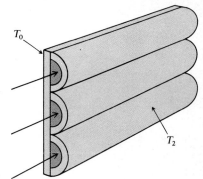

T_0

T_2

FIGURE 3-17
Cylindrical configuration.

3-14   Results for Cylindrical Coordinates

As an example of a two-dimensional problem with $T = T(r, \phi)$, consider the configuration shown in Fig. 3-17. A thin vertical surface is maintained at a temperature T_0, and semicircular tubes are attached to the right side of the flat surface. The outer surface temperature of the tubes is maintained constant at T_2. It is desired to determine the temperature distribution within the tube walls for various fluids flowing through the tubes. We will assume that the bulk temperature of the fluids is equal to the temperature of the vertical surface. Isothermal boundary conditions can be used along each surface of a tube except for the inner tube surface at $r = r_1$. Here convection from the fluid flowing inside the pipe will occur. The convection coefficient is a function of the physical properties of the fluid and the Reynolds number associated with the flow. Evaluation of convection coefficients for internal flows is discussed in Chaps. 6 and 7. The boundary conditions are indicated in Fig. 3-18.

   The finite difference equations needed are Eq. (3-75) for the interior nodes and Eq. (3-80) for the inner surface of the tube. Isothermal boundary conditions apply along $\phi = 0$ and $\phi = \pi$ for $R_1 \le r \le R_2$ and at $r = R_2$ for $0 < \phi < \pi$, with $\theta = (T - T_B)/(T_w - T_B)$, $\bar{r} = r/R_2$, $\bar{\delta} = \delta/R_2$, and $\mathrm{Bi} = \bar{h}R_2/k$.

   A program CYLIN used to obtain a solution to this problem is given in Listing 3-2. A list of variables appears in Table 3-2. It can be seen that 16 nodes were used along the angular direction and six nodes along the radial direction (Statements 100 and 120). Statements 450 through 458 employ Eq. (3-75). The isothermal boundary conditions are assigned in Statements 480, 496, and 498. The nondimensional term, $\mathrm{N1} = q'''R_2{}^2/k(T_2 - T_B)$, and the nondimensional radius ratio, $\mathrm{R0} = R_1/R_2$, are set in Statement 50. The value of the Biot number $\mathrm{B1} = \bar{h}R_2/k$ is set in Statement 60. We let D denote the nondimensional radial increment $\bar{\delta}$. Since δ is given by $(R_2 - R_1)/5$ for a grid with six nodes in the radial direction ($\mathrm{R1} = 6$), it follows that $\bar{\delta} = (1 - \mathrm{R0})/5$.

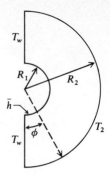

FIGURE 3-18
Cylindrical boundary conditions.

In general we write $\bar{\delta} = (1 - R0)/(R1 - 1.0)$. This value is calculated in Statement 140. Correspondingly, the number of angular increments in the sixteen by six grid representing the semicircular region is 15, each with a value of $\pi/15$. In general the value of each angular increment is $P = \pi/(A1 - 1.0)$ as calculated in Statement 160.

In the finite difference equation for conduction in cylindrical coordinates, the value of the radius which specifies the location of each grid appears. [See Eq. (3-80).] This variable, the nondimensional radius value for $R_1/R_2 \le r/R_2 \le 1.0$, is called R in the program and is calculated in Statement 434. As the value of $A = J$ (Statement 432) varies from 1.0 to $R1 = 6.0$, the value of R varies from $R0 = R_1/R_2$ to 1.0. The effect of internal generation, geometry, and convective boundary conditions can be studied by varying N1, R0, and B1, respectively. These values are specified in the data Statements 1040 and 1045.

Table 3-2   VARIABLES FOR PROGRAM CYLIN

Variable	Definition
N1	$q'' R_2^2/k(T_2 - T_B)$
R0	R_1/R_2
B1	$\bar{h} R_2/k$
A1	$A1 - 1 =$ number of angular increments $\Delta\phi$ between $0 \le \phi \le \pi$
R1	$R1 - 1 =$ number of radial increments δ between $R_1 < r < R_2$
D	$\bar{\delta} = \delta/R_2 = [(R_2 - R_1)/R_2]/(R_1 - 1)$
M, P1, N	Counting variables
E1	Convergence criterion
T(I, J)	Current nodal temperatures
Y(I, J)	Difference in nodal temperatures
U	Number of nodes not satisfying convergence criterion
R	Value of $\bar{r} = r/R_2$ for $R_1/R_2 \le r/R_2 \le 1.0$
X(I, J)	Newly calculated nodal temperatures
Z(I, J)	Truncated nodal temperatures

Listing 3-2   CYLIN

```
10    PRINT"TYPE 1 IF YOU WANT A PROGRAM EXPLANATION."
12    PRINT
13    PRINT"OTHERWISE, TYPE 2."
14    INPUTL
16    IF L=1 THEN 20
18    IF L=2 THEN 40
20    PRINT
21    PRINT"THIS PROGRAM SOLVES THE TWO-DIMENSIONAL LAPLACE OR"
22    PRINT"POISSON EQUATIONS FOR STEADY STATE HEAT CONDUCTION IN A SOLID"
23    PRINT"IN CYLINDRICAL COORDINATES SPECIFIED BY A 16X6 GRID."
24    PRINT
25    PRINT"THE GOVERNING EQUATION IS EXPRESSED BY USE OF A"
26    PRINT"FINITE DIFFERENCE APPROXIMATION."
27    PRINT
28    PRINT"THE OUTSIDE SURFACE OF THE CYLINDER IS ISOTHERMAL."
29    PRINT
30    PRINT"THE INSIDE SURFACE OF THE CYLINDER HAS A CONVECTIVE"
31    PRINT"BOUNDARY CONDITION."
32    PRINT
33    PRINT"THE TEMPERATURE IS A FUNCTION OF RADIUS AND CIRCUMFERENCE."
34    PRINT
35    PRINT"A UNIFORM TEMPERATURE DISTRIBUTION IN THE SOLID"
36    PRINT"IS ASSIGNED INITIALLY."
37    PRINT
38    PRINT"THIS DISTRIBUTION IS MODIFIED UNTIL ALL BOUNDARY"
39    PRINT"CONDITIONS AND THE DIFFERENTIAL EQUATION ARE SATISFIED."
40    DIM T(16,6),X(16,6),Y(16,6),Z(16,6)
45    REM   INPUT NON-DIMENSIONAL PARAMETERS
50    READ N1,RO
60    READ B1
95    REM ASSIGN GRID SIZE
100   LETA1=16
120   LETR1=6
140   LETD=(1-RO)/(R1-1.0)
160   LETP=3.14159/(A1-1.0)
200   LETM=100
202   LETP1=0
210   LETE1=0.001
290   REM ASSIGN INITIAL TEMPERATURE DISTRIBUTION
300   FOR I=1 TO A1
310   FOR J=1 TO R1
330   LETT(I,J)=1.0
340   NEXT J
350   NEXT I
355   PRINT
360   PRINT"DO YOU WANT A PRINT OUT OF THE INITIALLY ASSIGNED"
362   PRINT"LINEAR TEMP DISTRIBUTION?"
364   PRINT
366   PRINT"IF YES TYPE 1. IF NO TYPE 2"
370   INPUTB6
372   IF B6=1 THEN 380
374   IF B6=2 THEN 400
380   PRINT"THE ASSIGNED TEMPERATURE DISTRIBUTION IS:"
382   FOR I=1 TO A1 STEP 3
383   FOR J=1 TO R1
384   LETY(I,J)=INT(T(I,J)*10^2+0.5)/10^2
385   PRINT Y(I,J)
386   NEXT J
387   PRINT
388   NEXT I
400   PRINT
402   PRINT"THE ITERATION OF THE FINITE DIFFERENCE EQNS BEGINS NOW."
404   REM INITIALIZE COUNTING VARIABLES
405   LETN=0
410   LETU=0
```

Listing 3–2   CYLIN (*Continued*)

```
420   FOR J=1 TO R1
430   FOR I=1 TO A1
432   LETA=J
434   LETR=R0+(A-1)*D
440   IF I=1 THEN 496
442   IF I=A1 THEN 498
444   IF J=1 THEN 460
446   IF J=R1 THEN 480
450   LETZ1=2*(R^2)*(P^2)*(T(I,J+1)+T(I,J-1))
452   LETZ2=R*D*(P^2)*(T(I,J+1)-T(I,J-1))
454   LETZ3=2*(D^2)*(T(I+1,J)+T(I-1,J))
456   LETZ4=2*(R^2)*(P^2)*(D^2)*N1
458   LETX(I,J)=(Z1+Z2+Z3+Z4)/(4*(R^2)*(P^2)+4*(D^2))
459   GOTO 570
460   LETS1=(D^2)*(T(I-1,J)+T(I+1,J))
462   LETS2=2*((R^2)+R*D/2)*(P^2)*T(I,J+1)
464   LETS3=(R^2)*(P^2)*(D^2)*N1
466   LETS4=2*(D^2)+2*((R^2)+R*D/2)*(P^2)+2*D*(R^2)*(P^2)*B1
468   LETX(I,J)=(S1+S2+S3)/S4
469   GOTO 570
480   LETX(I,J)=1.0
490   GOTO 570
496   LETX(I,J)=0
497   GOTO 570
498   LETX(I,J)=0
499   GOTO 570
570   LETY(I,J)=X(I,J)-T(I,J)
580   IF ABS(Y(I,J))<E1 THEN 600
590   LETU=U+1
600   LETT(I,J)=X(I,J)
610   NEXT I
620   NEXT J
630   LETN=N+1
640   IF N>=M THEN 700
650   IF U=0 THEN 700
652   IF P1>0 THEN 660
653   PRINT"AFTER EACH ITERATION THE U VALUE INDICATES THE NUMBER"
654   PRINT"OF GRID POINTS WHERE A SOLUTION HAS NOT YET BEEN"
655   PRINT"OBTAINED TO SUFFICIENT ACCURACY."
656   LETP1=P1+1
657   PRINT
658   PRINT"IT. NO.      U VALUE"
659   PRINT
660   PRINT N,U
670   PRINT
690   IF U<>0 THEN 410
698   PRINT
700   PRINT N,U
710   PRINT
715   PRINT"THE FINAL SOLUTION IS"
716   PRINT
720   FOR I=1 TO A1
722   FOR J=1 TO R1
724   LETZ(I,J)=INT(T(I,J)*10^3+0.5)/10^3
728   NEXT J
732   NEXT I
740   FOR I=1 TO A1
742   PRINT Z(I,1);Z(I,2);Z(I,3);Z(I,4);Z(I,5);Z(I,6)
744   PRINT
746   NEXT I
1000  PRINT
1032  PRINT"R0="R0,"N1="N1,"B1="B1
1033  PRINT
1040  DATA 0,.333
1045  DATA 1.5
4000  END
```

Results for a Biot number, $\bar{h}R/k = 1.50$, N1 $= 0$, and R0 $= 0.333$ are given in Fig. 3-19. Since $T_w = T_B$ as indicated in Fig. 3-19, the inner surface temperature at $r = R_1$ is very close to T_B. Larger values of the Biot number make the difference in temperature $(T - T_B)_{r=R_1}$ even smaller. For a fixed tube size and tube material, the Biot number increases as a result of increasing the convection coefficient $\bar{h}$. This increase can occur by increasing the velocity of the internal flow or changing the type of fluid. The convective heat transfer per unit area is equal to the product $\bar{h}(T - T_B)_{r=R_1}$. In general, the inner surface temperature T at $r = R_1$ is a function of both ϕ and Bi. The conductive heat transfer across the surfaces at $\phi = 0$ and $\phi = \pi$ for $R_1 \leq r \leq R_2$ depends on the temperature gradients normal to these surfaces.

It is possible to improve the accuracy somewhat by choosing a smaller grid. The improvement depends on the value of the Biot number and the isothermal boundary conditions. The effect of the grid size was discussed when rectangular coordinates were considered, and the same conclusions apply to cylindrical and spherical co-ordinates.

The nondimensional parameters and boundary conditions of this problem can be varied to obtain a desired solution. To obtain a desired heat transfer rate from the fluid, a change in the geometry may be needed. If the external flow conditions are changed, the problem may need to be analyzed with convective boundary conditions over the outer surface. If a desired Biot number cannot be attained by varying the value of the convection heat transfer coefficient, perhaps a solid material with a dif-ferent thermal conductivity can be used. This type of interaction with the mathematical model provides a powerful tool in computer-aided analysis.

3-15 Conclusion

The numerical technique used in this chapter is called the *Gauss-Seidel method*. The Gauss-Seidel method was convenient for solving the Laplace and Poisson equations since the equations resulting from a finite difference approximation had dominant coefficients in each nodal equation and each dominant coefficient was associated with a different unknown nodal temperature in each separate equation. When the absolute value of this dominant coefficient is larger than the sum of the other coefficients in the equation, the set of equations is called a *diagonal system*, and convergence can be obtained by using the Gauss-Seidel method. This method is discussed in detail in Ref. 3-7.

It is not intended to imply that the Gauss-Seidel iteration scheme used in this chapter for solving two-dimensional heat conduction problems is always the most ef-ficient way of obtaining a solution. When the governing equations are written in finite difference form, the problem is basically one of obtaining the solution to a set of linear equations. Other direct methods such as matrix inversion or Gaussian elimination may be used.

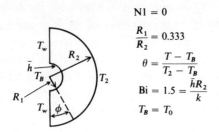

$$N1 = 0$$

$$\frac{R_1}{R_2} = 0.333$$

$$\theta = \frac{T - T_B}{T_2 - T_B}$$

$$Bi = 1.5 = \frac{\bar{h}R_2}{k}$$

$$T_B = T_0$$

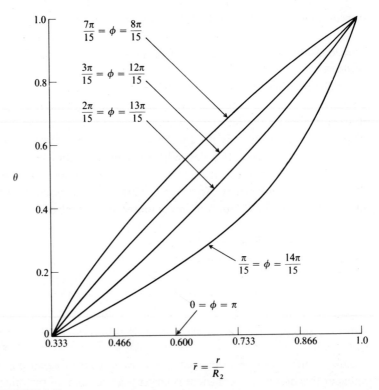

FIGURE 3-19
Conduction in a cylindrical shell with isothermal and convective boundary
conditions.

Considerations useful for choosing a numerical scheme are discussed in Ref. 3-6.
The reader can use any one of several numerical techniques. For example, Dusenberre
(Ref. 3-5) has presented numerical techniques for several different types of heat transfer
problems. Regardless of the numerical technique chosen, the analysis of the boundary
conditions, grid size, limiting assumptions, mathematical model, and accuracy desired

must accompany the use of any numerical technique if meaningful results are to be obtained. The major portion of Ref. 3-8 is concerned with the numerical solution of the Laplace equation, and many modifications to the numerical methods given in this chapter are suggested. Another alternate method is given in Ref. 3-9, where the application of orthonormalization methods is applied to heat conduction problems with arbitrary shape and arbitrary thermal boundary conditions.

In this chapter we have not considered heat conduction in a material undergoing a change of phase. A computer solution for the non-steady-state heat conduction in a solidifying alloy is given in Ref. 3-13. The governing equation given by

$$\frac{\partial^2 T}{\partial x^2} = \frac{\rho}{k}(c_p + \alpha \, \Delta H)\frac{\partial T}{\partial t} \qquad (3\text{-}94)$$

is solved in dimensional form. The term ΔH is the latent heat of fusion and $\alpha = \alpha(T)$ is a function of temperature which can be determined from a phase equilibrium diagram. The fact that α is a function of the dependent variable, temperature, makes the governing equation a nonlinear partial differential equation. However, the objective of a simple finite difference representation is to approximate a nonlinear partial differential equation by a linear algebraic equation. References 3-10 to 3-12 are suggested for further study.

Another method, called the *finite element method*, has proven to be advantageous when solving three-dimensional problems with irregular boundary conditions and nonhomogeneous properties. Whereas the finite difference method approximates the governing differential equation, the finite element method approximates the solution. This approximation results from a minimization process based upon the theories of variational calculus. Many current papers deal with this technique.

References

3-1   ZEMANSKY, N. W., and H. C. VAN NESS: "Basic Engineering Thermodynamics," McGraw-Hill, New York, 1966.

3-2   CARSLAW, H. S., and J. C. JAEGER: "Conduction of Heat in Solids," Oxford, London, 1947.

3-3   ARPACI, V. S.: "Conduction Heat Transfer," Addison-Wesley, Reading, Mass., 1966.

3-4   OZISIK, M. N.: "Boundary Value Problems of Heat Conduction," International Textbook, Scranton, Pa., 1968.

3-5   DUSENBERRE, G. M.: "Heat Transfer Calculations by Finite Differences," International Textbook, Scranton, Pa., 1961.

3-6   SMITH, GORDON D.: "Numerical Solutions of Partial Differential Equations with Exercises and Worked Solutions," Oxford, London, 1965.

3-7 JAMES, M. L., G. M. SMITH, and J. C. WOLFORD: "Analog and Digital Computer Methods in Engineering Analysis," International Textbook, Scranton, Pa., 1965.

3-8 THOM, A., and C. J. APELT: "Field Computations in Engineering and Physics," Van Nostrand, London, 1961.

3-9 SPARROW, E. M., and A. HAJI-SHEIKH: Transient and Steady Heat Conduction in Arbitrary Bodies with Arbitrary Boundary and Initial Conditions, J. Heat Transfer, ser. C, vol. 90, pp. 103–108, 1968.

3-10 MURRAY, W. D., and F. LANDIS: Numerical and Machine Solutions of Transient Heat Conduction Problems Involving Melting and Freezing, pt. I, J. Heat Transfer, ser. C, vol. 81, pp. 106–112, 1959.

3-11 SUNDERLAND, J. E., and R. J. GROSH: Transient Temperature in a Melting Solid, J. Heat Transfer, ser. C, vol. 83, pp. 409–414, November 1961.

3-12 CITRON, S. J.: Heat Conduction in a Melting Slab, Institute of Aeronautical Sciences Rept. No. 59–68, January 1959.

3-13 CARNAHAN, B., H. A. LUTHER, and J. O. WILKES: "Applied Numerical Methods," Wiley, New York, 1969.

3-14 CHAPMAN, A. J.: "Heat Transfer," Macmillan, New York, 1960.

Problems[1]

3-1 The nondimensional form of the two-dimensional Poisson equation is given by Eq. (3-9).

(a) What is the equation for one-dimensional, steady-state conduction in the x direction without internal generation?

(b) Integrate this equation, and apply the following boundary conditions:

$$at \quad \bar{x}=0 \quad \theta=1$$
$$at \quad \bar{x}=1 \quad \theta=0$$

(c) What is the nondimensional temperature gradient within the solid? What is the dimensional temperature gradient within the solid?

(d) What is the nondimensional governing equation for one-dimensional, steady-state conduction with uniform internal generation? Integrate this equation and apply the following boundary conditions.

$$at \quad \bar{x}=0 \quad \theta=1$$
$$at \quad \bar{x}=1 \quad \frac{d\theta}{d\bar{x}}=0$$

(e) What is the nondimensional temperature gradient for part (d) expressed in terms of $\bar{x}$?

[1] Problems whose numbers are followed by a superscript italic c should be analyzed by obtaining solutions on a digital computer.

3-2   Fig. 3-2P shows the cross section of a fin which can be assumed to be infinite in the
      x direction. The governing equation for the steady-state, two-dimensional conduction
      is given by the solution to the Laplace equation:

$$\frac{\partial^2\theta}{\partial x^2} + \frac{\partial^2\theta}{\partial y^2} = 0$$

where $\theta = (T - T_\infty)/(T_w - T_\infty)$.

The boundary conditions indicated in the figure may be written

$$\theta(0, y) = 1$$
$$\theta(\infty, y) = 0$$
$$\theta(x, 0) = 0$$
$$\theta(x, l) = 0$$

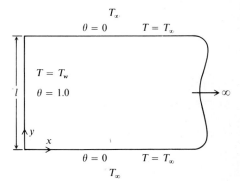

FIGURE 3-2P
Infinite fin cross section.

(a) Apply the separation of variables technique, and show that the product form of
the solution which satisfies all three homogeneous boundary conditions can be
expressed by

$$\theta(x, y) = \sum_{n=1}^{\infty} C_n e^{-\lambda_n x} \sin \lambda_n y$$

where $\lambda_n = n\pi/l$ and C_n is determined by applying the nonhomogeneous boundary
condition.

(b) Applying the nonhomogeneous boundary condition gives

$$1.0 = \sum_{n=1}^{\infty} C_n \sin \lambda_n y$$

where C_n is a coefficient in a Fourier sine series expansion given by

$$C_n = \frac{2}{l} \int_0^l \sin \lambda_n y \, dy$$

Show that this leads to the analytical solution given by

$$\theta(x, y) = 2 \sum_{n=1}^{\infty} \frac{1 - (-1)^n}{n\pi} e^{-n\pi x/l} \sin (n\pi y/l)$$

Compare this solution to the solution for the finite region given in Sec. 3-4.

3-3   The solution for the infinite fin in Prob. 3-2 was obtained by using isothermal boundary conditions on the top and bottom surfaces. Convective boundary conditions are usually more appropriate at these surfaces. Use a coordinate system which corresponds to the coordinate system used for the analysis of fins in Chap. 2. (See Fig. 3-3P.)

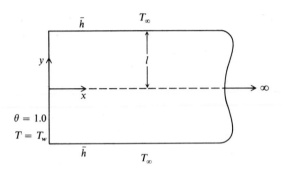

FIGURE 3-3P
Infinite fin with convective boundary
conditions.

(a) Show that the solution to the problem with convective boundary conditions depends on the solution to the following equations:

$$\frac{\partial^2 \theta}{\partial x^2} + \frac{\partial^2 \theta}{\partial y^2} = 0$$

$$\theta(0, y) = 1.0$$

$$\theta(\infty, y) = 0$$

$$\frac{\partial \theta(x, 0)}{\partial y} = 0$$

$$-k \frac{\partial \theta(x, l)}{\partial y} = \bar{h}\theta(x, l)$$

(b) Use the separation of variables technique, and show that the assumed product form is expressed by

$$\theta(x, y) = \sum_{n=1}^{\infty} C_n e^{-\lambda_n x} \cos \lambda_n y$$

where the values of λ_n (the eigenvalues) are given by the roots of the following transcendental equation

$$\cot \lambda_n l = \frac{1}{\text{Bi}} \lambda_n l$$

where $\text{Bi} = \bar{h}l/k$. The values of C_n are determined by applying the nonhomogeneous boundary condition.

(c) Numerical values for λ_n are given in Appendix IV of Ref. 3-2. Write a short computer program to generate the first three eigenvalues as a function of the Biot number.

(d) Use the nonhomogeneous boundary condition, and show that

$$1.0 = \sum_{n=1}^{\infty} C_n \cos \lambda_n y$$

where the coefficients of the Fourier cosine series expansion are given by

$$C_n = \frac{2 \sin \lambda_n l}{\lambda_n l + \sin \lambda_n l \cos \lambda_n l}$$

(e) Show that the final analytical solution is given by

$$\theta(x, y) = 2 \sum_{n=1}^{\infty} \frac{\sin \lambda_n l}{\lambda_n l + \sin \lambda_n l \cos \lambda_n l} e^{-\lambda_n x} \cos \lambda_n y$$

(f) When $\text{Bi} \to \infty$, why should the boundary conditions approach isothermal boundary conditions? What form does the analytical solution take for $\text{Bi} \to \infty$?

(g) Compare the two-dimensional solution in part (e) with the one-dimensional solution for the temperature distribution in an infinite fin. When do the two solutions agree?

3-4 Show that the analytical solution to the Laplace equation given by Eq. (3-31) agrees with the numerical solution given in Fig. 3-8. How many terms in the infinite series are required to obtain accurate values of local temperatures?

3-5 Rederive the finite difference equation for a near-surface node, Eq. (3-53), in terms of a nondimensional temperature $\theta = (T - T_1)/(T_0 - T_1)$. The values of T_0 and T_1 are reference temperatures. Suggest values that could appropriately be used as the two reference temperatures for various surface boundary conditions. Repeat the above procedure for the nodal equation for a surface node on a curved boundary, Eq. (3-55).

3-6 Define a rectangular grid with $\Delta y = 2\Delta x$. Start with an energy balance, and derive the nodal equations for surface nodes with convection and interior nodes as given by Eqs. (3-62) to (3-64). Consider a situation in which $T_{\infty, R} \neq T_{\infty, L}$ and $\bar{h}_R \neq \bar{h}_L$. Define $\theta = (T - T_{\infty, L})/(T_w - T_{\infty, L})$, and derive the nondimensional equation for the right surface node with convection. Show that when $T_{\infty, R} = T_{\infty, L}$ and $\bar{h}_R = \bar{h}_L$, the equation reduces to Eq. (3-63).

3-7 (a) Using the method of finite differences discussed in Sec. 3-5, derive the interior nodal equation for the node shown in Fig. 3-14. The results should agree with Eq. (3-74).

(b) Derive the finite difference equation for a node on the outer surface of the cylindrical element shown in Fig. 3-14. The node loses energy by convection to an environment at a temperature T_∞.

(c) By using finite difference approximations for the derivatives, express the governing equation for two-dimensional conduction in a sphere [Eq. (3-91)] in finite difference form.

3-8   The top surface of a nine by four rectangle with $l = 4.5$ cm and $w = 2.0$ cm is maintained at $T_2 = 100°C$, and the lower isothermal surface is $T_1 = 200°C$. The two-dimensional temperature distribution is governed by the Laplace equation. The two vertical surfaces have boundary conditions which produce the results presented in Fig. 3-8 and 3-9.

(a) Using these results, determine the temperature (°C) at the center of the rectangle in each figure.

(b) What is the approximate temperature gradient $\partial T/\partial y(°C/cm)$ at $x/w = \frac{1}{2}$ and $y/l = \frac{5}{18}$ in each figure?

(c) What is the approximate temperature gradient $\partial T/\partial x$ at $x/w = 0$ and $y/l = \frac{1}{9}$ in each figure?

(d) What is the magnitude and direction of the local heat transfer per unit area at $x/w = \frac{1}{4}$ and $y/l = \frac{5}{18}$ in each figure? The value of the thermal conductivity of the solid material is $k = 150$ W/m-°C.

3-9   Define $\bar{r} = r/R$ and $\theta = (T - T_\infty)/(T_1 - T_\infty)$.

(a) Express the governing equation for steady-state, two-dimensional conduction in cylindrical coordinates with uniform internal generation, Eq. (3-73), in nondimensional form.

(b) Identify the nondimensional parameters which result.

3-10 (a) Show that the governing equation for steady-state, one-dimensional, radial conduction in a cylinder without internal generation can be expressed by

$$\frac{d}{dr}\left(r\,\frac{dT}{dr}\right) = 0$$

(b) Integrate this equation using the boundary conditions

$$\text{at} \quad r = r_1 \quad T = T_1$$
$$\text{at} \quad r = r_2 \quad T = T_2$$

Express the resulting temperature distribution in nondimensional form.

(c) Use Fourier's law, and derive an expression for the radial heat transfer.

3-11 Consider steady-state, radial heat conduction in a cylinder with uniform internal generation.

(a) Obtain an analytical solution for the temperature distribution by integrating the governing equation and applying the following boundary conditions:

$$\text{at} \quad r = 0 \quad dT/dr = 0$$
$$\text{at} \quad r = R \quad T = T_w$$

(b) Define $\theta = (T - T_w)/(T_{max} - T_w)$, where T_{max} is the temperature at the center of the cylinder. Express the temperature distribution θ in terms of $\bar{r} = r/R$.

(c) Using Fourier's law, obtain an expression for the heat transfer by conduction within the cylinder.

(d) Discuss the changes to the analytical solution when the isothermal boundary condition at $r = R$ is replaced by a convective boundary condition at $r = R$.

3-12 A nuclear moderator is heated because of the nuclear radiation from the fuel element. The moderator is in the form of a hollow cylinder. It can be assumed that the governing equation for the temperature distribution is given by

$$\frac{d^2T}{dr^2} + \frac{1}{r}\frac{dT}{dr} + \frac{q'''}{k} = 0$$

The boundary conditions are given by

$$\text{at} \quad r = r_i \qquad T = T_i$$

$$\text{at} \quad r = r_0 \qquad \frac{dT}{dr} = 0$$

(a) Obtain the temperature distribution within the moderator.

(b) Express this temperature distribution in nondimensional form.

(c) Determine an expression for the heat transfer per unit area that must of necessity occur across the surface at $r = r_i$ in order for steady-state conditions to be maintained.

3-13 By plotting both results on a single graph, compare the two-dimensional solution for the temperature distribution in the rectangular region shown in Fig. 3-13P with the corresponding temperature distribution given by a one-dimensional fin analysis with isothermal ends. The parameters are chosen such that Fig. 3-10 gives the two-dimensional solution for this problem. The convection coefficient on both vertical surfaces is equal to 20.0 W/m²-K, and the thermal conductivity is 10.0 W/m-K.

The rectangular region exists in a cryogenic environment with $T_2 = 20.0$ K and $T_1 = T_\infty = 10.0$ K. At $y/l = \frac{8}{9}$, what is the difference between the temperature at $x = 0$ and the temperature at $x/w = \frac{1}{2}$? Answer the same question for $y/l = \frac{1}{3}$. At $\bar{y} = \frac{2}{3}$ and $\bar{x} = 0$, what is the ratio of q''_x to q''_y?

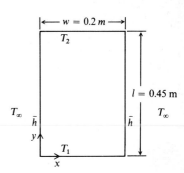

FIGURE 3-13P
Rectangular region.

3-14   Calculate the steady-state temperature distribution in a new material that is proposed for use in the dental profession for capping teeth. Study a model of rectangular shape, $l = 0.9$ cm and $w = 0.4$ cm. The boundary conditions are chosen such that Fig. 3-11 is a solution to the problem when $\text{Bi} = 10.0$. Use $\bar{h} = 300$ W/m²-K, $k = 0.27$ W/m-K, and temperatures $T_\infty = 10°C$, $T_2 = 20°C$, and $T_1 = 30°C$.

(a) Show that this problem does indeed fit the solution given in Fig. 3-11.

(b) What is the lowest temperature in the material, and where does it occur? What is the temperature at the center of the material?

(c) Calculate the heat transfer per unit area crossing the top surface, $\bar{y} = 0$.

3-15   What is the governing equation for steady-state, one-dimensional axial conduction in a cylinder with uniform generation? Discuss physical situations where one-dimensional axial conduction will occur. Compare the governing equation and resulting temperature profiles with those for steady-state, one-dimensional conduction in rectangular coordinates.

3-16   The governing equation for steady-state, two-dimensional conduction in a hollow sphere is given by Eq. (3-91).

(a) Show that the governing equation for one-dimensional, radial conduction without internal generation can be written

$$\frac{d}{dr}\left(r^2 \frac{dT}{dr}\right) = 0$$

(b) Integrate this equation, and obtain the temperature distribution for the following boundary conditions:

$$\begin{array}{lll} \text{at} & r = r_1 & T = T_1 \\ \text{at} & r = r_2 & T = T_2 \end{array}$$

(c) Define $\theta = (T - T_2)/(T_1 - T_2)$ and $\bar{r} = r/r_1$. Express the temperature profile in nondimensional form.

(d) Derive the expression for the heat transfer by conduction at the surface $r = r_1$.

3-17   A spherical nuclear fuel element has a variable internal generation rate given by

$$q''' = q_0''' C\left(\frac{r}{R}\right)^2 \qquad 0 \le r \le R$$

where C and q_0''' are constants.

(a) For one-dimensional, radial conduction, what is the governing nondimensional differential equation for the steady-state temperature distribution?

(b) Obtain the nondimensional temperature distribution for the following boundary conditions:

$$\begin{array}{lll} \text{at} & \bar{r} = 0 & \theta < \infty \\ \text{at} & \bar{r} = 1 & \theta = 1 \end{array}$$

(c) What is the expression for the heat transfer per unit area within the sphere?

3-18   The results in Fig. 3-19 show that the temperature at $r = R_1$ is approximately equal to T_B since $\theta \approx 0$. This means that an isothermal boundary condition $T = T_B$ could have been specified at this surface. Choose various values of R_2 and k, and for each set of these values calculate the unit surface resistance $1/\bar{h}$ and the inner radius R_1 which give the results given in Fig. 3-19. Refer to the computer program for this solution. What changes would be necessary if $T_1 = 2T_B$? What changes would be necessary if the boundary condition $d\theta/d\bar{r} = 0$ at $\bar{r} = 0.333$ were applied? What changes would be necessary if a convective boundary condition at $\bar{r} = 1.0$ were applied? What changes would be necessary if $R_1/R_2 = 0.10$?

3-19 For steady-state, one-dimensional, radial conduction in cylindrical coordinates without internal generation, the governing equation is

$$\frac{d}{dr}\left(r\frac{dT}{dr}\right) = 0$$

Consider the composite cylinders made up of materials 1 and 2 shown in Fig. 3-19P.

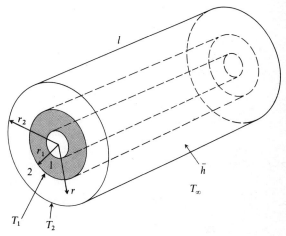

FIGURE 3-19P
Composite cylinders.

(a) Show that the heat transfer by conduction at $r = r_2$ is given by

$$q = \frac{2\pi k_2 l(T_1 - T_2)}{\ln(r_2/r_1)}$$

(b) The heat transfer by convection from the surface is given by Newton's law of cooling as follows:

$$q = 2\pi r_2 \, l\bar{h}(T_2 - T_\infty)$$

Show that the equation for the steady-state heat transfer can also be written

$$q = \frac{2\pi k_2 l(T_1 - T_\infty)}{\ln(r_2/r_1) + k/\bar{h}r_2}$$

(c) Show that the heat transfer as a function of radius r_2 for a fixed value of r_1 will be a maximum when $r_2 = k_2/\bar{h}$. (Note: If material 2 is insulation covering a wire or pipe, the value of r_2 which gives a maximum heat transfer is called the *critical radius* which specifies the *critical thickness* of insulation given by $r_2 - r_1$.)

(d) What would the critical radius be if material 2 were a material with a uniform internal generation q'''?

3-20ᶜ The solution for the two-dimensional, nondimensional temperature distribution in a nine by four rectangle with no internal generation is given in Figs. 3-8 and 3-9 for various boundary conditions.

(a) Study the effect of internal generation for various boundary conditions by varying the value of the parameter N1. (This compares solutions of the Laplace equation with solutions of the Poisson equation.)

(b) How large must the value of N1 be before a significant effect on the temperature distribution is noticed? What is the corresponding value of the internal generation rate for chosen values of length, temperature at the isothermal surfaces, and thermal conductivity?

3-21ᶜ A corrosion study is to be made on a new metal ($k = 170$ Btu/h-ft-°R) for use in an oceanographic vessel. A long test specimen which is 0.32 ft thick and 0.72 ft high is to be installed between two pipes which carry heated liquids as shown in Fig. 3-21P. The upper surface of the test specimen is in contact with a pipe surface maintained at 200°F, and the lower surface is in contact with a pipe surface maintained at 100°F. The outside is exposed to sea water at 32°F, and the inside surface is exposed to air at 70°F. The values of the convection coefficients are given in the figure. An electric current which will generate a uniform internal generation of $q''' = 25,000$ Btu/h-ft³ will be passed through the test specimen during the corrosion test.

(a) During the corrosion test it is desirable that all parts of the specimen remain at or above 100°F. Determine if this will occur.

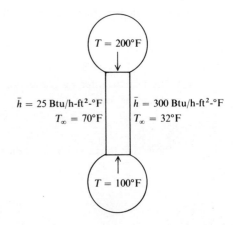

FIGURE 3-21P
Corrosion test specimen.

(b) It is also desirable that the maximum temperature difference between the inside surface and the outside surface along a line of $y = $ constant does not exceed 30°F. Will this be achieved?

(c) How much thermal energy will be convected away from the test specimen?

3-22ᶜ A portable steam generator which utilizes nuclear energy to generate steam is used to generate steam in remote desert locations. The supporting legs protrude through the outer shell as shown in Fig. 3-22P to provide supporting structure for internal components. During operation the internal section of the legs is in contact with saturated vapor, and the internal exposed surface may be assumed to be isothermal at 110°C. All sections of the legs outside the shell are insulated except for the bottom surface. The unit is normally placed directly on the hot desert sand which may be assumed to be at a temperature of 40°C. Calculate the two-dimensional temperature distribution in a leg and the heat conduction through a leg during steady-state operation. The thermal conductivity is 20.0 W/m-K.

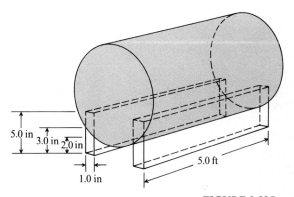

FIGURE 3-22P
Steam generator.

3-23ᶜ One end of a structural member of square cross section supports a combustion chamber, and the other end is attached to an insulated wall and may be assumed adiabatic. The

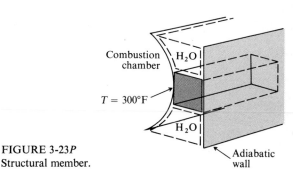

FIGURE 3-23P
Structural member.

cross section of the member is 0.25 ft by 0.25 ft, and the thermal conductivity is 10.0 Btu/h-ft-°R. In a new design it is proposed to install a water chamber between the combustion chamber and the wall (the region indicated by the dashed lines). When this chamber is flooded, the convection coefficient between the member and the surroundings is 250 Btu/h-ft²-°R on both top and bottom surfaces. The combustion chamber wall is at a temperature of 300°F, and the water flows through the chamber at a mean temperature of 60°F. Find the heat transfer rate from the structural member. When the chamber is not flooded, the ambient air will be at 100°F, and the convection coefficient will be 5.0 Btu/h-ft²-°R. Compare the temperature profile for this situation with the flooded situation. How much error is made by using a one-dimensional fin analysis with an insulated end boundary condition?

TRANSIENT CONDUCTION

4-1 Introduction

Chapters 2 and 3 were concerned with steady-state heat transfer by conduction. The temperature distribution in the solid materials was a function of the spatial coordinates but was independent of time. The one-dimensional problems in Chap. 2 were formulated in terms of ordinary differential equations, and the two-dimensional problems in Chap. 3 were formulated in terms of elliptic partial differential equations. This chapter treats transient heat conduction problems. The temperature profile in the solid materials is a function of both the spatial coordinates and time. The analysis leads to partial differential equations that are parabolic rather than elliptic.

The classification of partial differential equations is based on the sign of the discriminate $b^2 - ac$ in the general second-order partial differential equation

$$az_{xx} + 2bz_{xy} + cz_{yy} = H(x, y, z, z_x, z_y)$$

where a, b, and c are real continuous functions. The subscripts indicate partial differentiation of the dependent variable z with respect to the indicated independent variable (x or y), and H is a continuous function of the indicated arguments.

When $b^2 - ac < 0$, the equation is said to be *elliptic*. The Laplace and Poisson equations analyzed in Chap. 3 are elliptic with $b = 0$ and $a = c = 1$. When $(b^2 - ac) = 0$, the equation is said to be *parabolic*. Equation (4-1) in the next section is parabolic since $b = c = 0$. When $b^2 - ac > 0$, the equation is said to be *hyperbolic*. An example of this type of equation is the wave equation. This terminology is based upon the similarity to the equation for a conic given by

$$ax^2 + 2bxy + cy^2 = H$$

for constant a, b, c, and H. This equation is for an ellipse, parabola, or hyperbola, depending upon the sign of $b^2 - ac$.

The difference between elliptic and parabolic equations can also be explained in terms of physical behavior. A thermal disturbance in a field specified by an elliptic equation will propagate in all directions over the entire field. The propagation in a field specified by a parabolic equation will only occur in a semi-infinite region, such as the region specified by time greater than zero. The propagation of a disturbance in a field governed by a hyperbolic equation occurs in a characteristic direction or along characteristic lines.

In this chapter solutions to the governing parabolic partial differential equations are obtained. We limit ourselves to problems which can be specified in terms of one or two independent spatial variables and time. One additional complication that must be considered when applying finite difference techniques to parabolic equations is convergence. This item is discussed in Sec. 4-3 along with the derivation of the finite difference equations.

This chapter begins with a derivation of an analytical solution for transient heat conduction. Analytical solutions are available in the literature for many boundary and initial conditions (cf. Refs. 4-1 to 4-3). Charts based on these analytical solutions are useful when the assumptions inherent in the analytical solutions are applicable (cf. Refs. 4-4 to 4-6). Next, numerical techniques which can be used to complement and supplement the existing analytical solutions are presented. Computer programs which may be used to obtain solutions to the numerical models are discussed, and the results of some numerical solutions are analyzed.

4-2   Analytical Analysis

The method of separation of variables used in Chap. 3 can be used to obtain analytical solutions to some transient conduction problems. Consider transient, one-dimensional conduction in rectangular coordinates. When constant thermal conductivity is assumed and there is no internal generation, the governing equation (3-4) reduces to the following form for one-dimensional, transient conduction:

$$\frac{\partial^2 T}{\partial x^2} = \frac{1}{\alpha} \frac{\partial T}{\partial t} \qquad \text{(4-1)}$$

This equation is called the one-dimensional *Fourier field equation*. If we define $\Delta T = T - T_r$, where T_r is a constant reference temperature, we can write $\partial^2(\Delta T)/\partial x^2 = (1/\alpha)\partial([\Delta T)/\partial t]$.

The first step is to assume that the solution can be expressed as the product of two functions which are each in turn a function of only one independent variable. Mathematically we write

$$\Delta T(x, t) = X(x)\tau(t) \qquad (4\text{-}2)$$

Substituting Eq. (4-2) into Eq. (4-1) results in

$$\frac{1}{\alpha\tau}\frac{d\tau}{dt} = \frac{1}{X}\frac{d^2X}{dx^2} = -\lambda^2 \qquad (4\text{-}3)$$

Since the variables have been separated, one side of the equation is only a function of time and the other only a function of x. Thus, for the equality to hold for all x and t, each side of the equation must be equal to the same constant $-\lambda^2$. The negative sign insures an exponentially decaying solution in terms of the independent variable time. The two ordinary differential equations in Eq. (4-3) can be written as

$$\frac{d\tau}{dt} + \alpha\lambda^2\tau = 0$$

and

$$\frac{d^2X}{dx^2} + \lambda^2 X = 0$$

The corresponding solutions are

$$\tau = C_1 e^{-\alpha\lambda^2 t} \qquad X = C_2 \sin \lambda x + C_3 \cos \lambda x$$

where C_1, C_2, and C_3 are constants of integration to be determined from the boundary conditions. We now see how the choice of a negative constant in Eq. (4-3) results in a solution which decays exponentially with time. The assumed form of the solution given by Eq. (4-2) can now be written

$$\Delta T = (B_1 \sin \lambda x + B_2 \cos \lambda x)\, e^{-\alpha\lambda^2 t} \qquad (4\text{-}4)$$

where $B_1 = C_1 C_2$ and $B_2 = C_1 C_3$.

To specify boundary conditions, consider the infinite slab shown in Fig. 4-1. The initial temperature distribution through the slab is assumed to be known. The two surface temperatures at $x = 0$ and $x = l$ are T_1 and T_2, respectively. Assume that at time $t = 0$ the temperature of both surfaces is suddenly reduced to $T = T_0$ by exposure to a constant temperature heat sink. We now choose the reference

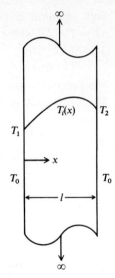

FIGURE 4-1
Infinite slab.

temperature T_r to be T_0. This will produce a homogeneous boundary condition for ΔT at $x = 0$ and $x = l$ for time equal to or greater than zero. Thus,

$$\Delta T = \Delta T_i(x) \qquad 0 \leq x \leq l \qquad t < 0 \qquad (4\text{-}5a)$$

$$\Delta T = 0 \qquad x = 0 \qquad t \geq 0 \qquad (4\text{-}5b)$$

$$\Delta T = 0 \qquad x = l \qquad t \geq 0 \qquad (4\text{-}5c)$$

To satisfy Eq. (4-5b), B_2 must equal zero. To satisfy Eq. (4-5c), $\sin \lambda x$ must equal zero when $x = l$. This condition is satisfied for

$$\lambda_n = \frac{n\pi}{l} \qquad n = 1, 2, 3, \ldots$$

The assumed solution can then be written

$$\Delta T = (B_n \sin \lambda_n x)\, e^{-\lambda_n^2 \alpha t} \qquad n = 1, 2, 3, \ldots$$

This notation indicates that there is a different solution to the differential equation for each consecutive integer, for n equal one to infinity. Also, a different integration constant B_n can be associated with each solution. If one uses the superposition theorem for the solution of a linear equation, one can write that the sum of these solutions is also a solution:

$$\Delta T = \sum_{n=1}^{\infty} (B_n \sin \lambda_n x)\, e^{-\lambda_n^2 \alpha t} \qquad n = 1, 2, 3, \ldots$$

The initial condition at $t = 0$ imposes the further requirement that

$$\Delta T_i(x) = \sum_{n=1}^{\infty} B_n \sin \lambda_n x$$

The constants B_n can now be chosen as the constants in the Fourier series representation of $\Delta T_i(x)$. The Fourier coefficients are based on orthogonal functions as discussed in Chap. 3. For $f(x) = \Delta T_i(x)$ and $g_n(x) = \sin \lambda_n x$,

$$B_n = \frac{\int_0^l \Delta T_i(x) \sin \lambda_n x \, dx}{\int_0^l \sin^2 \lambda_n x \, dx} \qquad n = 1, 2, 3, \ldots$$

Using trigonometric substitution to evaluate the integral in the denominator yields

$$\int_0^l \sin^2 \lambda_n x \, dx = \left(-\frac{1}{2\lambda_n} \cos \lambda_n x \sin \lambda_n x + \frac{1}{2\lambda_n} \lambda_n x \right)_0^l = \frac{l}{2}$$

The expression for the Fourier coefficients can then be written

$$B_n = \frac{2}{l} \int_0^l \Delta T_i(x) \sin \lambda_n x \, dx \qquad n = 1, 2, 3, \ldots$$

If one uses these values for B_n, the resulting solution satisfies the differential equation, the boundary conditions, and the initial condition. Thus, the assumed product form of the solution is valid. The solution is written

$$\Delta T = \frac{2}{l} \sum_{n=1}^{\infty} e^{-(n\pi/l)^2 \alpha t} \sin \frac{n\pi x}{l} \int_0^l \Delta T_i(x) \sin \frac{n\pi x}{l} \, dx \qquad n = 1, 2, 3, \ldots \qquad (4\text{-}6)$$

When the initial temperature distribution is a constant, $T - T_1 = \Delta T_i(x) = \Delta T_1$, and the integration can be readily carried out. The solution for this special initial condition is

$$\theta = \frac{\Delta T}{\Delta T_1} = 2 \sum_{n=1}^{\infty} e^{-(n\pi/l)^2 \alpha t} \sin \frac{n\pi x}{l} \frac{1 - (-1)^n}{n\pi} \qquad n = 1, 2, 3, \ldots$$

or

$$\theta = \frac{4}{\pi} \sum_{n=1}^{\infty} \frac{e^{-(n\pi/l)^2 \alpha t}}{n} \sin \frac{n\pi x}{l} \qquad n = 1, 3, 5, \ldots \qquad (4\text{-}7)$$

The exponential term contains $\alpha t/l^2$. This ratio is defined as the *Fourier number*. Its significance is discussed throughout this chapter.

Many of the solutions that appear in graphical form, such as those given in Refs. 4-4, 4-5, and 4-6, are limited by the assumption of a uniform initial temperature distribution. In many practical problems this initial condition does not apply. However, in a numerical analysis, a nonuniform initial condition can be treated almost as easily as a uniform initial condition.

Once the temperature distribution as given by Eq. (4-7) is known, it is possible to calculate the heat transfer by using Fourier's law. The instantaneous heat transfer rate per unit area is $q/A = -k(\partial\theta/\partial x)(\partial T/\partial\theta)$, where $\theta = (T - T_0)/(T_1 - T_0)$. Thus,

$$\frac{q}{A} = \frac{4k(T_1 - T_0)}{l} \sum_{n=1}^{\infty} e^{-(n\pi/l)^2\alpha t} \cos\frac{n\pi x}{l} \qquad n = 1, 3, 5, \ldots$$

The total heat transfer per unit area up to time t is

$$\frac{Q}{A} = \int_0^t \frac{4k(T_1 - T_0)}{l} \sum_{n=1}^{\infty} e^{-(n\pi/l)^2\alpha t} \cos\frac{n\pi x}{l} \, dt$$

$$\frac{Q}{A} = \frac{4k(T_1 - T_0)l}{\pi^2\alpha} \sum_{n=1}^{\infty} \frac{1}{n^2}[1 - e^{-(n\pi/l)^2\alpha t}] \cos\frac{n\pi x}{l} \qquad n = 1, 3, 5, \ldots$$

(4-8)

These calculations show how the temperature distribution controls the heat transfer in transient conduction. As in the previous chapters, the solution to a heat transfer problem depends upon obtaining the unique solution to the temperature distribution.

Other analytical methods such as Laplace transforms may be used to advantage for some problems in transient conduction. Reference 4-8 gives further discussion of the application of Laplace transforms to transient heat conduction.

The governing equation for one-dimensional, transient conduction with internal generation is Eq. (3-4):

$$\frac{\partial^2 T}{\partial x^2} + \frac{q'''}{k} = \frac{1}{\alpha}\frac{\partial T}{\partial t}$$

It is instructive to nondimensionalize this equation and its associated boundary conditions. For a convective boundary condition the following nondimensional variables are used:

$$\theta = \frac{T - T_\infty}{T_i - T_\infty} \qquad \bar{x} = \frac{x}{l} \qquad \bar{t} = \frac{\alpha t}{l^2} \equiv \text{Fo}$$

For now, we treat the initial temperature T_i and the environmental temperature T_∞ as constants. The nondimensional time variable defined above is very useful in transient analysis. It is the same Fourier number which appeared in the analytical solution [cf. Eq. (4-7)]. Physically the Fourier number is the ratio of the rate of heat transfer by conduction to the rate of energy storage in a reference volume. This fact becomes clear when we write

$$\text{Fo} = \frac{\alpha t}{l^2} = \frac{kl^2(\Delta T/l)}{\rho c l^3(\Delta T/t)} \sim \frac{kA(dT/dx)}{mc(dT/dt)} \sim \frac{\text{heat transfer by conduction}}{\text{rate of energy storage}}$$

Introducing the above nondimensional variables we write

$$\frac{\partial T}{\partial x} = \frac{\partial\theta}{\partial\bar{x}}\frac{\partial\bar{x}}{\partial x}\frac{\partial T}{\partial\theta} = \frac{T_i - T_\infty}{l}\frac{\partial\theta}{\partial\bar{x}}$$

$$\frac{\partial^2 T}{\partial x^2} = \frac{\partial}{\partial \bar{x}} \left(\frac{\partial T}{\partial x} \right) \frac{\partial \bar{x}}{\partial x} = \frac{\partial}{\partial \bar{x}} \left(\frac{T_i - T_\infty}{l} \frac{\partial \theta}{\partial \bar{x}} \right) \frac{\partial \bar{x}}{\partial x}$$

$$\frac{\partial^2 T}{\partial x^2} = \frac{T_i - T_\infty}{l^2} \frac{\partial^2 \theta}{\partial \bar{x}^2}$$

$$\frac{\partial T}{\partial t} = \frac{\partial \theta}{\partial \text{Fo}} \frac{\partial \text{Fo}}{\partial t} \frac{\partial T}{\partial \theta} = \alpha \frac{T_i - T_\infty}{l^2} \frac{\partial \theta}{\partial \text{Fo}}$$

Substituting these derivatives into the governing partial differential equation gives

$$\frac{\partial^2 \theta}{\partial \bar{x}^2} + \frac{q''' l^2}{k(T_i - T_\infty)} = \frac{\partial \theta}{\partial \text{Fo}}$$

The nondimensional internal generation term is identical to that used in the previous chapters. Recall that

$$\frac{q''' l^2}{k(T_i - T_\infty)} = \frac{q''' l^3}{kl^2(T_i - T_\infty)/l} \sim \frac{\text{energy generation rate}}{\text{heat transfer by conduction}}$$

Now consider a convective boundary condition at $x = 0$ for $t \geq 0$. Mathematically we express this by

$$\bar{h}[T(0, t) - T_\infty] = k \frac{\partial [T(0, t)]}{\partial x}$$

If we use the previously defined nondimensional variables, the right side of this equation can be written as

$$\frac{k(T_i - T_\infty)}{l} \frac{\partial \theta(0, \text{Fo})}{\partial \bar{x}}$$

and the left side can be written

$$(T_i - T_\infty)\bar{h} \frac{T(0, t) - T_\infty}{T_i - T_\infty}$$

Equating these two expressions gives

$$\frac{\bar{h}l}{k} \frac{T(0, t) - T_\infty}{T_i - T_\infty} = \frac{\partial \theta(0, \text{Fo})}{\partial \bar{x}}$$

Finally one can write

$$(\text{Bi})\theta(0, \text{Fo}) = \frac{\partial \theta(0, \text{Fo})}{\partial \bar{x}}$$

This nondimensional equation specifies the convective boundary condition on the surface at $\bar{x} = 0$. This equation contains the nondimensional Biot number in addition to the Fourier number. Thus when convective boundary conditions occur, the nondimensional temperature distribution is a function of the nondimensional internal generation term as well as both the Biot and Fourier numbers. These parameters are the important governing ones that should be varied to analyze transient heat conduction.

4-3   Finite Difference Formulation

In Chap. 3 it was shown that the finite difference equations for steady-state heat transfer can be applied to a relatively large grid network. Then the size of the network can be reduced to see if a finer grid size would substantially improve the accuracy of the results. There was no danger of selecting a grid size so large that convergence of the solution became a problem. The only complicating factor was the determination of the appropriate nodal equations for surface or near-surface nodes. In the finite difference approximation to parabolic partial differential equations, the problems of convergence and stability are more critical. It is no longer possible to arbitrarily select a large grid network and always stay within the stability and convergence limits. There is also a choice of methods that can be used to obtain the finite difference approximations. These methods are classified as either explicit or implicit. They will be discussed separately.

Explicit methods   Consider transient, one-dimensional conduction with constant properties. The governing equation is given by Eq. (4-1). The second derivative with respect to the spatial coordinate can be expressed by a central difference approximation as shown in Chap. 3. Since we initially consider only one independent space variable, only one subscript i is needed to indicate an increment in a spatial variable. A time interval is denoted by a superscript n. We then write the finite difference approximation to the second derivative as follows:

$$\frac{\partial^2 T}{\partial x^2} = \frac{T_{i+1}^n - 2T_i^n + T_{i-1}^n}{(\Delta x)^2} \qquad (4-9)$$

The superscripts indicate that the three nodal temperatures are evaluated at the same instant.

The time derivative can be expressed as a forward difference approximation

$$\frac{\partial T}{\partial t} = \frac{T_i^{n+1} - T_i^n}{\Delta t}$$

Here we have indicated the temperature change of a single node during one increment in time divided by the time increment. To clarify this notation, consider the space-time grid in Fig. 4-2. The increment $n+1$ refers to an increment in time and the increment $i+1$ refers to an increment in space. The nodes T_i^n, T_{i-1}^n, and T_i^{n+1} are indicated.

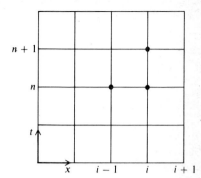

FIGURE 4-2
Space-time grid.

Using these finite difference approximations in Eq. (4-1), we obtain

$$\frac{T_i^{n+1} - T_i^n}{\Delta t} = \alpha \frac{T_{i+1}^n - 2T_i^n + T_{i-1}^n}{(\Delta x)^2} \qquad (4\text{-}10)$$

An approximation associated with this finite difference equation is that while the temperature at T_i^n changes during a small time interval to a value T_i^{n+1}, the values of T_{i+1}^n and T_{i-1}^n are assumed to remain constant. In effect we are assuming that the internal energy change of the volume element with time can be calculated without considering any spatial temperature variation through the element. Such a spatial temperature variation might be caused by the temperature of adjacent nodes changing at different rates. This approximation infers that during the small time increment the heat conduction across the surface of the element is constant. In other words, we can say that the finite difference approximation assumes that the spatial variation of volume terms, such as internal energy, is negligible with respect to their variation with time, and the variation with time of surface terms, such as heat flux, is negligible with respect to their spatial variation.

Using Eq. (4-10) and solving for the temperature one time increment in the future yields

$$T_i^{n+1} = T_i^n + \frac{\alpha\,\Delta t}{(\Delta x)^2}\,(T_{i+1}^n - 2T_i^n + T_{i-1}^n) \qquad (4\text{-}11)$$

The superscripts indicate that all nodal temperatures on the right side of Eq. (4-11) are evaluated at the current time. Equation (4-11) can be nondimensionalized by defining $\theta = (T - T_0)/(T_1 - T_0)$. The resulting equation is

$$\theta_i^{n+1} = \theta_i^n + \frac{\alpha\,\Delta t}{(\Delta x)^2}\,(\theta_{i+1}^n - 2\theta_i^n + \theta_{i-1}^n) \qquad (4\text{-}12)$$

Introducing $\overline{\Delta x} = \Delta x/l$ we write $\alpha\,\Delta t/(\Delta x)^2 = [1/(\overline{\Delta x})^2]\mathrm{fo}$, where $\mathrm{fo} = \alpha\,\Delta t/l^2$ is a Fourier modulus based on a time increment Δt.

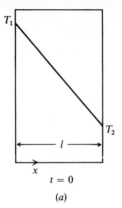
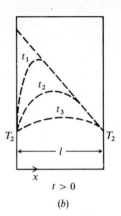

FIGURE 4-3
Transient and steady-state conduction.

We now turn to the analysis of transient conduction in a material with a non-uniform initial temperature distribution. To establish this initial temperature distribution, consider one-dimensional, constant-property, steady-state heat transfer without internal generation. Equation (4-1) reduces to $d^2T/dx^2 = 0$. The solution is $T = C_1 x + C_2$, a straight-line temperature distribution as shown in Fig. 4-3a. Applying the boundary conditions indicated in Fig. 4-3a leads to

$$T = \frac{T_2 - T_1}{l} x + T_1$$

Defining $\theta = (T - T_2)/(T_1 - T_2)$ and $\bar{x} = x/l$, we can write this equation as $\theta = 1 - \bar{x}$. Now suppose that at $t = 0$ the temperature at the left face is suddenly changed to equal T_2, the temperature at the right face. In other words, the value of θ at $\bar{x} = 0$ suddenly changes from one to zero. The governing equation for the temperature distribution in the slab for time greater than zero is Eq. (4-1). We expect a temperature response with time as indicated in Fig. 4-3b. Assume that the initial nondimensional temperature $\theta = (T - T_2)/(T_1 - T_2)$ at $t = 0$ is given as indicated in Table 4-1.

During the transient response the initial linear temperature distribution must approach the final steady-state distribution which is given by $\theta = \text{constant} = 0$. There is no physical way that the nondimensional temperature can be negative during this process. The governing finite difference equation for the transient temperature response, Eq. (4-12), can be written

$$\theta_i^{n+1} = \theta_i^n \left[1 - \frac{2\alpha\,\Delta t}{(\Delta x)^2} \right] + \frac{\alpha\,\Delta t}{(\Delta x)^2} (\theta_{i+1}^n + \theta_{i-1}^n) \quad (4\text{-}12)$$

Once a value of $\alpha\,\Delta t/(\Delta x)^2$ is chosen, θ_i^{n+1} can be calculated by using the initial condition of θ given in Table 4-1. The thermal diffusivity $\alpha = k/\rho c$ is a physical property of the material. The value Δt is the time increment required for θ_i^n to change to θ_i^{n+1}. The grid spacing Δx is the distance between the nodes used in the finite difference representation. Let us arbitrarily choose $\alpha \Delta t/(\Delta x)^2 = 1$. Using the finite difference equation to calculate θ_2^{n+1} for $n + 1 = 2$ yields

$$\theta_2{}^2 = 0.8(1 - 2) + 1(0.6 + 0) = -0.8 + 0.6 = -0.2$$

This answer is physically impossible. Examination of Eq. (4-12) shows that if we choose $\alpha \Delta t/(\Delta x)^2 = \frac{1}{2}$, no negative terms will appear in the equation. With this value for $\alpha \Delta t/(\Delta x)^2$ the equation reduces to

$$\theta_i^{n+1} = \tfrac{1}{2}(\theta_{i+1}^n + \theta_{i-1}^n)$$

Repeating the calculation for $\theta_2{}^2$ yields

$$\theta_2{}^2 = \tfrac{1}{2}(0.6 + 0) = 0.3$$

In a similar manner, $\theta_3{}^2 = 0.6$, $\theta_4{}^2 = 0.4$, and $\theta_5{}^2 = 0.2$. These values can now be used to calculate θ for $n + 1 = 3$. It can be shown mathematically (cf. Ref. 4-9, p. 93) that $\alpha \Delta t/(\Delta x)^2 = \frac{1}{2}$ is the maximum permissible value which allows the results obtained from the finite difference equation (4-12) to converge to the proper solution. Thus, once a value of Δx is chosen, the value of Δt is limited, and conversely choosing Δt limits Δx. Smaller values of $\alpha \Delta t/(\Delta x)^2$ improve the finite difference approximation somewhat, but a value of 0.5 usually gives acceptable accuracy.

The explicit method is characterized by the use of temperature values at the current time increment n to calculate the spatial derivative. One disadvantage of this method is obvious. Because of convergence considerations, the maximum time interval is limited once a space increment is chosen. If the solution requires results over long time periods, an excessive amount of computational time is necessary. One way to avoid this problem, at the expense of introducing another complication, is to use the implicit method.

Table 4-1 NODAL VALUES OF
NONDIMENSIONAL
TEMPERATURES
$\theta = (T - T_2)/(T_1 - T_2)$

i	$n = 1$	$n = 2$	$n = 3$	$n = 4$	$n = 5$
1	0	0	0	0	0
2	0.8				
3	0.6				
4	0.4				
5	0.2				
6	0	0	0	0	0

Implicit method   A fully implicit finite difference approximation to Eq. (4-1) is given by

$$\frac{T_i^{n+1} - T_i^n}{\Delta t} = \alpha \frac{T_{i+1}^{n+1} - 2T_i^{n+1} + T_{i-1}^{n+1}}{(\Delta x)^2} \tag{4-13}$$

Comparison of Eqs. (4-13) and (4-10) shows that the only difference in the implicit method is that the three nodal values on the right side of the equation used to approximate $\partial^2 T/\partial x^2$ are evaluated at the advanced time row $n + 1$ rather than the current time row n (cf. Fig. 4-2). Equation (4-13) has three unknown temperatures on the $n + 1$ time row, and N equations are obtained from Eq. (4-13), where N is the number of spatial nodes representing the solid. These must be solved simultaneously to obtain the nodal temperatures. Although the implicit method involves the simultaneous solution of a set of algebraic equations, the system of equations is stable for all values of $\alpha \Delta t/(\Delta x)^2$ (Ref. 4-9). Many modifications, such as the Crank-Nicolson method (Ref. 4-10), have been suggested as compromises between implicit and explicit methods. Other schemes are discussed in Refs. 4-11 and 4-16.

4-4   Boundary Conditions

The explicit method is suitable for many transient, one-dimensional problems in heat conduction. This method is used for the development of the boundary condition equations in finite difference form.

If we return to the explicit approximation of the governing differential equation (4-11), and define $\theta = (T - T_\infty)/(T_1 - T_\infty)$ and $\bar{x} = x/l$, the equation can be rewritten as

$$\theta_i^{n+1} = \left[1 - \frac{2\text{fo}}{(\Delta x)^2}\right]\theta_i^n + \frac{\text{fo}}{(\Delta x)^2}(\theta_{i+1}^n + \theta_{i-1}^n) \tag{4-14}$$

where $\text{fo} = \alpha \Delta t/l^2$. This equation applies for all interior nodes. The finite difference error due to truncating the Taylor series expansion is of the order $\alpha \Delta t + (\Delta x)^2$ (cf. Ref. 4-12, p. 395).

When convection occurs at a boundary, one can write

$$k\left(\frac{\partial T}{\partial x}\right)_s = h(T - T_\infty)_s \tag{4-15}$$

This equation states that the energy conducted to the surface is equal to the energy leaving the surface by convection, or vice versa. It is based on the application of the conservation of energy principle across the surface. Equation (4-15) is true only when radiation from the surface can be neglected. In general, one must equate the heat conduction at the surface to the heat transfer crossing the surface by both convection and radiation.

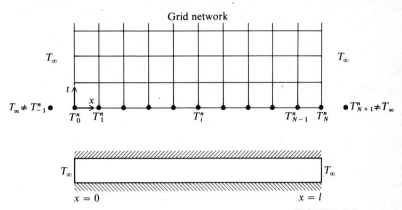

FIGURE 4-4
One-dimensional slab.

To illustrate the approximation of convective boundary conditions with an explicit finite difference method, consider a slab with transient one-dimensional conduction in the x direction. The slab has adiabatic boundary conditions along its length as shown in Fig. 4-4. Convection occurs at $x = 0$ and $x = l$, and the finite difference form of Eq. (4-15) is written for the surface nodes by using a forward difference approximation for dT/dx at $x = 0$. This method yields

$$\frac{T_1^n - T_0^n}{\Delta x} = \frac{\bar{h}}{k}(T_0^n - T_\infty) \qquad (4\text{-}16)$$

Solving for T_0^n gives

$$T_0^n = \frac{T_1^n + (\bar{h}\,\Delta x/k)T_\infty}{1 + \bar{h}\,\Delta x/k} \qquad (4\text{-}17)$$

This equation can be written

$$\theta_0^n = \frac{\theta_1^n}{1 + \overline{\Delta x}\,\text{Bi}} \qquad (4\text{-}18)$$

where $\text{Bi} = \bar{h}l/k$, $\overline{\Delta x} = \Delta x/l$, and $\theta = (T - T_\infty)/(T_1 - T_\infty)$. In a similar manner, using a backward finite difference approximation for dT/dx at $x = l$, one writes

$$\theta_N^n = \frac{\theta_{N-1}^n}{1 + \overline{\Delta x}\,\text{Bi}} \qquad (4\text{-}19)$$

The transient temperature distribution in the slab may be solved using Eq. (4-14) for $1 \le i \le N - 1$ along with Eq. (4-18) for $i = 0$ and Eq. (4-19) for $i = N$. To start the solution, the initial temperature distribution throughout the slab must be known at each node. In addition the convergence requirement that $\alpha \, \Delta t/(\Delta x)^2 \leq 0.5$ must be satisfied.

To write a central difference approximation for a convective boundary condition at a surface where $\bar{x} = 0$, it is necessary to use a fictitious node outside the solid at $i = -1$. Likewise, at a surface $\bar{x} = 1$ a fictitious node at $i = N + 1$ is used. The boundary condition at $\bar{x} = 0$ can then be written

$$\frac{T_1^n - T_{-1}^n}{2\Delta x} = \frac{\bar{h}}{k}(T_0^n - T_\infty) \tag{4-20}$$

or

$$\theta_1^n = \theta_{-1}^n + 2\overline{\Delta x} \, \text{Bi} \theta_0^n$$

From Eq. (4-14) the explicit finite difference form of the governing equation at $\bar{x} = 0$ is

$$\theta_0^{n+1} = \left[1 - \frac{2\text{fo}}{(\Delta x)^2}\right]\theta_0^n + \frac{\text{fo}}{(\Delta x)^2}(\theta_1^n + \theta_{-1}^n) \tag{4-21}$$

The fictitious term θ_{-1}^n can be eliminated by use of Eqs. (4-20) and (4-21) to obtain

$$\theta_0^{n+1} = \left[1 - \frac{2\text{fo}}{(\Delta x)^2} - \frac{2\text{fo Bi}}{\overline{\Delta x}}\right]\theta_0^n + \frac{2\text{fo}}{(\Delta x)^2}\theta_1^n \tag{4-22}$$

In a similar manner, at $\bar{x} = 1$ one can write

$$\theta_{N+1}^n = \theta_{N-1}^n + 2\overline{\Delta x} \, \text{Bi} \theta_N^n \tag{4-23}$$

and

$$\theta_N^{n+1} = \left[1 - \frac{2\text{fo}}{(\Delta x)^2}\right]\theta_N^n + \frac{\text{fo}}{(\Delta x)^2}(\theta_{N+1}^n + \theta_{N-1}^n)$$

The imaginary temperature θ_{N+1}^n can be eliminated between these two equations to obtain

$$\theta_N^{n+1} = \left[1 - \frac{2\text{fo}}{(\Delta x)^2} - \frac{2\text{fo Bi}}{(\overline{\Delta x})}\right]\theta_N^n + \frac{2\text{fo}}{(\Delta x)^2}\theta_{N-1}^n \tag{4-24}$$

Equations (4-22) and (4-24) can be used to give the values of θ_i^{n+1} for the nodes at $\bar{x} = 0$ and $\bar{x} = 1.0$, respectively. Equation (4-14) still applies for all other nodal values of θ_i^{n+1}. This set of explicit equations can be solved individually to obtain the values θ_i^{n+1} once the initial values of θ_i^n are known. The values for subsequent time intervals may be calculated in a like manner by using the previously calculated values as a starting point.

Chapter 3 of Ref. (4-9) proves that the limiting value of $\alpha \, \Delta t/(\Delta x)^2$ necessary to insure convergence when using a central difference approximation for convective boundary conditions is given by

$$\frac{\alpha \, \Delta t}{(\Delta x)^2} \le \frac{1}{2 + \bar{h} \, \Delta x/k}$$

If the convection heat transfer coefficient is different for each exposed surface, the surface with the highest coefficient will determine the maximum allowable value. The corresponding limitation on $\alpha \, \Delta t/(\Delta x)^2$ for the forward-backward finite difference scheme for convective boundary conditions discussed earlier is $\alpha \, \Delta t/(\Delta x)^2 < 0.5$, identical to that required for the interior nodes (cf. Ref. 4-9).

Let us review the significance of the appearance of the Biot and Fourier numbers in the finite difference equations. The Biot number occurred in the finite difference equations for surface nodes undergoing convection. The physical interpretation of the Biot number was discussed in Chap. 2. It indicates the relative magnitudes of the internal and external resistances to heat transfer by conduction and convection, respectively. Large Biot numbers indicate a relatively small resistance to convective heat transfer and that the surface temperature will be controlled by the environmental temperature of the solid. A very large Biot number produces results approaching an isothermal boundary condition $T = T_\infty$, and a very small Biot number produces results approaching an adiabatic boundary condition.

The Fourier number arose in the finite difference equations for the interior nodes. This number can be considered as a nondimensional time constant given by $t/(l^2/\alpha)$. It is a measure of the penetration depth of a thermal disturbance. A large value of the Fourier number indicates small temperature gradients within the solid or a large penetration depth of a temperature disturbance, and conversely. The larger the ratio α/l^2, the more rapid is the response of the temperature distribution to a sudden change in thermal conditions.

4-5   Numerical Analysis

The numerical analysis for transient conduction can now be performed. We first obtain a finite difference solution to the problem solved analytically in Sec. 4-2. Problems with a nonuniform initial temperature distribution are then analyzed.

Consider transient, one-dimensional heat conduction with constant properties and no internal generation. The initial temperature distribution $T_i(x)$ is uniform at T_1. At time $t = 0$ both exposed surfaces are suddenly brought to the same new temperature T_0. A transient symmetrical temperature profile occurs within the solid. The material under consideration has a thickness $2l$. Because of symmetry we consider only the half-thickness with the boundary conditions shown in Fig. 4-5.

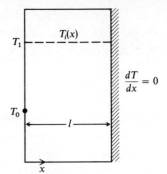

FIGURE 4-5
Left half of slab.

It is logical to define $\theta = (T - T_0)/(T_1 - T_0)$, $\bar{x} = x/l$, and Fo $= \alpha t/l^2$. The nondimensional governing equation corresponding to Eq. (4-11) is then

$$\frac{\partial^2 \theta}{\partial \bar{x}^2} = \frac{\partial \theta}{\partial \text{Fo}}$$

The initial uniform temperature distribution is specified by

$$\theta(\bar{x}, 0) = 1 \qquad \text{at} \qquad \text{Fo} = 0$$

The isothermal boundary condition at the left face is written

$$\theta(0, \text{Fo}) = 0 \qquad \text{for} \qquad \text{Fo} > 0$$

The insulated boundary condition is specified for all times greater than zero by writing

$$\frac{\partial \theta(1, \text{Fo})}{\partial \bar{x}} = 0 \qquad \text{for} \qquad \text{Fo} > 0$$

The finite difference form of the governing equation is given by Eq. (4-12) with the requirement that $\alpha \, \Delta t/(\Delta x)^2 \leq 0.5$.

We now must determine a method of obtaining the finite difference solution to the governing partial differential equation. A grid network is needed to form the nodal points which represent the solid region and its boundary. Recall that in the finite difference solutions for the Laplace and Fourier equations obtained in the previous chapter it was necessary to initially assign arbitrary temperatures to all nodes within the grid. Then the finite difference equations were applied to each node to calculate new values of the nodal temperatures. It was then necessary to repeat this process until the proper temperatures of all nodes were found. This again points out the behavior of an elliptic equation. Since a thermal disturbance will propagate throughout the solid in all directions, several iterations over the entire field are necessary to find the solution which satisfies the boundary conditions on all sides.

In the analysis of transient conduction we must accurately specify the initial condition of each node within the grid. Therefore it is not necessary to continually recalculate the initially assigned temperature values to reach the solution. Once a proper step size Δx and a time increment Δt are chosen, we apply the finite difference equations to each nodal point for each successive time increment. This numerical technique is a result of the behavior of a parabolic equation. A thermal disturbance propagates only at times greater than the initial disturbance which occurs at time $t = 0$.

The program TRAN, discussed below, has been written to obtain the solution to the transient temperature response for the initial conditions and boundary conditions indicated in Fig. 4-5. The variables and their physical meanings are given in Table 4-2.

Twelve space increments specified by 13 nodes between $0 \leq \bar{x} \leq 1$ are used as the grid network. This choice of grid fixes $\overline{\Delta x} = \Delta x/l = \frac{1}{12}$ and is assigned to the variable X in the program. The value of the convergence criterion $F = \alpha \, \Delta t/(\Delta x)^2$ is also read in Statement 100. For a given material with thermal diffusivity α, this choice of F and X fixes the time increment Δt for integration.

At time zero, the nodal temperature at $\bar{x} = 0$ is discontinuous. This occurs since we assume that the temperature of this face is instantaneously changed from the initial value $\theta = 1.0$ (Statement 230) to the value $\theta = 0.0$ for time greater than zero. Physically this condition can be approached when the surface is exposed to an environmental temperature T_0 with the existence of a very low resistance to convective heat transfer $R_c = 1/\bar{h}A$ between the surface and the environment. Since the "instantaneous" change in the surface temperature at $\bar{x} = 0$ actually requires a small finite time, the surface node at $\bar{x} = 0$ is assigned an average value between $\theta_1{}^0$ and $\theta_1{}^1$ for the calculation over the first nondimensional time increment. This is accomplished in Statement 510. All subsequent calculations are based on the boundary condition $\theta_1^n = 0$ at $\bar{x} = 0$ because of Statement 560. Equation (4-14) occurs in Statement 530,

Table 4-2 VARIABLES FOR THE TRAN PROGRAM

Variable	Definition
T(I)	Nondimensional temperature at time n
U(I)	Nondimensional temperature at time $n+1$
X(I)	Nondimensional truncated temperature used in the printout
P	Print interval
E	Nondimensional temperature value used to terminate calculations
F	Convergence criterion $= \alpha \, \Delta t/(\Delta x)^2$
FO	Fourier number $= \alpha \, t/l^2$
C	Counting variable
X	Nondimensional space increment $= \Delta x/l$

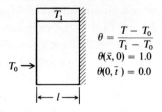

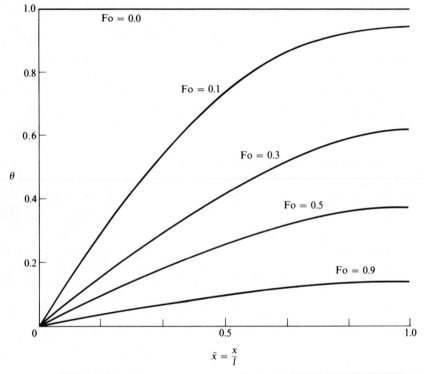

FIGURE 4-6
One-dimensional transient conduction with an isothermal boundary condition.

where the temperatures for all interior nodes are calculated. These temperatures, denoted by U(I), are the nodal values for each subsequent time increment. Throughout the program we make use of the equality $\alpha \, \Delta t/(\Delta x)^2 = \mathrm{fo}/(\overline{\Delta x})^2 = \mathrm{F}$.

An imaginary node outside of the specified wall boundary is used to satisfy the boundary condition $d\theta/d\bar{x} = 0$ at $\bar{x} = 1$. This node is denoted by U(14). Using a central finite difference approximation for the derivative at $\bar{x} = 1$, one can write

$$k \, \frac{\theta_{i+1}^{n+1} - \theta_{i-1}^{n+1}}{2\Delta x} = 0$$

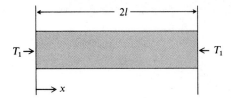

FIGURE 4-7
Rod with internal generation.

It follows that

$$\theta_{i+1}^{n+1} = \theta_{i-1}^{n+1}$$

The node at $\bar{x} = 1$ is node 13, and using the above equation we write

$$\theta_{14}^{n+1} = \theta_{12}^{n+1}$$

In terms of the computer variables for the time $n + 1$ we require U(14) = U(12). This condition is specified in Statement 550. As the final steady-state temperature distribution is reached, the change in the nodal temperatures during each successive time interval approaches zero. For the boundary conditions being considered, the values of θ_i^n also approach zero, the steady-state value. The computation is terminated (Statements 650 to 660) when the nondimensional temperature of an interior node is close to zero, $\theta_i^n < E$, where E is a small value (0.01). An alternate way to terminate the calculations is to specify a maximum Fourier number. During the integration the nondimensional temperatures for the odd-numbered nodes are printed at the nondimensional time intervals defined by P in Statement 200. A flow diagram for the following program is shown in Flow Chart 4-1.

Numerical results obtained from program TRAN are shown in Fig. 4-6. Excellent agreement is obtained with the analytical series solution, Eq. (4-7). Since the initial temperature distribution was uniform, the solution to this problem can also be obtained from the transient conduction charts presented in most elementary heat transfer texts.

In the following section numerical solutions are obtained for situations where the initial temperature distribution is not uniform. The results are obtained by making a few modifications to the program used in this section.

4-6 Initial Conditions of Nonuniform Temperature

An example is used to illustrate the analysis of a problem with a nonuniform initial condition. Consider the transient, one-dimensional temperature response of the solid cylindrical rod shown in Fig. 4-7. Initially the rod has a steady-state temperature distribution caused by uniform internal generation within the solid. Thermal insulation on the surface restricts the heat flow to the axial direction. The ends of the rod

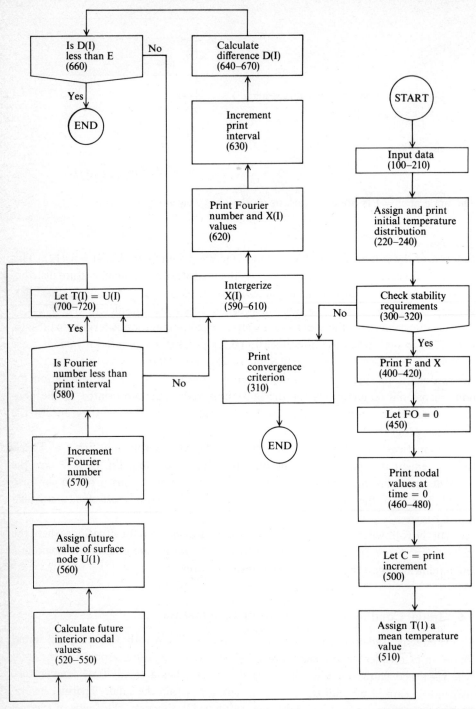

FLOW CHART 4-1

Listing 4-1   TRAN

```
10   PRINT"TYPE 1 IF YOU WANT A PROGRAM EXPLANATION."
12   PRINT
13   PRINT"OTHERWISE, TYPE 2."
14   INPUTL
16   IF L=1 THEN 20
18   IF L=2 THEN 40
20   PRINT
21   PRINT"THIS PROGRAM SOLVES THE PARABOLIC, PARTIAL DIFFERENTIAL EQN."
22   PRINT"DESCRIBING THE TEMPERATURE DISTRIBUTION FOR TRANSIENT,"
23   PRINT"ONE-DIMENSIONAL HEAT CONDUCTION IN RECTANGULAR COORDINATES."
24   PRINT
25   PRINT"AN ISOTHERMAL BOUNDARY CONDITION APPLIES AT X=0."
26   PRINT
27   PRINT"IT IS ASSUMED THAT THE LEFT FACE SUDDENLY BECOMES EQUAL"
28   PRINT"TO A NEW CONSTANT TEMPERATURE AT TIME=0."
29   PRINT
30   PRINT"THE INITIAL TEMPERATURE DISTRIBUTION MUST BE KNOWN."
31   PRINT
32   PRINT"IT IS ASSUMED THAT THE INITIAL TEMPERATURE DISTRIBUTION"
33   PRINT"IS UNIFORM THROUGHOUT THE SOLID."
34   PRINT
35   PRINT"AN EXPLICIT, FINITE DIFFERENCE APPROXIMATION IS USED"
36   PRINT"TO EXPRESS THE GOVERNING EQUATION."
37   PRINT
40   PRINT
100  READF,X
150  DIM T(14),U(14),D(14),X(14)
160  REM INPUT CONTROL VARIABLES
200  LETP=0.1
210  LETE=0.01
215  REM ASSIGN INITIAL TEMPERATURE DISTRIBUTION
220  FOR I=1 TO 14
230  LETT(I)=1.0
240  NEXT I
300  IF F<=0.5 THEN 400
310  PRINT"CONVERGENCE CRITERION ="F        "TOO LARGE"
320  GOTO 800
400  PRINT"CONVERGENCE CRITERION="F
410  PRINT
420  PRINT"DELTA X BAR="X
430  PRINT
450  LETFO=0
460  PRINT"FOURIER NO. 1     3     5     7     9     11     13"
470  PRINT
480  PRINTFO;T(1);T(3);T(5);T(7);T(9);T(11);T(13)
490  PRINT
500  LETC=P
505  REM SURFACE NODAL EQUATION FOR INITIAL TIME INCREMENT
510  LETT(1)=T(2)/2
515  REM APPLY EQUATION FOR INTERIOR NODES
520  FOR I=2 TO 13
530  LETU(I)=(1-2*F)*T(I)+F*(T(I+1)+T(I-1))
540  NEXT I
545  REM APPLY ADIABATIC NODAL EQUATION
550  LETU(14)=U(12)
555  REM SURFACE NODAL EQUATION FOR ALL SUBSEQUENT TIME INCREMENTS
560  LETU(1)=0
565  REM INCREMENT NON-DIMENSIONAL TIME
570  LETFO=FO+(X^2)*F
580  IF FO<C THEN 700
585  REM INTEGERIZE THE NODAL TEMPERATURE VALUES
590  FOR I=1 TO 13 STEP 2
600  LETX(I)=INT(U(I)*10^3 +0.5)/10^3
610  NEXT I
```

Listing 4-1   TRAN (*continued*)

```
620 PRINTFO;X(1);X(3);X(5);X(7);X(9);X(11);X(13)
625 REM INCREMENT PRINT INTERVAL
630 LETC=C+P
635 REM CHECK FOR CONVERGENCE
640 FOR I=2 TO 13
650 LETD(I)=U(I)-E
660 IF ABS(D(I))<E THEN 800
670 NEXT I
690 REM UPDATE MOST RECENT NODAL VALUES
700 FOR I=1 TO 14
710 LETT(I)=U(I)
720 NEXT I
730 GOTO 520
750 DATA 0.144,0.08333
800 END
```

Computer Results 4-1   TRAN

```
TYPE 1 IF YOU WANT A PROGRAM EXPLANATION.

OTHERWISE, TYPE 2.
? 2

CONVERGENCE CRITERION= 0.144

DELTA X BAR= 0.08333

FOURIER NO. 1    3    5    7    9    11     13

 0   1   1   1   1   1   1   1

0.100992  0   0.289  0.542  0.733  0.859  0.927  0.948
0.200984  0   0.204  0.393  0.552  0.671  0.745  0.77
0.300976  0   0.157  0.303  0.429  0.525  0.585  0.605
0.400968  0   0.123  0.237  0.335  0.41   0.457  0.473
0.50096   0   0.096  0.185  0.262  0.32   0.357  0.37
0.600952  0   0.075  0.145  0.204  0.25   0.279  0.289
0.700944  0   0.058  0.113  0.16   0.196  0.218  0.226
0.800936  0   0.046  0.088  0.125  0.153  0.17   0.176
0.900928  0   0.036  0.069  0.098  0.119  0.133  0.138
```

are maintained at a constant temperature T_1. Under these circumstances the governing equation for the steady-state temperature distribution, Eq. (3-4), becomes

$$\frac{d^2T}{dx^2} = \frac{-q'''}{k}$$

Two integrations yield

$$T = -\frac{q'''x^2}{2k} + C_1 x + C_2 \qquad 0 \le x \le l$$

The stated isothermal boundary condition is written

$$\text{at} \qquad x = 0 \qquad T = T_1$$

Because of symmetry we can also write

$$\text{at} \quad x = l \quad \frac{dT}{dx} = 0$$

If one uses the boundary conditions to evaluate the constants of integration, the steady-state temperature distribution is given by

$$T - T_1 = -\frac{q'''x^2}{2k} + \frac{q'''lx}{k} \qquad 0 \le x \le l \qquad (4\text{-}26)$$

It is desired to find the transient temperature response within the cylinder under the following situation: Assume that at time $t = 0$, the power source fails ($q''' = 0$). At the same time the temperature at the two exposed ends is suddenly reduced to $T = T_0$ by exposing them to a flowing coolant. This problem is similar to the one analyzed in Sec. 4-5 with the important difference that the initial temperature distribution is nonuniform. Thus, the analytical solution obtained in Sec. 4-2 is not valid. The final form of an analytical solution would have to be obtained from Eq. (4-6). This fact requires evaluation of the integral $\int_0^l \Delta T_i(x) \sin (n\pi x/l) \, dx$ by either analytical or numerical methods. The only required modification to the numerical procedure is to assign the proper initial temperature distribution to the nodes within the rod. For consistency we introduce a nondimensional dependent variable equal to the one in Fig. 4-6. In addition we introduce $\bar{x} = x/l$. Equation (4-26) can then be written

$$\theta - 1 = N1\left(\bar{x} - \frac{\bar{x}^2}{2}\right) \qquad 0 \le \bar{x} \le 1 \qquad (4\text{-}26a)$$

where $\theta = (T - T_0)/(T_1 - T_0)$ and $N1 = q'''l^2/k(T_1 - T_0)$.

A solution for $N1 = 2.0$ is given in Fig. 4-8. The program modifications to TRAN required to specify the initial temperature distribution and obtain the solution are

```
 50  LET N1 = 2.0
 75  LET X0 = 0
220  FOR I = 1 to 13
230  LET T(I) = N1 * (X0 − (X0 ↑ 2)/2) + 1.0
235  LET X0 = X0 + 1.0/12
240  NEXT I
440  PRINT "N1 ="N1
445  PRINT
```

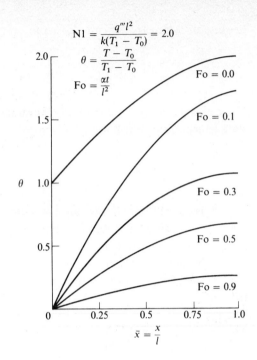

FIGURE 4-8
One-dimensional transient conduction
with an initial parabolic temperature
distribution.

It is interesting to compare Figs. 4-6 and 4-8. Figure 4-8 indicates values of
$\theta > 1.0$ at Fo $= 0.3$ for $\bar{x} > 0.75$. The maximum value of θ for Fo $= 0.3$ in Fig. 4-6
is 0.6. The effect of the nonuniform initial condition is not as noticeable at larger
values of the Fourier number. As an example, for $\alpha = 0.001$ ft^2/s (0.93 cm^2/s) and a
half-thickness l of 1.0 ft (30.48 cm), the value Fo $= 0.3$ corresponds to a time of 5.0
min. The initial conditions, boundary conditions, material properties, or material
geometry may be varied to study their effects on the cooling rate.

We are now in a position to analyze transient conduction for a variety of nonuniform initial conditions. Consider an initial steady-state, one-dimensional temperature distribution specified by Eq. (3-4). Assume variable internal generation given
by $q''' = q_0''' e^{-mx}$, where $q_0''' = $ constant. Equation (3-4) then becomes

$$\frac{d^2T}{dx^2} + \frac{q_0''' e^{-mx}}{k} = 0 \qquad (4\text{-}27)$$

The solution to this governing equation gives a good approximation to the temperature
distribution within a solid with thermal radiation absorbed by the surface at $x = 0$.
The constant term q_0''' is the effective value of the internal generation at $x = 0$. The
value of m is a function of the radiation properties of the surface and the wavelength

of the incoming radiation. We assume that m is a known constant for this calculation. The solution to Eq. (4-27) is the sum of a homogeneous and particular solution given by

$$T = C_1 + C_2 x - \frac{q_0''' e^{-mx}}{km^2} \qquad 0 \le x \le l$$

For the boundary conditions

$$x = 0 \qquad T = T_1$$

$$x = l \qquad \frac{dT}{dx} = 0$$

the resulting initial steady-state temperature distribution within the solid is

$$T = T_1 + \frac{q_0'''}{km^2}(1 - e^{-mx}) - \frac{q_0'''}{km} e^{-ml} x \qquad (4\text{-}28)$$

In nondimensional form with $\theta = (T - T_0)/(T_1 - T_0)$ and $\bar{x} = x/l$, one can write

$$\theta - 1 = \frac{q_0'''(1 - e^{-ml\bar{x}})}{km^2(T_1 - T_0)} - \frac{q_0''' l e^{-ml}\bar{x}}{km(T_1 - T_0)} \qquad (4\text{-}28a)$$

We can identify a nondimensional internal generation parameter $\mathrm{N1} = q_0''' l^2/k(T_1 - T_0)$ and a nondimensional radiation parameter $\mathrm{R0} = ml$. These are the parameters that should be varied when analyzing the effect of various initial steady-state temperature profiles in the solid. Equation (4-28a) can be written in the alternate form

$$\theta - 1 = \frac{\mathrm{N1}}{\mathrm{R0}^2}(1 - e^{-\mathrm{R0}\bar{x}}) - \frac{\mathrm{N1}}{\mathrm{R0}}\bar{x}e^{-\mathrm{R0}} \qquad (4\text{-}28b)$$

where the defined nondimensional parameters have been used.

The program used to obtain the results presented in Fig. 4-8 can be used to study the transient response in a wall with an initial temperature distribution given by Eq. (4-28b). The only change required is to put this equation in Statement 230 and read in the desired value of R0. One could then study the transient behavior caused by suddenly removing the incoming radiation at time $t = 0$.

Now let us consider a plane wall with a linear, steady-state temperature distribution as shown in Fig. 4-9. The boundary at $x = l$ is held constant at T_2, and the temperature at the surface $x = 0$ is suddenly changed at time $t = 0$. From Eq. (3-4) the governing equation for the initial temperature distribution within the wall is given by $d^2T/dx^2 = 0$. This equation can be integrated to show that the initial temperature profile is given by

$$T = -\frac{T_1 - T_2}{l} x + T_1 \qquad 0 \le x \le l \qquad (4\text{-}29)$$

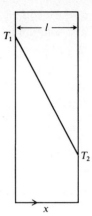

FIGURE 4-9
Plane wall with $(dT/dx)_i = C$.

Using the established procedure for nondimensionalizing this equation and defining $\theta = (T - T_0)/(T_1 - T_0)$, we obtain

$$\theta - 1 = -\frac{T_1 - T_2}{T_1 - T_0}\,\bar{x} \qquad (4.29a)$$

This particular form of θ is defined based on the boundary condition that will be used for the transient response caused by suddenly changing the surface temperature at $\bar{x} = 0$ from T_1 to T_0 at time $t = 0$. Here we see that the important nondimensional parameter for the steady-state distribution is a nondimensional temperature ratio. We will define N8 $= (T_1 - T_2)/(T_1 - T_0)$. It is interesting to determine how long it will take steady-state conditions to be reestablished if the nondimensional temperature of the left face at $\bar{x} = 0$ suddenly becomes zero at time $t = 0$. Results are given in Fig. 4-10 for N8 $= \frac{2}{3}$. The following modifications to the TRAN program were used to obtain the results given in Fig. 4-10 for an initial linear temperature distribution:

```
 50   LET N8 = 2/3
 75   LET X0 = 0
220   FOR I = 1 to 13
225   LET N8 = 0.667
230   LET T(I) = − N8*X0 + 1.0
235   LET X0 = X0 + 1.0/12
240   NEXT I
440   PRINT "N8 = "N8
445   PRINT
520   FOR I = 2 to 12
550   LET U(13) = T(13)
```

640   (REMOVE)
650   (REMOVE)
660   IF FO > 2.0 THEN 800
670   (REMOVE)
700   FOR I = 1 to 13

     The results in Fig. 4-10 show that the nondimensional time Fo required to reach the final linear, steady-state temperature profile is 0.8. If $\alpha = 0.001$ ft^2/s (0.93 cm^2/s) and $l = 1.0$ ft (30.48 cm), the time required to reach steady state is approximately 800 s. Another observation is that it takes 200 s before the nondimensional temperature gradient at $\bar{x} = 1.0$ changes sign. This information would be useful in designing control systems which use thermal sensing elements within the material.

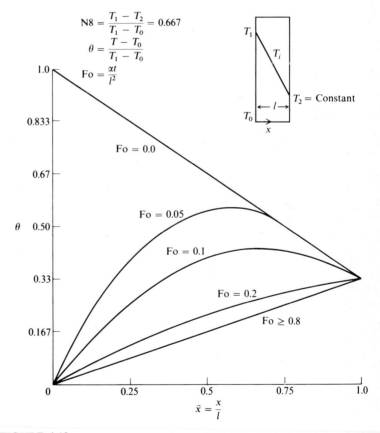

FIGURE 4-10
One-dimensional transient conduction with an initial linear temperature distribution.

4-7   Convective Boundary Conditions

Often when exposed surfaces of a body with transient conduction are experiencing heat transfer through the surfaces, a convective rather than an isothermal boundary condition is more appropriate. When this is true, the finite difference Biot number is introduced as a result of using the nodal equations at the surface. The proper equations (4-18) to (4-24) have been derived in Sec. 4-4.

Consider the configuration shown in Fig. 4-5, but assume that the surface is suddenly exposed to an environmental temperature T_∞ at time $t = 0$. During the transient response there will be a temperature difference between the surface temperature at $x = 0$, across which convection occurs, and the environmental temperature T_∞. The nodal equation for the surface node U(1) is given by Eq. (4-22) which is based on an explicit scheme using a central finite difference approximation to the spatial temperature gradient. The corresponding stability requirement given in Sec. 4-4 is

$$\frac{\alpha\,\Delta t}{(\Delta x)^2} \le \frac{1}{2 + \bar{h}\,\Delta x/k}$$

The following modifications to the program TRAN are required to analyze a convective boundary condition rather than an isothermal boundary condition at $\bar{x} = 0$. Note that $\bar{h}\,\Delta x/k = \text{Bi}\,\Delta\bar{x}$.

```
 20   PRINT "CONVECTIVE BOUNDARY CONDITIONS AT X = 0"
100   READ F, X, B1
300   IF F < = 1/(2 + B1*X) THEN 400
440   PRINT "BIOT NUMBER = " B1
445   PRINT
560   LET U(1) = (1 − 2*F − 2*F*X*B1)*T(1) + 2*F*T(2)
640   FOR I = 1 to 13
750   DATA 0.1, 0.08333, 1.0
```

Results for two different Biot numbers are given in Fig. 4-11. The temperature profiles are significantly different from those shown in Fig. 4-6. The only difference between these two problems is the boundary condition at $\bar{x} = 0$. Figure 4-6 is based on an isothermal boundary condition at $\bar{x} = 0$ for Fo > 0, and Fig. 4-11 is based on a convective boundary condition at $\bar{x} = 0$ for Fo > 0. Since the initial temperature distribution is uniform, the results in Fig. 4-11 can be checked by using transient conduction charts which graphically display analytical solutions. (See, for example, p. 149, Ref. 4-13.) For clarity, only Biot numbers of 1.0 and 0.1 are presented in Fig.

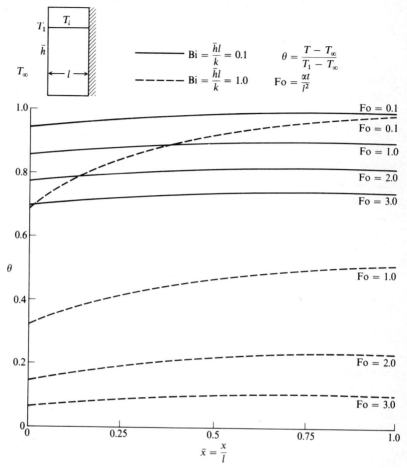

FIGURE 4-11
One-dimensional transient conduction with convective boundary conditions.

4-11. Results for larger Biot numbers would more closely resemble the isothermal boundary condition results given in Fig. 4-6. Figure 4-11 clearly shows that the surface temperature at $\bar{x} = 0$ is predominantly controlled by the interior temperatures for small values of the Biot number.

4-8 Negligible Internal Resistance

The results for $\text{Bi} = 0.1$ in Fig. 4-11 show that the temperature at any time (any Fourier number) is almost uniform throughout the solid material. This fact indicates that the internal resistance to heat conduction is much smaller than the external resistance to convection. However, severe temperature gradients occur within the solid for a Biot number of 1.0, especially during the initial transient response. This fact indicates an appreciable resistance to conduction within the solid. As the value of the Biot number approaches zero, the temperature gradients within the solid approach zero. A value $Bi = 0$ occurs when all surfaces are adiabatic. If the temperature gradients within a body are small at each instant of time, one can treat the entire solid as a closed system in thermal equilibrium at each instant when applying the first law of thermodynamics. This treatment leads to an analysis based on a lumped mathematical model as discussed in Chap. 1. No work crosses the boundary of this closed system. Thus, the first law of thermodynamics is given by $\delta Q = dU$. In heat transfer analysis we write this energy balance as an instantaneous rate equation as was done in the previous chapters. Because of the assumption of negligible temperature gradients within the solid, we cannot use Fourier's law of conduction to express the heat transfer rate. Instead, Newton's law of convection can be used since the total heat transfer is controlled by the external convective heat transfer resistance at the surface. We denote the total surface area on both sides of the wall by A_s. According to the conservation of energy principle, we equate the heat transfer crossing the system boundary to the time rate of change of internal energy within the sytem. This gives

$$-\bar{h}A_s(T - T_\infty) = \rho V c \frac{dT}{dt} \qquad (4\text{-}30)$$

Note that when the temperature T of the solid is greater than the temperature T_∞ of the surroundings, heat transfer will occur from the solid. This transfer must be accompanied by a decrease in internal energy. The negative sign in Eq. (4-30) is necessary to satisfy this requirement. One can now separate the variables in Eq. (4-30) and integrate. The initial temperature of the solid at time $t = 0$ is denoted by T_1.

$$\int_{T_1}^{T} \frac{dT}{T - T_\infty} = \int_{0}^{t} \frac{-\bar{h}A_s}{\rho V c} \, dt$$

$$\frac{T - T_\infty}{T_1 - T_\infty} = e^{-\bar{h}A_s t / \rho V c} \qquad (4\text{-}31)$$

The nondimensional exponent can be rearranged to form the product of the Biot and Fourier numbers based on the characteristic length L_c:

$$\frac{\bar{h}A_s t}{\rho V c} = \frac{\bar{h}L_c}{k} \frac{kt}{\rho c L_c^2} = Bi_{L_c} Fo_{L_c}$$

The *characteristic length* is defined as the total volume of the solid divided by the surface area across which heat transfer occurs by convection, $L_c = V/A_s$. This characteristic length is normally used only when the internal resistance to transient, conductive heat transfer is negligible, and Eq. (4-30) is used as the mathematical model. The solution for the transient temperature response within the solid, Eq. (4-31), can then be written

$$\theta = \frac{T - T_\infty}{T_1 - T_\infty} = e^{-\text{Bi}_{L_c}\,\text{Fo}_{L_c}} \qquad (4\text{-}32)$$

If the transient temperature response within the geometry shown in Fig. 4-5 is analyzed by using Eq. (4-32), the temperature within the solid must be nearly uniform at any instant in time. The value of L_c for this analysis is $V/A_s = A_s l/A_s = l$, where l is the *total* thickness as shown in Fig. 4-1. On the other hand, the results shown in Fig. 4-11 are not based on a negligible internal resistance analysis but are obtained from a finite difference solution to Eq. (4-1). The chosen characteristic length was the *half*-thickness which was denoted as l in Fig. 4-5. Thus the Biot and Fourier numbers used in the solution of a negligible internal resistance problem are not the same as those used previously. The following equalities apply: $\text{Bi}_{L_c} = 2\text{Bi}$, and $\text{Fo}_{L_c} = \frac{1}{4}\text{Fo}$, where Bi and Fo are defined in Fig. 4-11. We can now compare the results for an approximate, negligible internal resistance solution to the numerical results given in Fig. 4-11. For $\text{Bi} = 0.1$ and $\text{Fo} = 2.0$ we calculate $\text{Bi}_{L_c} = 2(0.1) = 0.2$, $\text{Fo}_{L_c} = \frac{1}{4}(2.0) = 0.5$. Using Eq. (4-32) we then obtain $\theta = e^{-(0.2)(0.5)} = 0.905$. The average value of θ for $\text{Bi} = 0.1$ and $\text{Fo} = 2.0$ is approximately 0.8 as shown in Fig. 4-11, substantially different from the value (0.905) predicted by Eq. (4-32). When the Biot number Bi_{L_c} based on $L_c = V/A_s$ is less than 0.1, it is suggested in most textbooks that a sufficiently accurate answer can be obtained by using Eq. (4-32) to calculate the temperature as a function of time. To check this approximation, we let $\text{Bi} = 0.05$ and $\text{Fo} = 2.0$. The numerical results give a value of $\theta \approx 0.89$ and Eq. (4-32) gives a value of $\theta = 0.951$. As the Biot number decreases, the accuracy of Eq. (4-32) is improved. Figure 4-12 compares the negligible internal resistance solutions to corresponding numerical solutions obtained from Eq. (4-1) for several values of the Biot number. The solutions given are for the same geometry and boundary conditions used to obtain the results in Fig. 4-11. Figure 4-12 clearly shows the error involved when Eq. (4-32) is incorrectly used to calculate transient temperature profiles for values of Biot numbers Bi_{L_c} greater than 0.1. When the value of the Biot number is greater than 0.1, the analysis of one-dimensional, transient conduction in rectangular coordinates must be based on the analytical or numerical solution to the Fourier field equation as given by Eq. (4-1).

The preceding analysis for negligible internal resistance contained two important limiting assumptions, in addition to the assumption of a small Biot number. It was assumed that at time $t = 0$ both surfaces were exposed to the same boundary condition. It was also assumed that the environmental temperature was not a function of time.

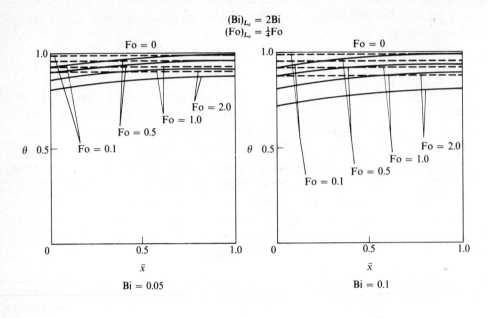

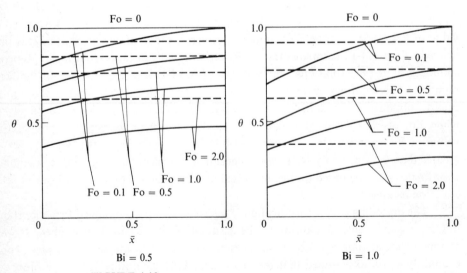

FIGURE 4-12
Transient conduction solutions for negligible and nonnegligible internal resistance.

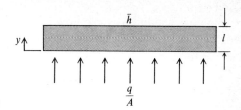

FIGURE 4-13
Slab with negligible internal resistance.

Consider the situation shown in Fig. 4-13. The cross-sectional area, which is equal to the top surface or bottom surface, is denoted by A. At time $t = 0$ the solid surface is suddenly exposed to a constant heat transfer per unit area on the lower surface and convection to an environment at T_∞ on the upper surface. (This model could represent the placing of a skillet over a fire.) We treat the problem as one with negligible internal resistance by assuming that $\mathrm{Bi}_{L_c} < 0.1$. Equating the heat transfer crossing the system boundary to the time rate of change of internal energy within the system gives

$$q - \bar{h}A(T - T_\infty) = \rho V c \frac{dT}{dt}$$

Using $V = Al$, where V is the volume and l is the thickness, this equality reduces to

$$\frac{dT}{dt} + \frac{\bar{h}(T - T_\infty)}{\rho c l} = \frac{q/A}{\rho c l} \qquad (4\text{-}33)$$

When the environmental temperature is constant, we let $\Delta T = T - T_\infty$ and write

$$\frac{d\,\Delta T}{dt} + \bar{h}\frac{\Delta T}{\rho c l} = \frac{q/A}{\rho c l}$$

The homogeneous solution obtained by setting the left side equal to zero is

$$\Delta T = Be^{-\bar{h}t/\rho c l}$$

where B is a constant of integration. By inspection, a particular solution is

$$\Delta T = \frac{q}{\bar{h}A}$$

Direct substitution of this equation into Eq. (4-33) satisfies the equality since both q and $\bar{h}$ are independent of time. The sum of the homogeneous and particular solutions yields

$$\Delta T = \frac{q}{\bar{h}A} + Be^{-\bar{h}t/\rho c l} \qquad (4\text{-}34)$$

An initial condition is required to evaluate the constant of integration. We take $\Delta T = 0$ at $t = 0$. This condition gives $B = -q/\bar{h}A$. The final solution in nondimensional form is

$$\frac{\Delta T}{q/\bar{h}A} = 1 - e^{-\bar{h}t/\rho cl} \qquad (4\text{-}35)$$

The nondimensional exponent is a Fourier-Biot product, where $\text{BiFo} = (\bar{h}l/k)(kt/\rho cl^2) = \bar{h}t/\rho cl$. As $t \to \infty$, the temperature difference $T - T_\infty$ approaches the steady-state value given by $q/\bar{h}A$.

After steady state is reached, assume that the constant heat flux boundary condition at the lower surface is suddenly replaced by an adiabatic boundary condition. The equation for the transient temperature response is then given by

$$\frac{d(\Delta T)}{dt} + \frac{\bar{h}(\Delta T)}{\rho cl} = 0 \qquad (4\text{-}36)$$

The initial condition for this equation is now the steady-state temperature distribution given by Eq. (4-35) for $t \to \infty$. The solution to Eq. (4-36) is

$$\frac{\Delta T}{q/\bar{h}A} = e^{-\bar{h}t/\rho cl} \qquad (4\text{-}37)$$

Now suppose that the environmental temperature increases linearly with time as given by $T_\infty = T_1 + bt$. The initial environmental temperature is T_1, and b is a constant value to indicate the constant rate of increase in T_∞. From Eq. (4-33) the governing equation can be written as

$$\frac{dT}{dt} + \frac{\bar{h}}{\rho cl}[T - (T_1 + bt)] = \frac{q/A}{\rho cl}$$

We define $s = \bar{h}/\rho lc$ for convenience and write

$$\frac{dT}{dt} + sT = sT_1 + sbt + \frac{q/A}{\rho cl} \qquad (4\text{-}38)$$

The homogeneous solution is $T = c_1 e^{-st}$. The particular solution is $T = c_2 + c_3 t$. Therefore

$$T = c_2 + c_3 t + c_1 e^{-st} \qquad (4\text{-}39)$$

Substituting Eq. (4-39) into Eq. (4-38) gives

$$c_3 + sc_2 + sc_3 t = sT_1 + \frac{q}{A\rho cl} + sbt$$

Matching coefficients requires that

$$c_3 + sc_2 = sT_1 + \frac{q}{A\rho cl}$$

and

$$sc_3 = sb$$

This condition in turn gives

$$c_3 = b \qquad c_2 = T_1 + \frac{q}{A\rho cls} - \frac{b}{s}$$

The solution is then

$$T = T_1 + \frac{q}{A\rho cls} - \frac{b}{s} + bt + c_1 e^{-st}$$

The initial condition is $T = T_\infty = T_1$ at $t = 0$. This condition gives

$$c_1 = \frac{b}{s} - \frac{q}{A\rho cls}$$

The final temperature distribution is then

$$T = T_1 + \frac{q}{A\rho cls} - \frac{b}{s} + bt + \left(\frac{b}{s} - \frac{q}{A\rho cls}\right) e^{-st} \qquad (4\text{-}40)$$

We can define $\theta = (T - T_1)/(T_\infty - T_1)$ and $\tau = st = \text{BiFo}$ and write

$$\theta = \frac{q}{A\rho cls(T_\infty - T_1)} - \frac{b}{s(T_\infty - T_1)} + \frac{b\tau}{s(T_\infty - T_1)} + \left[\frac{b}{s(T_\infty - T_1)} - \frac{q}{A\rho cls(T_\infty - T_1)}\right] e^{-\tau}$$

The important nondimensional parameters can be identified as $q/A\rho cls(T_\infty - T_1)$ and $b/s(T_\infty - T_1)$. The first parameter can also be written $q/\bar{h}A(T_\infty - T_1)$, and it is obviously a nondimensional ratio of incoming heat transfer (q) to heat transfer by convection.

Once again the value of the Biot number plays an important role in analysis. For Biot numbers ($\bar{h}L_c/k$) less than 0.1, the temperature distribution can be approximated by considering a function of only one independent variable—time. The resulting lumped analysis leads to an ordinary differential equation. For larger Biot numbers two independent variables, distance and time, are required in an analysis which leads to a partial differential equation. When energy crosses a surface by both radiation and convection, an effective heat transfer coefficient $\bar{h}_T$ [see Eq. (1-61)] should be used in calculating the Biot number to determine when a lumped analysis is valid. In Chap. 1 a lumped analysis was used entirely. We are now in a position to intelligently choose between a lumped analysis and a differential analysis. It is also clear why the analysis in Chap. 1 was applied to thin bodies with high thermal conductivity.

4-9   Transient Conduction in a Fin

The numerical analysis for steady-state, one-dimensional conduction in fins of various profiles was considered in Chap. 2. Now we consider a cylindrical fin of constant area with uniform internal generation shown in Fig. 4-14. The steady-state temperature distribution is governed by the equation

$$\frac{d^2T}{dx^2} - \frac{\bar{h}p}{kA}(T - T_\infty) = \frac{-q'''}{k} \qquad 0 \le x \le l$$

This equation was derived in Chap. 2. Letting $\theta = (T - T_\infty)/(T_w - T_\infty)$, $\bar{x} = x/l$, and $m^2 = \bar{h}p/kA$ we write

$$\frac{d^2\theta}{d\bar{x}^2} - m^2 l^2 \theta = \frac{-q''' l^2}{k(T_w - T_\infty)} \qquad 0 \le \bar{x} \le 1.0 \qquad (4\text{-}41)$$

The analytical solution to this equation expressed as the sum of a homogeneous and a particular solution is

$$\theta = C_1 e^{ml\bar{x}} + C_2 e^{-ml\bar{x}} + \frac{q''' A}{\bar{h}p(T_w - T_\infty)} \qquad (4\text{-}42)$$

Using the boundary conditions

$$x = 0 \qquad \theta = 1.0$$

$$\bar{x} = 1.0 \qquad \frac{d\theta}{d\bar{x}} = 0$$

gives

$$C_1 = \frac{e^{-2ml}}{1 + e^{-2ml}}\left[1 - \frac{q''' A}{\bar{h}p(T_w - T_\infty)}\right]$$

$$C_2 = \frac{1 - q''' A/\bar{h}p(T_w - T_\infty)}{1 + e^{-2ml}}.$$

and the steady-state temperature distribution is determined. Note that the ratio $q''' A/\bar{h}p(T_w - T_\infty)$ can be written as $\frac{1}{4}N1/Bi$, where we have defined $N1 = q''' D^2/k(T_w - T_\infty)$ and $Bi = \bar{h}D/k$.

To express Eq. (4-42) in nondimensional form, we can write $ml = (\bar{h}p/kA)^{1/2}l = (4\bar{h}/kD)^{1/2}l = 2(\bar{h}D/k)^{1/2}(l/D) = 2Bi^{1/2}L1$, where the nondimensional parameter $L1 = l/D$. The steady-state solution is then

$$\theta = C_1 e^{2L1(Bi)^{1/2}\bar{x}} + C_2 e^{-2L1(Bi)^{1/2}\bar{x}} + \frac{1}{4}\frac{N1}{Bi} \qquad (4\text{-}42a)$$

where

$$C_1 = \frac{e^{-4L1(Bi)^{1/2}}}{1 + e^{-4L1(Bi)^{1/2}}}\left(1 - \frac{1}{4}\frac{N1}{Bi}\right)$$

and

$$C_2 = \frac{1 - \frac{1}{4}N1/Bi}{1 + e^{-4L1(Bi)^{1/2}}}$$

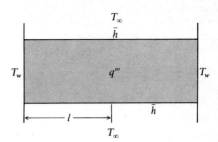

FIGURE 4-14
Fin with internal generation.

Suppose that at time $t = 0$ the internal generation term becomes zero, and it is desired to determine the transient temperature response of the fin. After a sufficient time we know that the temperature distribution will again reach the steady-state value given in Sec. 2-4 for a fin without internal generation. That is,

$$\theta = \frac{T - T_\infty}{T_w - T_\infty} = \frac{\cosh ml(1 - \bar{x})}{\cosh ml} = \frac{\cosh 2L1Bi^{1/2}(1 - \bar{x})}{\cosh 2L1Bi^{1/2}} \qquad (4\text{-}43)$$

The differential equation governing the transient temperature distribution in a fin can be derived by considering a one-dimensional model as was done in Chap. 2. The results of applying the conservation of energy principle to a properly chosen differential element with $q''' = 0$ gives

$$\frac{\partial^2 T}{\partial x^2} - \frac{\bar{h}p(T - T_\infty)}{kA} = \frac{1}{\alpha}\frac{\partial T}{\partial t} \qquad (4\text{-}44)$$

In finite difference form, Eq. (4-44) becomes

$$\frac{T_{i+1}^n - 2T_i^n + T_{i-1}^n}{(\Delta x)^2} - \frac{\bar{h}p(T_i^n - T_\infty)}{kA} = \frac{1}{\alpha}\frac{T_i^{n+1} - T_i^n}{\Delta t}$$

Solving for T_i^{n+1} yields

$$T_i^{n+1} = T_i^n\left[1 - \frac{2\alpha\,\Delta t}{(\Delta x)^2} - \frac{\alpha\,\Delta t\bar{h}p}{kA}\right] + \frac{\alpha\,\Delta t}{(\Delta x)^2}(T_{i+1}^n + T_{i-1}^n) + \frac{\alpha\,\Delta t\bar{h}pT_\infty}{kA} \qquad (4\text{-}45)$$

To eliminate the T_i^n term, let

$$\frac{\alpha\,\Delta t}{(\Delta x)^2} = \frac{1}{2}\left(1 - \frac{\alpha\,\Delta t\bar{h}p}{kA}\right) \qquad (4\text{-}46)$$

This choice of $\alpha\,\Delta t/(\Delta x)^2$ eliminates the negative terms in Eq. (4-45) and assures solutions which are physically realistic. From Eq. (4-46) the time increment is given by

$\Delta t = (\Delta x)^2/\alpha[2 + (\Delta x)^2 \bar{h}p/kA]$, and it follows that the nondimensional time increment must satisfy the relation

$$\frac{\alpha \, \Delta t}{(\Delta x)^2} \leq \frac{1}{2 + (\Delta x)^2 \bar{h}p/kA} \qquad (4\text{-}47)$$

to insure convergence, based on the discussion in Sec. 4-3. In terms of the nondimensional parameters, Eq. (4-47) becomes

$$F = \frac{\alpha \, \Delta t}{(\Delta x)^2} \leq \frac{1}{2 + 4(\overline{\Delta x})^2 L1^2 Bi} \qquad (4\text{-}48)$$

If we introduce $\theta = (T - T_\infty)/(T_w - T_\infty)$, where T_w is the temperature at $\bar{x} = 0$, the nondimensional form of the finite difference equation (4-45) becomes

$$\theta_i^{n+1} = [1 - 2F - 4FBiL1^2(\overline{\Delta x})^2]\theta_i^n + F(\theta_{i+1}^n + \theta_{i-1}^n) \qquad (4\text{-}49)$$

Note that $BiL1^2$, where $Bi = \bar{h}D/k$ and $L1 = l/D$, can also be written $BiL1$, where $Bi = \bar{h}l/k$.

The computer program TFIN was used to obtain the solutions given in Fig. 4-15 for the transient temperature response in a fin. A listing of the entire program appears below. The variables used are identified in Table 4-3. The initial steady-state temperature distribution is assigned in Statement 230 as given by the analytical solution, Eq. (4-42). The finite difference equation for the interior nodes, Eq. (4-49), is employed in Statement 530. The program uses the same logic as that described for the TRAN program.

Table 4-3   VARIABLES FOR THE TFIN PROGRAM

Variable	Definition
X0	Nondimensional increment x/l used to specify initial temperature distribution
F	Convergence criterion $= \alpha \, \Delta t/(\Delta x)^2$
B1	Biot number $= \bar{h}D/k$
N1	Internal generation parameter $= q'' D^2/[k(T_w - T_\infty)]$
L1	Length-to-diameter ratio
T(I)	Nondimensional temperatures at time n
U(I)	Nondimensional temperatures at time $n + 1$
X(I)	Nondimensional truncated temperatures used in the printout
P	Print interval
FO	Fourier number $= \alpha t/D^2$
C	Counting variable
X	Nondimensional increment $\Delta x/l$ which is fixed by the grid size

The curves in Fig. 4-15 (page 216) show the transient response within a fin for Bi $= \bar{h}L/k = 0.02$, N1 $= 0.1$, and L1 $= 10.0$. Note that an analysis which assumes negligible internal resistance to conduction is not appropriate for the analysis of a fin, even for Biot numbers less than 0.1. This is because the isothermal boundary condition at the fin base causes a temperature gradient within the fin material. It can be seen that the temperature profile approaches the final known steady-state value at Fo $= \alpha t/D^2 \geq 50$. An interesting observation is the fact that inflection points occur in the profile during the initial cooling period. This is caused by the specified boundary condition $d\theta/d\bar{x} = 0$ at $\bar{x} = 1$. Note also that a thermal disturbance diffuses through the fin as cooling takes place. A parametric analysis is made by varying N1, B1, and L1 to determine the transient temperature response and heat transfer rates for a wide variety of conditions.

One can also analyze transient conduction in a fin of arbitrary cross-sectional area by using the proper finite difference form of the governing equation. Other boundary conditions at $\bar{x} = 1$ may also be needed for certain problems. The required initial condition represented by the steady-state temperature distribution within a fin of arbitrary profile may be determined by using the methods in Chap. 2.

4-10   Cylindrical and Spherical Coordinates

The governing equations for conduction in cylindrical and spherical coordinates were given in Sec. 3-10. For transient, one-dimensional, radial conduction in a cylinder the governing equation is

$$\frac{\partial^2 T}{\partial r^2} + \frac{1}{r}\frac{\partial T}{\partial r} = \frac{1}{\alpha}\frac{\partial T}{\partial t} \qquad (4\text{-}50)$$

The corresponding equation in spherical coordinates is

$$\frac{1}{r^2}\frac{\partial^2(rT)}{\partial r^2} = \frac{1}{\alpha}\frac{\partial T}{\partial t} \qquad (4\text{-}51)$$

The method of expressing the left side of these equations in finite difference form was discussed in Sec. 3-10. The finite difference expression for the time derivative is identical to the form used in this chapter for transient conduction in rectangular coordinates. Finite difference expressions for boundary conditions on cylindrical or spherical surfaces are obtained by applying the conservation of energy principle to surface nodes. We again expect the nondimensional temperature distribution to be a function of the Biot and Fourier numbers which are based on an appropriate characteristic length, such as a radius.

Listing 4-2   TFIN

```
10   PRINT"TYPE 1 IF YOU WANT A PROGRAM EXPLANATION."
12   PRINT
13   PRINT"OTHERWISE, TYPE 2."
14   INPUTL
16   IF L=1 THEN 20
18   IF L=2 THEN 40
20   PRINT
21   PRINT"THIS PROGRAM SOLVES THE PARABOLIC, PARTIAL DIFFERENTIAL EQN"
22   PRINT"DESCRIBING THE TEMPERATURE DISTRIBUTION FOR TRANSIENT,"
23   PRINT"ONE-DIMENSIONAL HEAT CONDUCTION IN A FIN."
24   PRINT
25   PRINT"AN ADIABATIC BOUNDARY CONDITION APPLIES AT X=1.0."
26   PRINT
27   PRINT" IT IS ASSUMED THAT THE INTERNAL GENERATION SUDDENLY"
28   PRINT"BECOMES ZERO AT TIME=0."
29   PRINT
30   PRINT"THE INITIAL TEMPERATURE DISTRIBUTION IS GIVEN BY"
31   PRINT"THE STEADY STATE SOLUTION FOR ONE-DIMENSIONAL HEAT"
32   PRINT"CONDUCTION IN A FIN."
33   PRINT
34   PRINT"AN EXPLICIT, FINITE DIFFERENCE APPROXIMATION IS USED"
35   PRINT"TO EXPRESS THE GOVERNING EQUATION."
40   PRINT
75   LETX0=0
100  READ F,X,B1,N1,L1
150  DIM T(14),U(14),X(14),Y(14)
155  REM CALCULATE CONSTANTS
160  LETC2=(1-0.25*(N1/B1))/(1+EXP(-4*L1*(B1^.5)))
170  LETC1=C2*EXP(-4*L1*(B1^.5))
180  LETZ1=0.25*(N1/B1)
200  LETP=0.5
210  REM CALCULATE INITIAL STEADY STATE TEMPERATURE DISTRIBUTION
220  FOR I=1 TO 13
230  LETT(I)=C1*EXP((2*L1*(B1^.5))*X0)+C2*EXP((-2*L1*(B1^.5))*X0)+Z1
235  LETX0=X0+1/12
240  NEXT I
300  IF F<=1.0/(2.0+4*(X^2)*(L1^2)*B1) THEN 400
310  PRINT"CONVERGENCE CRITERION="F          "TOO LARGE"
320  GOTO 800
400  PRINT"CONVERGENCE CRITERION="F
410  PRINT
420  PRINT"DELTA X BAR="X,"L1=L/D="L1
430  PRINT
440  PRINT"B1="B1, "N1="N1
445  PRINT
450  LETF0=0
460  PRINT"FOURIER NO  1    3    5    7    9    11    13"
470  PRINT
480  PRINT FO   T(1)   T(3)   T(5)   T(7)   T(9)   T(11)   T(13)
490  PRINT
500  LETC=P
510  REM APPLY EQUATION FOR INTERIOR NODES
520  FOR I=2 TO 13
530  LETU(I)=(1-2*F-4*F*B1*(L1^2)*(X^2))*T(I) + F*(T(I+1)+T(I-1))
540  NEXT I
545  REM APPLY ADIABATIC NODAL EQUATION
550  LETU(14)=U(12)
555  REM SURFACE NODAL EQUATION FOR ALL SUBSEQUENT TIME INCREMENTS
560  LETU(1)=T(1)
565  REM INCREMENT NON-DIMENSIONAL TIME
570  LETFO=F0+(L1^2)*(X^2)*F
580  IF FO<C THEN 700
585  REM INTEGERIZE THE NODAL TEMPERATURE VALUES
590  FOR I=1 TO 13 STEP 2
```

LISTING 4-2   TFIN (*continued*)

```
600 LETX(I)=INT(U(I)*10^3+0.5)/10^3
610 NEXT I
620 PRINT FO   X(1)   X(3)   X(5)   X(7)   X(9)   X(11)   X(13)
625 REM INCREMENT PRINT INTERVAL
630 LETC=C+P
635 REM CHECK FOR CONVERGENCE
640 IF FO>100 THEN 800
690 REM UPDATE MOST RECENT NODAL VALUES
700 FOR I=1 TO 14
710 LETT(I)=U(I)
720 NEXT I
730 GOTO 520
750 DATA 0.4,0.08333,0.02,0.1,10
800 END
```

Computer Results 4-2   TFIN

TYPE 1 IF YOU WANT A PROGRAM EXPLANATION.

OTHERWISE, TYPE 2.
? 2

CONVERGENCE CRITERION= 0.4

DELTA X BAR= 0.08333   L1=L/D= 10

B1= 0.02   N1= 0.1

FOURIER NO 1   3   5   7   9   11   13

 0 1. 1.09312 1.15072 1.18585 1.20046 1.21722 1.22055

0.555511	1	1.038	1.096	1.131	1.152	1.162	1.079
1.11102	1	0.996	1.043	1.078	1.099	1.068	1.027
1.66653	1	0.962	0.995	1.028	1.038	1.011	0.987
2.22204	1	0.933	0.951	0.978	0.983	0.964	0.95
2.77756	1	0.909	0.911	0.931	0.935	0.922	0.913
3.05531	1	0.897	0.892	0.908	0.911	0.901	0.893
3.61082	1	0.877	0.857	0.866	0.869	0.862	0.857
4.16633	1	0.858	0.825	0.826	0.829	0.824	0.821
4.72184	1	0.842	0.795	0.79	0.791	0.788	0.787
5.27736	1	0.827	0.768	0.757	0.756	0.754	0.753
5.55511	1	0.819	0.756	0.741	0.739	0.737	0.736
6.11062	1	0.806	0.732	0.711	0.706	0.705	0.704
6.66613	1	0.794	0.71	0.683	0.676	0.674	0.674
7.22164	1	0.783	0.69	0.657	0.647	0.645	0.644
7.77716	1	0.773	0.671	0.632	0.62	0.617	0.616
8.05491	1	0.768	0.662	0.621	0.607	0.603	0.603
8.61042	1	0.759	0.646	0.599	0.582	0.577	0.576
9.16593	1	0.75	0.63	0.578	0.559	0.553	0.551
9.72144	1	0.743	0.616	0.559	0.537	0.529	0.528
10.277	1	0.735	0.602	0.541	0.516	0.507	0.505
10.5547	1	0.732	0.596	0.533	0.506	0.497	0.494

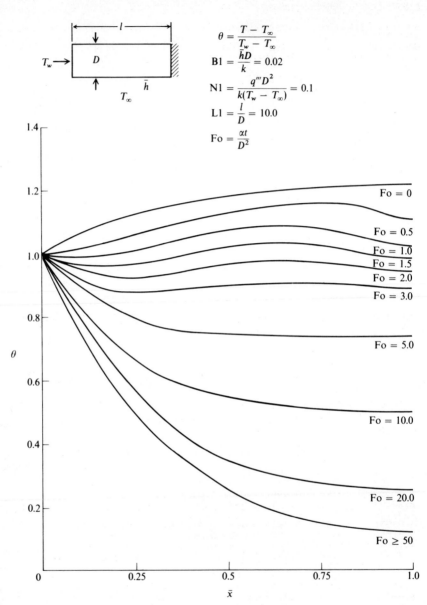

FIGURE 4-15
Transient conduction in a fin.

Initial conditions for a transient conduction analysis are often given by a steady-state, one-dimensional, analytical solution. In cylindrical coordinates, the steady-state, radial temperature distribution is given by a solution to Eq. (4-50) when the time derivative is zero. This model results in the equation

$$\frac{\partial^2 T}{\partial r^2} + \frac{1}{r}\frac{\partial T}{\partial r} = 0 \qquad (4\text{-}52)$$

This equation can be written

$$\frac{d}{dr}\left(r\frac{dT}{dr}\right) = 0$$

Two integrations yield

$$T = C_1 \ln r + C_2 \qquad (4\text{-}53)$$

If we specify the following boundary conditions

$$r = r_1 \qquad T = T_1$$

$$r = r_2 \qquad T = T_2$$

we obtain $C_1 = (T_2 - T_1)/\ln (r_2/r_1)$ and $C_2 = T_1 - (T_2 - T_1)\ln r_1/\ln (r_2/r_1)$. If we define $\theta = (T - T_1)/(T_2 - T_1)$, we obtain

$$\theta = \frac{1}{\ln (r_2/r_1)} \ln \frac{r}{r_1} \qquad (4\text{-}54)$$

This relationship gives the steady-state temperature distribution $\theta = \theta(r)$.

The heat transfer for one-dimensional, radial conduction is obtained from Fourier's law

$$q = kA_r \frac{dT}{dr} \qquad (4\text{-}55)$$

where A_r is the area normal to the radial heat conduction and is given by $A_r = 2\pi r L$ for a cylinder of length L. Combining Eqs. (4-54) and (4-55) gives

$$q = \frac{2\pi k L(T_2 - T_1)}{\ln (r_2/r_1)} \qquad (4\text{-}56)$$

This equation can be rewritten as $q = (T_2 - T_1)/R_c$, where R_c is defined as the resistance to radial heat conduction and is given by $R_c = \ln (r_2/r_1)/2\pi k L$. The corresponding term which represents the resistance to one-dimensional heat conduction in rectangular coordinates is given by $R_c = l/kA$. Both of these definitions of resistance to conduction are limited to one-dimensional heat conduction in a material with constant properties.

In spherical coordinates the governing equation for the steady-state, radial temperature distribution in a material with constant properties and no internal generation is obtained from Eq. (4-51) and can be written

$$\frac{\partial^2(rT)}{\partial r^2} = 0 \qquad (4\text{-}57)$$

This equation can be integrated twice to obtain

$$T = C_1 + \frac{C_2}{r} \qquad (4\text{-}58)$$

The two constants of integration can be determined by specifying the following boundary conditions

$$r = r_1 \qquad T = T_1$$
$$r = r_2 \qquad T = T_2$$

Applying the boundary conditions and defining $\theta = (T - T_2)/(T_1 - T_2)$ and $\bar{r} = r/r_2$ leads to

$$\theta = \frac{r_1}{r_2 - r_1}\left(\frac{1}{\bar{r}} - 1\right) \qquad (4\text{-}59)$$

which is the steady-state, nondimensional, radial temperature distribution.

Fourier's law for one-dimensional heat conduction in spherical coordinates is written

$$q = kA_r \frac{dT}{dr} \qquad (4\text{-}60)$$

where the area is given by $A_r = 4\pi r^2$. Combining Eqs. (4-59) and (4-60) gives

$$q = \frac{4\pi k r_1 r_2 (T_1 - T_2)}{r_1 - r_2} \qquad (4\text{-}61)$$

The resistance to one-dimensional, spherical conduction in a material when the temperature distribution is governed by Eq. (4-57) is found by equating Eq. (4-61) to $q = (T_1 - T_2)/R_c$. This move gives $R_c = (1/r_2 - 1/r_1)/4\pi k$.

A transient analysis is required when one of the boundary conditions which specifies the initial steady-state temperature distribution is suddenly changed at time $t = 0$. Equations (4-54) and (4-59) can then be used to specify the initial temperature distribution in cylindrical and spherical coordinates, respectively. Note that in each case the initial temperature distribution is not uniform but is a function of the radius.

Analytical solutions for transient, one-dimensional conduction in cylindrical and spherical coordinates can be found in the literature. (See Refs. 4-1, 4-3, and 4-8 for examples.) Numerical solutions may be obtained by using the methods presented in this chapter. It is necessary to specify the grid size to fit the chosen coordinate system, to express the finite difference equation for the governing equation in terms of the chosen coordinate system, and to determine the limitation of the convergence criterion $\alpha \, \Delta t / (\Delta r)^2$ as described in Sec. 4-3.

4-11   Two-dimensional Transient Conduction

The governing equation for transient, two-dimensional conduction with constant properties and no internal generation follows directly from Eq. (3-4). In dimensional form the equation is

$$\frac{\partial^2 T}{\partial x^2} + \frac{\partial^2 T}{\partial y^2} = \frac{1}{\alpha} \frac{\partial T}{\partial t} \tag{4-62}$$

Let us define $\theta = (T - T_1)/(T_i - T_1)$, where T_1 is an isothermal boundary and T_i is the initial temperature. This equation can be put in nondimensional form using the method discussed in Sec. 4-2. Using the same nondimensional independent variables and applying the chain rules of differentiation gives

$$\frac{\partial^2 \theta}{\partial \bar{x}^2} + \frac{\partial^2 \theta}{\partial \bar{y}^2} = \frac{\partial \theta}{\partial \mathrm{Fo}} \tag{4-63}$$

We will use this nondimensional form of the governing equation as the starting point in obtaining a nondimensional finite difference equation.

The finite difference approximations for the spatial derivatives in Eq. (4-63) were derived in Sec. 3-4. The finite difference approximations for the time derivative were given in Sec. 4-3. Substituting these finite difference expressions into Eq. (4-63) yields

$$\frac{1}{(\overline{\Delta x})^2} (\theta^n_{i+1, j} - 2\theta^n_{i, j} + \theta^n_{i-1, j}) + \frac{1}{(\overline{\Delta y})^2} (\theta^n_{i, j+1} - 2\theta^n_{i, j} + \theta^n_{i, j-1})$$

$$= \frac{\theta^{n+1}_{i, j} - \theta^n_{i, j}}{\overline{\Delta \mathrm{Fo}}} \tag{4-64}$$

The associated convergence requirement (cf. Ref. 4-14) is

$$\frac{\overline{\Delta t}}{(\overline{\Delta x})^2} + \frac{\overline{\Delta t}}{(\overline{\Delta y})^2} \leq 0.5$$

When $\overline{\Delta x} = \overline{\Delta y} = \bar{\delta}^2$, this expression reduces to $\overline{\Delta t}/\bar{\delta}^2 \leq 0.25$. This requirement restricts the size of the time increment more severely than in one-dimensional problems.

The corresponding implicit equation is obtained by using the nondimensional temperatures on the left side of Eq. (4-64) at time $n+1$ rather than at time n. The implicit equation is stable but requires simultaneous solution of a large number of equations. As in the case of one-dimensional, transient conduction, many methods

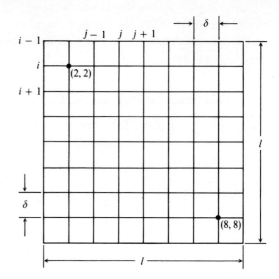

FIGURE 4-16
Two-dimensional transient conduction grid.

have been suggested which are combinations of the implicit and explicit methods. (See Refs. 4-14 and 4-15.)

Consider the two-dimensional, square region shown in Fig. 4-16. Assume that the initial temperature distribution is uniform and at time $t = 0$ all four surfaces are suddenly brought to an isothermal value T_1 and held constant. The mathematical boundary conditions which represent this situation are

$$\theta(\bar{x}, \bar{y}, 0) = 1.0$$
$$\theta(\bar{x}, 0, \bar{t}) = 0$$
$$\theta(0, \bar{y}, \bar{t}) = 0 \qquad (4\text{-}65)$$
$$\theta(1, \bar{y}, \bar{t}) = 0$$
$$\theta(\bar{x}, 1, \bar{t}) = 0$$

where $\theta = (T - T_1)/(T_i - T_1)$.

Solving for $\theta_{i,j}^{n+1}$ from Eq. (4-64) yields

$$\theta_{i,j}^{n+1} = \theta_{i,j}^n \left(1 - \frac{4\,\overline{\Delta t}}{\bar{\delta}^2}\right) + \frac{\overline{\Delta t}}{\bar{\delta}^2}(\theta_{i+1,j}^n + \theta_{i,j+1}^n + \theta_{i-1,j}^n + \theta_{i,j-1}^n) \qquad (4\text{-}66)$$

For convergence we choose $\overline{\Delta t}/\bar{\delta}^2 \le 0.25$ where

$$\frac{\overline{\Delta t}}{\bar{\delta}^2} = \frac{\overline{\Delta\mathrm{Fo}}}{\bar{\delta}^2} = \frac{\alpha\,\Delta t/l^2}{\delta^2/l^2} = \frac{\alpha\,\Delta t}{\delta^2} = \mathrm{F}$$

Equation (4-66) is used to calculate the temperature of the interior nodes on the $n + 1$ time row. The computer program TRAN may be modified to obtain a solution to

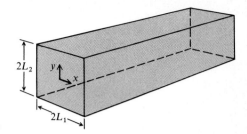

FIGURE 4-17
Two-dimensional bar.

this two-dimensional transient conduction problem. One possibility is shown in the following program called TWOD. (See Listing 4-3.) The variables are identified in Table 4-4. A nine by nine grid is used to simulate the square region. When the grid size was reduced by a factor of 2, the numerical results obtained from the two different grid sizes agreed to within the second decimal place.

Since the initial temperature distribution was uniform, the numerical results can be checked by using the analytical solutions presented in graphical form on page 149 of Ref. 4-13. This reference shows how solutions to transient two-dimensional problems can be obtained from one-dimensional results by using the relation

$$\left(\frac{T - T_1}{T_i - T_1}\right)_{2D} = \left(\frac{T - T_1}{T_i - T_1}\right)_{L_1} \left(\frac{T - T_1}{T_i - T_1}\right)_{L_2} \qquad (4\text{-}67)$$

The nondimensional temperature $[(T - T_1)/(T_i - T_1)]_{L_1}$ is the solution for one-dimensional transient conduction in the x direction across a thickness $2L_1$, as shown in Fig. 4-17. The nondimensional temperature $[(T - T_1)/(T_i - T_1)]_{L_2}$ is the solution for one-dimensional, transient conduction in the y direction across a thickness $2L_2$ as shown in Fig. 4-17. The product of these two nondimensional temperatures at a

Table 4-4   VARIABLES FOR THE TWOD PROGRAM

Variable	Definition
M	Number of increments in Δx and Δy
F	Convergence criterion $= \alpha \, \Delta t/(\Delta x)^2$
X	Nondimensional increment $\Delta x/l = \Delta y/l =$ fixed grid size
T(I, J)	Nondimensional temperatures at time n
U(I, J)	Nondimensional temperatures at time $n + 1$
D(I, J)	Difference U(I, J) − T(I, J)
X(I, J)	Nondimensional truncated temperatures used in the printout
P	Print interval
E	Minimum value of D(I, J) used to terminate the computation
FO	Fourier number $\alpha t/l^2$
C	Counting variable

Listing 4-3 TWOD

```
10    PRINT"TYPE 1 IF YOU WANT A PROGRAM EXPLANATION."
12    PRINT
13    PRINT"OTHERWISE, TYPE 2."
14    INPUT L
16    IF L=1 THEN 20
18    IF L=2 THEN 40
20    PRINT
21    PRINT"THIS PROGRAM SOLVES THE PARABOLIC, PARTIAL DIFFERENTIAL EQN"
22    PRINT"DESCRIBING THE TEMPERATURE DISTRIBUTION FOR TRANSIENT,"
23    PRINT"TWO-DIMENSIONAL HEAT CONDUCTION IN A BAR."
24    PRINT
25    PRINT"IT IS ASSUMED THAT ALL SURFACES SUDDENLY BECOME EQUAL"
26    PRINT"TO A NEW, CONSTANT TEMPERATURE AT TIME=0."
27    PRINT
28    PRINT"THE INITIAL TEMPERATURE DISTRIBUTION IS UNIFORM"
29    PRINT"THROUGHOUT THE BAR."
30    PRINT
31    PRINT"AN EXPLICIT, FINITE DIFFERENCE APPROXIMATION IS USED"
32    PRINT"TO EXPRESS THE GOVERNING EQUATION."
33    PRINT
34    PRINT"THE BAR HAS A SQUARE CROSS SECTION WHICH IS"
35    PRINT"REPRESENTED BY A 9X9 GRID."
40    PRINT
75    LETM=9
100   READ F,X
150   DIM T(50,50),U(50,50),D(50,50),X(50,50)
190   REM ASSIGN CONTROL CONSTANTS
200   LETP=0.02
210   LETE=0.0001
215   REM ASSIGN INITIAL TEMPERATURE DISTRIBUTION
220   FOR I=1 TO M
230   FOR J=1 TO M
240   LETT(I,J)=1.0
250   NEXT J
260   NEXT I
300   IF 2*F<=0.5 THEN 400
310   PRINT"CONVERGENCE CRITERION="F      "TOO LARGE"
320   GOTO 2000
400   PRINT"CONVERGENCE CRITERION="F
410   PRINT
420   PRINT"DELTA X BAR="X
430   PRINT
440   REM INITIALIZE NON-DIMENSIONAL TIME
450   LETFO=0
500   LETC=P
505   REM SURFACE NODAL EQUATIONS FOR INITIAL TIME INCREMENT
510   FOR J=1 TO M
520   LETT(1,J)=T(2,2)/2
530   NEXT J
540   FOR J=1 TO M
550   LETT(M,J)=T(2,2)/2
560   NEXT J
570   FOR I=1 TO M
580   LETT(I,1)=T(2,2)/2
590   NEXT I
600   FOR I=1 TO M
610   LETT(I,M)=T(2,2)/2
620   NEXT I
625   REM APPLY EQUATION FOR INTERIOR NODES
630   FOR I=2 TO M-1
640   FOR J=2 TO M-1
```

Listing 4-3 TWOD (*continued*)

```
650  LETU(I,J)=T(I,J)*(1-4*F)+F*(T(I+1,J)+T(I,J+1)+T(I-1,J)+T(I,J-1))
660  NEXT J
670  NEXT I
675  REM SURFACE NODAL EQUATIONS FOR ALL SUBSEQUENT TIME INCREMENTS
680  FOR J=1 TO M
690  LETU(1,J)=0
700  NEXT J
710  FOR J=1 TO M
720  LETU(M,J)=0
730  NEXT J
740  FOR I=1 TO M
750  LETU(I,1)=0
760  NEXT I
770  FOR I=1 TO M
780  LETU(I,M)=0
790  NEXT I
795  REM INCREMENT NON-DIMENSIONAL TIME
800  LETFO=FO+(X^2)*F
810  IF FO<C THEN 1000
820  PRINT"FOURIER NUMBER="FO
830  PRINT
835  REM INTEGERIZE THE NODAL TEMPERATURE VALUES
840  FOR I=1 TO M STEP 2
850  FOR J=1 TO M STEP 2
852  LETX(I,J)=INT(U(I,J)*10^4+0.5)/10^4
854  NEXT J
856  NEXT I
860  PRINT X(1,1),X(1,3),X(1,5),X(1,7),X(1,9)
870  PRINT X(3,1),X(3,3),X(3,5),X(3,7),X(3,9)
880  PRINT X(5,1),X(5,3),X(5,5),X(5,7),X(5,9)
890  PRINT X(7,1),X(7,3),X(7,5),X(7,7),X(7,9)
900  PRINT X(9,1),X(9,3),X(9,5),X(9,7),X(9,9)
902  REM INCREMENT  PRINT INTERVAL
905  LETC=C+P
909  REM CHECK FOR CONVERGENCE
910  FOR I=2 TO M-1
920  FOR J=2 TO M-1
930  LETD(I,J)=U(I,J)-T(I,J)
940  IF ABS(D(I,J)) <E THEN 2000
950  NEXT J
960  NEXT I
970  IF FO>1.0 THEN 2000
990  REM UPDATE MOST RECENT NODAL VALUES
1000 FOR I=1 TO M
1010 FOR J=1 TO M
1020 LETT(I,J)=U(I,J)
1030 NEXT J
1040 NEXT I
1050 GOTO 630
1500 DATA 0.1,0.125
2000 END
```

Computer Results 4-3

TYPE 1 IF YOU WANT A PROGRAM EXPLANATION.

OTHERWISE, TYPE 2.
? 2

CONVERGENCE CRITERION= 0.1

DELTA X BAR= 0.125

FOURIER NUMBER= 2.03125 E-2

0	0	0	0	0
0	0.6086	0.7561	0.6086	0
0	0.7561	0.9368	0.7561	0
0	0.6086	0.7561	0.6086	0
0	0	0	0	0

FOURIER NUMBER= 0.040625

0	0	0	0	0
0	0.3695	0.5077	0.3695	0
0	0.5077	0.6973	0.5077	0
0	0.3695	0.5077	0.3695	0
0	0	0	0	0

FOURIER NUMBER= 6.09375 E-2

0	0	0	0	0
0	0.2417	0.3397	0.2417	0
0	0.3397	0.4774	0.3397	0
0	0.2417	0.3397	0.2417	0
0	0	0	0	0

FOURIER NUMBER= 0.08125

0	0	0	0	0
0	0.1609	0.2272	0.1609	0
0	0.2272	0.3209	0.2272	0
0	0.1609	0.2272	0.1609	0
0	0	0	0	0

FOURIER NUMBER= 0.1

0	0	0	0	0
0	0.1109	0.1568	0.1109	0
0	0.1568	0.2217	0.1568	0
0	0.1109	0.1568	0.1109	0
0	0	0	0	0

FOURIER NUMBER= 0.120313

0	0	0	0	0
0	0.0742	0.1049	0.0742	0
0	0.1049	0.1483	0.1049	0
0	0.0742	0.1049	0.0742	0
0	0	0	0	0

given value of the Fourier number gives the nondimensional temperature for the two-dimensional solution at the corresponding Fourier number. The Biot and Fourier numbers for each of the one-dimensional solutions are based upon the characteristic lengths L_1 and L_2, respectively.

To check the above numerical results obtained using the TWOD computer program, we need results for the two corresponding one-dimensional solutions.

Since an isothermal boundary condition is used, the effective value of the convection heat transfer coefficient at the surface is infinite, resulting in $k/\bar{h}l = 0$. At the center of the bar ($\bar{x} = 0$) and for Fo $= 0.4$, the value of $[(T - T_1)/(T_i - T_1)]_{L_1} = [(T - T_1)/(T_i - T_1)]_{L_2} = 0.48$ is read from Fig. 4-8 of Ref. 4-13. Using Eq. (4-67), $[(T - T_1)/(T_i - T_1)]_{2-D} = \theta = (0.48)(0.48) = 0.23$.

Care must be taken when comparing analytical and numerical solutions. The characteristic length used to plot the results in Fig. 4-8, Ref. 4-13, is the slab half-thickness L_1. This length is not the same characteristic length used in the numerical solution. The value of $\overline{\Delta x}$ (variable X in program TWOD) used for the nine by nine grid is one-eighth. (See Statement 1500.) Thus $l = \Delta x/\overline{\Delta x} = 8\,\Delta x$, which is the total length of one side of the square grid. Therefore, since $l = 2L_1$, the Fourier numbers used in Statement 800 (Fo $= \alpha t/l^2$) are equal to one-fourth the Fourier numbers (Fo $= \alpha t/L_1^2$) plotted in Fig. 4-8 of the cited reference. The Fourier number corresponding to 0.4 in the numerical solution is then 0.1. The numerical results shown in Computer Results 4-3 give a value of $\theta = 0.222$ at the center of the plate for FO $= 0.1$, as compared to 0.23 obtained from the curve in Ref. 4-13.

Minor modifications can be made to the computer program TWOD to study the effects of nonuniform initial conditions and various boundary conditions on transient, two-dimensional heat conduction. The modifications required are very similar to those made to the TRAN program to study the effects of nonuniform initial conditions and boundary conditions on transient, one-dimensional conduction.

The explicit method of finite difference representation of partial differential equations has proved very useful in this chapter. A wide variety of transient heat conduction problems can be analyzed using this method. As mentioned earlier, if results for a transient temperature distribution are desired over a long time period, the explicit method of finite difference representation can lead to excessive computational time, and an implicit method may be more appropriate. A computer program which uses an implicit method to obtain a solution to the nondimensional form of Eq. (4-62) is discussed in Ref. 4-16.

4-12   Conclusion

This chapter has indicated when a model based upon negligible internal resistance can be used in a transient analysis of conduction heat transfer. An equally important consideration is deciding when a transient analysis should be used rather than a steady-state analysis. This depends upon the time constant $t_c = L^2/\alpha$, where L is a characteristic length for the thermal system. For a solid metal wall with $\alpha = 5 \times 10^{-3}$ cm^2/s and thickness 10 cm, $t_c = 20,000$ s $= 5.56$ h. For an insulating material 10 cm thick and $\alpha = 5 \times 10^{-4}$ cm^2/s, the value $t_c = 55.6$ h. These large values of time constants indicate that the transient effects on heat conduction are important and must be considered in the mathematical model, unless interest is focused on a time span much larger than the time constant. Then steady-state conditions can be reached.

For convective heat transfer the time constants are much shorter. This shortening is due to the diffusion mechanism in convection which leads to larger diffusivities, especially in turbulent flow. For example, for liquids flowing turbulently through a pipe of 10-cm diameter, the time constant is of the order of magnitude of 10 s. For gas flow, one can expect a time constant of the order of 100 s. Thus, the transient nature of convection need be considered only when interest is focused on a very short time span. For example, to study the transient response of a heated pipe suddenly exposed to the internal flow of a liquid coolant for a period of 10 min, a steady-state value of the convection heat transfer coefficient can be used to specify the boundary condition when analyzing the transient temperature distribution within the pipe wall. This type of approximation was made throughout this chapter to obtain transient temperature distributions in solids.

In the first four chapters the average steady-state values of convection heat transfer coefficients were specified without considering their local variation over a surface. We now turn our attention to the steady-state flow fields and temperature fields that occur near a solid surface during heat transfer by convection. These fields specify the mass, momentum, and energy transfer at a surface. The fundamental laws of fluid mechanics are used to formulate mathematical models which are based on the conservation principles of mass, momentum, and energy. The solution to these models allows the determination of local steady-state convection heat transfer coefficients.

References

4-1 SCHNEIDER, P. J.: "Conduction Heat Transfer," Addison-Wesley, Reading, Mass., 1957.

4-2 JAKOB, M.: "Heat Transfer," vol. I, Wiley, New York, 1949.

4-3 CARSLAW, H. S., and J. G. JAEGER: "Conduction of Heat in Solids," Clarendon Press, Oxford, 1959.

4-4 GRÖBER, H., and S. ERK: "Die Grundgesetze der Warmeubertragung," Springer-Verlag, Berlin, 1933.

4-5 HEISLER, M. P.: Temperature Charts for Induction and Constant Temperature Heating, *Trans. ASME*, vol. 69, pp. 227–236, 1947.

4-6 SCHNEIDER, P. J.: "Temperature Response Charts," Wiley, New York, 1963.

4-7 CHAPMAN, A. J.: "Heat Transfer," 2d ed., Macmillan, New York, 1968.

4-8 ARPACI, V. S.: "Conduction Heat Transfer," Addison-Wesley, Reading, Mass., 1966.

4-9 SMITH, G. D.: "Numerical Solution of Partial Differential Equations," Oxford, London, 1965.

4-10 CRANK, J., and P. NICOLSON: A Practical Method for Numerical Evaluation of Solutions of Partial Differential Equations of the Heat Conduction Type, *Proc. Cambridge Phil. Soc.*, vol. 43, pp. 50–67, 1947.

4-11 RICHTMEYER, R. D.: "Difference Methods for Initial Value Problems," Interscience, New York, 1957.

4-12 OZISIK, M. N.: "Boundary Value Problems of Heat Conduction," International Textbook, Scranton, Pa., 1968.

4-13 KREITH, F.: "Principles of Heat Transfer," 2d ed., International Textbook, Scranton, Pa., 1966.

4-14 BARAKAT, H. Z., and J. A. CLARK: On the Solution of the Diffusion Equations by Numerical Methods, *J. Heat Transfer*, vol. 88, ser. C, pp. 421–427, 1966.

4-15 LARKIN, B.: Some Finite Difference Methods for Problems in Transient Heat Flow, *AIChE J.*, preprint 16, August 1964.

4-16 CARNAHAN, B., H. A. LUTHER, and J. O. WILKES: "Applied Numerical Methods," Wiley, New York, 1969.

4-17 JAKOB, M., and G. HAWKINS: "Elements of Heat Transfer," 3d ed., Wiley, New York, 1957.

Problems[1]

4-1 Consider the one-dimensional, transient temperature distribution in a solid wall. When the initial temperature distribution is uniform and suddenly undergoes a stepwise change in surface temperature at time $t = 0$, an analytical solution is given by Eq. (4-7). Because of symmetry, an alternate form of the solution can be obtained by using the coordinates shown in Fig. 4-1P along with the boundary conditions given below. We define $\theta = (T - T_\infty)/(T_i - T_\infty)$.

$$\frac{\partial^2 \theta}{\partial x^2} = \frac{1}{\alpha} \frac{\partial \theta}{\partial t}$$

initial condition: $\quad \theta(x, 0) = 1$

boundary condition: $\quad \theta(l, t) = 0$

boundary condition: $\quad \dfrac{\partial \theta(0, t)}{\partial x} = 0$

where l is the slab half-thickness.

(*a*) Assume a product solution, and show that the two boundary conditions lead to a solution of the form

$$\theta(x, t) = \sum_{n=0}^{\infty} C_n e^{-\alpha \lambda_n^2 t} \cos \lambda_n x$$

where C_n is determined by the initial condition, and $\lambda_n = (2n + 1)\pi/2l$.

(*b*) Show that each C_n is a coefficient of a Fourier cosine series expansion of unity over the interval $0 \leq x \leq l$ and each is given by (See Ref. 4-6, Chap. 5):

$$C_n = (-1)^n \frac{2}{\lambda_n l}$$

(*c*) What is the final expression for $\theta(x, t)$? Compare this result with Eq. (4-7).

---

[1] Problems whose numbers are followed by a superscript italic *c* should be analyzed by obtaining solutions on a digital computer.

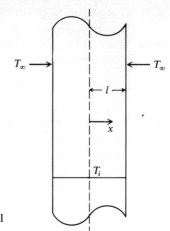

FIGURE 4-1P
Transient conduction, isothermal boundary conditions.

4-2   Many times a surface convective boundary condition is more appropriate than an isothermal boundary condition. Consider the solid wall analyzed in Prob. 4-1. Replace the isothermal boundary condition $\theta(l, t) = 0$ with the convective boundary condition

$$-k\,\frac{\partial\theta(l, t)}{\partial x} = \bar{h}\theta(l, t)$$

(a) Assume a product solution, and show that the eigenvalues λ_n are determined by the equality

$$\lambda_n l \sin \lambda_n l = \mathrm{Bi} \cos \lambda_n l$$

where $\mathrm{Bi} = \bar{h}l/k$.

(b) Show that the product form of the solution can be written

$$\theta(x, t) = \sum_{n=1}^{\infty} C_n e^{-\alpha\lambda_n^2 t}\cos\lambda_n x$$

where the orthogonality of the trigonometric functions allows one to write

$$C_n = \frac{2\sin\lambda_n l}{\lambda_n l + \sin\lambda_n l\cos\lambda_n l}$$

4-3   Consider transient one-dimensional conduction with uniform internal generation. The governing nondimensional equation is

$$\frac{\partial^2\theta}{\partial\bar{x}^2} + \frac{q'''l^2}{k(T_i - T_\infty)} = \frac{\partial\theta}{\partial\mathrm{Fo}}$$

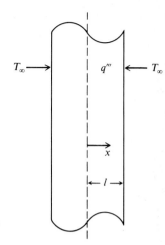

FIGURE 4-3P
Transient conduction with internal generation.

where $\bar{x} = x/l$, $\text{Fo} = \alpha t/l^2$, and $\theta = (T - T_\infty)/(T_i - T_\infty)$. We define $\text{N1} = q'''l^2/k(T_i - T_\infty)$. The coordinate system is shown in Fig. 4-3P. The initial condition and boundary conditions are

$$\theta(\bar{x}, 0) = 0$$

$$\theta(1, \text{Fo}) = 0$$

$$\frac{\partial \theta(0, \text{Fo})}{\partial \bar{x}} = 0$$

The solution to this type of problem is discussed in Ref. 4-8, Sec. 4-7. The temperature can be written as the sum of two temperatures:

$$\theta(\bar{x}, \text{Fo}) = \psi(\bar{x}, \text{Fo}) + \phi(\bar{x})$$

where $\phi(\bar{x})$ satisfies the equation

$$\frac{d^2\phi}{d\bar{x}^2} + \text{N1} = 0$$

and boundary conditions

$$\frac{d\phi(0)}{d\bar{x}} = 0$$

$$\phi(1) = 0$$

(a) Show that the solution for $\phi(\bar{x})$ is

$$\phi(\bar{x}) = \frac{\text{N1}}{2}(1 - \bar{x}^2)$$

(b) The function $\psi(\bar{x}, \text{Fo})$ is obtained from the solution to

$$\frac{\partial^2 \psi}{\partial \bar{x}^2} = \frac{\partial \theta}{\partial \text{Fo}}$$

$$\text{initial condition:} \quad \psi(\bar{x}, 0) = -\phi(\bar{x})$$

$$\text{boundary condition:} \quad \psi(1, t) = 0$$

$$\text{boundary condition:} \quad \frac{\partial \psi(0, \text{Fo})}{\partial \bar{x}} = 0$$

Show that a product solution leads to

$$\psi(\bar{x}, \text{Fo}) = \sum_{n=0}^{\infty} C_n e^{-(l\lambda_n)^2 \text{Fo}} \cos l\lambda_n \bar{x}$$

where C_n is determined by the initial condition.

(c) Show that the application of the initial condition leads to

$$\frac{\text{N1}}{2}(\bar{x}^2 - 1) = \sum_{n=0}^{\infty} C_n \cos l\lambda_n \bar{x}$$

where the coefficients C_n are given by

$$C_n = \frac{\int_1^0 -\phi(\bar{x}) \cos (l\lambda_n \bar{x}) \, d\bar{x}}{\int_1^0 \cos^2 (l\lambda_n \bar{x}) \, d\bar{x}}$$

or

$$C_n = -2\frac{(-1)^n \text{N1}}{(l\lambda_n)^3}$$

(d) Write the expression for the nondimensional analytical solution for $\theta(\bar{x}, \text{Fo})$.

4-4   Use the results given in Fig. 4-6 to obtain an approximate solution for the following problem. A slab as shown in Fig. 4-1 has a total thickness of 2.0 ft. The initial uniform temperature distribution is 100°F. At time $t = 0$, both faces are suddenly brought to a temperature of 50°F and held constant. The slab has a thermal diffusivity of 0.001 ft²/s.
(a) How long does it take the center of the slab to reach 70°F?
(b) If the slab is only 1.0 ft thick, how long does it take the center to reach 70°F?
(c) If the slab material has a thermal diffusivity of 0.002 ft²/s and a thickness of 1.0 ft, how long does it take the center to reach 70°F?
(d) For $\alpha = 0.001$ ft²/s and a thickness of 2.0 ft, how long does it take the center of the slab to reach 70°F if both faces are suddenly brought to a temperature of 0.0°F at time $t = 0$?

4-5   The analytical solution for the results given in Fig. 4-6 is given by Eq. (4-7). Evaluate the infinite series for certain values of $\bar{x} = x/l$ and $\bar{t} = \alpha t/l^2$, and show that the analytical results agree with the numerical results.

4-6   Consider the plane wall shown in Fig. 4-9 and the numerical results given in Fig. 4-10. If T_2 is held constant at 100°C and $T_1 = 300$°C, the solution can be applied to a problem with $T_0 = 0.0°C$. The temperature T_0 is the temperature at the left surface for time greater than zero. The thickness of the wall is 3.0 cm.

(a) It is desirable to have a wall that will allow heat conduction in the negative x direction at the right face, $\bar{x} = 1.0$, within 60 s after the left face, $\bar{x} = 0$, is exposed to T_0. What must be the value of the thermal diffusivity α (cm²/s) to meet this requirement?

(b) What is the maximum temperature (°C) within the wall in part (a) 45.0 s after the left face is exposed to T_0?

(c) How long does it take to reach steady-state conditions for the material chosen in part (a)?

4-7 Consider one-dimensional, transient conduction in a cylinder where $T = f(r, t)$. The governing equation for the temperature distribution is given by

$$\frac{\partial^2 T}{\partial r^2} + \frac{1}{r}\frac{\partial T}{\partial r} = \frac{1}{\alpha}\frac{\partial T}{\partial t}$$

(a) Choose appropriate nondimensional dependent and independent variables, and write the governing equation in nondimensional form.

(b) Convection occurs from the outer surface of the cylinder at $r = R$. Express this boundary condition in nondimensional form. Identify the important nondimensional variable.

4-8 The following two problems illustrate the general application of nondimensional solutions:

(a) The surface material of a space vehicle has a high temperature after reentering the earth's atmosphere. Assume that the initial temperature is uniform at 1,000°F and is insulated on the back side. The thickness is 0.2 ft, $\alpha = 0.16$ ft²/h, and $k = 10.0$ Btu/h-ft-°R. At time $t = 0$ the vehicle lands in the ocean where $T = 40°F$, and the convection coefficient is 50 Btu/h-ft²-°R. What is the surface temperature 1.5 min after the vehicle lands in the ocean? Use the results given in Fig. 4-11. What is the maximum temperature in the material after 1.5 min?

(b) A baker has a large flat piece of dough which is 0.1 ft thick. The dough is initially at a uniform temperature of 70°F and at time $t = 0$ it is placed in a hot oven where both sides are exposed to a temperature of 200°F. The thermal diffusivity is 0.003 ft²/h, $k = 0.20$ Btu/h-ft-°R, and $\bar{h} = 4.0$ Btu/h-ft²-°R. What is the surface temperature 5.0 min after the dough is placed in the oven? Use the results given in Fig. 4-11. What is the maximum temperature in the dough after 5.0 min?

4-9 A cylindrical fin with internal generation due to dissipation of electric current is to be used as a timer. It is planned to place a temperature sensor in the fin and measure $T - T_\infty$ as a function of time after the electric current is shut off at $t = 0$. The required initial condition of the fin is given by the curve for Fo $= 0$ in Fig. 4-14. The fin must be designed such that the indicated reading $T - T_\infty$ will equal zero 30.0 s after the current is shut off and $N1 = 0$. Suggest three different designs to meet these requirements. In each design specify

1 the location of the sensor in the fin
2 the values of l, D, and α
3 materials which approximately meet the required values of α
4 the corresponding value of the convection coefficient $\bar{h}$ required to make the solution in Fig. 4-14 valid
5 the value of q''' required to meet the required initial conditions as a function of $T_w - T_\infty$

4-10 A large slab of aluminum is 2.0 in thick. It is initially at a uniform temperature of 450°F, and both faces are suddenly exposed to an environmental temperature of 150°F. The edges of the slab are adiabatic, so heat transfer only occurs across the two exposed faces where an average convection heat transfer coefficient of $\bar{h} = 100$ Btu/h-ft²-°R exists. The properties of the aluminum are $k = 120$ Btu/h-ft-°R, $\rho = 169$ lb$_m$/ft³, and $c = 0.21$ Btu/lb$_m$-°R.

(*a*) Show that this problem can be analyzed by assuming negligible internal resistance.

(*b*) Calculate the temperature of the slab 1 min after it is exposed to the new environment.

(*c*) How much energy per unit surface area in the form of heat transfer has been removed from the slab in 1 min?

4-11 For a semi-infinite solid whose surface temperature is suddenly lowered and maintained constant at a temperature T_1, the transient temperature profile is obtained by solving the equation

$$\frac{\partial^2 T}{\partial x^2} = \frac{1}{\alpha}\frac{\partial T}{\partial t}$$

subject to the initial condition $T(x, 0) = T_i$ and the boundary condition $T(0, t) = T_1$. It can be shown (Ref. 4-1) that the solution to this problem is

$$\theta = \frac{2}{\sqrt{\pi}}\int_0^{x/2\sqrt{\alpha t}} e^{-n^2}\,d\eta = \text{erf}\left(\frac{x}{2\sqrt{\alpha t}}\right)$$

where $\theta = (T - T_1)/(T_i - T_1)$ and $\text{erf}(x/2\sqrt{\alpha t})$ is the error function. (Tabulated values of the error function may be found in standard math tables.)

(*a*) Derive the expression for the instantaneous heat transfer per unit area within the solid by using Fourier's law.

(*b*) What is the expression for the instantaneous heat transfer per unit area at the surface $(x = 0)$?

(*c*) Integrate the expression obtained in part (*b*) to obtain the expression for the total heat transfer per unit area crossing the surface as a function of time.

4-12 On a hot day, a concrete highway reaches a temperature of 45°C. A sudden rainstorm reduces the surface temperature to 17°C. How long does it take to cool the concrete to 28°C at a depth 2.0 cm below the surface? Properties are $k = 1.04$ W/m-K, $\rho = 2.5$

kg/m³, and $C_p = 418$ kJ/kg-K. The solution to the governing equation is given in Prob. 4-11. Tabulated values of the error function should be used to obtain a solution.

4-13 A spherical ball bearing, with its diameter $= 1.0$ in, $k = 23$ Btu/h-ft-°R, and $\alpha = 0.66$ ft²/h is initially at a uniform temperature of 1,000°F. It is suddenly placed in an oil bath ($T_\infty = 200°F$). Determine the time required for the center to cool to 500°F if $\bar{h} = 120$ Btu/h-ft²-°R. Show that the assumption of negligible internal resistance is valid for this problem.

4-14 The spherical ball bearing in Prob. 4-13 is initially at a nonuniform temperature, and it is suddenly placed in a boiling fluid ($T_\infty = 200°F$) where the average convection coefficient is $\bar{h} = 1,200$ Btu/h-ft²-°R. Discuss how you would obtain a solution for the transient temperature distribution within the solid sphere.

4-15 A long solid cylinder with a length much greater than the square of its radius is initially at a uniform temperature of 300°C. It is suddenly exposed to an environment where $T_\infty = 10°C$. The diameter of the cylinder is 0.3 m, and the properties are $k = 43$ W/m-K, $\alpha = 0.2$ m²/h, and $\bar{h} = 55$ W/m²-K. Estimate how long it will take the temperature at the center of the cylinder to reach 100°C.

4-16 The steel pipe shown in Fig. 4-16P is initially at a uniform temperature of 400°F after having transported liquid sodium to a nuclear reactor. At time $t = 0$, helium gas begins to flow through the pipe. The internal convection heat transfer coefficient for the flow of helium is $\bar{h} = 10.0$ Btu/h-ft²-°R. During a 30-min period, the average bulk temperature of the gas decreases linearly with time from 260°F to 230°F. The outer surface is perfectly insulated.

(a) Find an analytical solution for the temperature in the pipe at the end of 30 min assuming negligible internal resistance in the pipe.

(b) If the flow rate of the helium is adjusted to maintain a constant helium bulk temperature of 260°F during a 30-min period, find the analytical solution for the temperature of the pipe at the end of the 30-min period.

(c)ᶜ Repeat the problem if oil flows through the pipe instead of helium, and the internal convection heat transfer coefficient is $\bar{h} = 200.0$ Btu/h-ft²-°R. The problem should now be formulated as a finite difference solution.

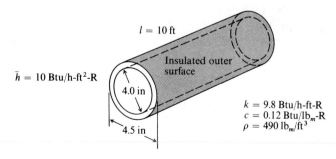

$l = 10$ ft

Insulated outer surface

$\bar{h} = 10$ Btu/h-ft²-R

4.0 in

4.5 in

$k = 9.8$ Btu/h-ft-R
$c = 0.12$ Btu/lb$_m$-R
$\rho = 490$ lb$_m$/ft³

FIGURE 4-16P
Reactor pipe.

4-17^c A new baking process required in the making of fine china is being considered to improve the aesthetic and artistic qualities of the product. The clay plates are supported in the oven by small legs and baked in a horizontal position. Infrared radiation falls uniformly on the top surface of a plate at a rate which can vary between 1,500 W/m² and 4,500 W/m². The new oven will also control the velocity and temperature of the air which flows over the top and bottom surfaces of the plates. The plates transfer energy by convection from both the top and bottom surfaces. The convection heat transfer coefficient is equal over both surfaces and will vary between 25.0 W/m²-K and 150 W/m²-K depending upon the Reynolds number for the forced convection over the plates.

It is necessary to know the rate of temperature increase within the china and the uniformity of the temperature throughout the plate at any instant in order to design the belt drives for the oven. Obtain this information for various operating conditions of radiation and convection. The plate is initially at room temperature when it enters the oven. Analyze a plate of radius 0.15 m and thickness 0.5 cm. The properties of the china are $k = 0.35$ W/m-K, $c = 0.30$ kcal$_{IT}$/kg-K, $\rho = 1{,}900$ kg/m³.

4-18 A boiler tube has an outside surface temperature of 1,000°F. Saturated steam flows inside the tube under steady-state conditions, and the inside surface temperature is 220°F. At time $t = 0$ the supply of steam is suddenly shut off, and a vacuum is generated inside the tube. This event produces an adiabatic boundary condition at the interior surface.

(a) Calculate the initial steady-state, one-dimensional, radial temperature distribution in the tube.

(b)^c After the tube is evacuated, how long does it take for the inner surface temperature to be increased by 300°F?

(c)^c If at $t = 0$ the boundary condition at the outer surface becomes one of convection to an environment of 800°F through a convection heat transfer coefficient of 5.0 Btu/h-ft²-°F, what is the maximum temperature that will be reached at the inner surface after the supply of steam is shut off? How does this answer depend upon the radius ratio r_o/r_i? Use the same initial temperature distribution as that calculated in part (a).

4-19 A piston in a new engine design receives a thermal heat flux at the lower surface from a radiation source at the end of the cylinder. It loses energy by convection from the upper surface with $\bar{h} = 35.0$ Btu/h-ft²-°F. The piston is stainless steel ($k = 8.0$ Btu/h-ft-°F, $c = 0.11$ Btu/lb$_m$-°F, $\rho = 488$ lb$_m$/ft³, and $\alpha = 0.15$ ft²/h). The radiation absorbed at the lower surface is a function of time and is given by $q'' = q_0'' \cos \omega t$, where $q_0'' = 10^4$ Btu/h-ft² and $\omega = 2\pi/P$, with P the period.

(a) If the internal resistance to conduction is neglected, show that the governing equation for the transient temperature distribution in the piston can be written

$$\frac{d\theta}{dt} + m\theta = n \cos \omega t$$

where $m = \bar{h}/\rho c l$ and $n = q_0''/\rho c l(T_i - T_\infty)$.

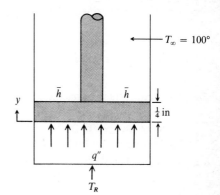

FIGURE 4-19P
Piston-cylinder.

(b) Show that the solution to this equation for the initial condition of $\theta = 0$ at $t = 0$ is given by

$$\frac{T - T_\infty}{q_0''/\bar{h}} = \frac{m\cos(\omega t - \alpha)}{(m^2 + \omega^2)^{1/2}} - \frac{m^2 e^{-mt}}{m^2 + \omega^2} \qquad \text{where } \alpha = \tan^{-1}(\omega/m).$$

(c)^c Solve the same problem by obtaining a finite difference solution to the governing partial differential equation for different values of the Biot number.

$$\frac{\partial^2 T}{\partial x^2} = \frac{1}{\alpha}\frac{\partial T}{\partial t} \qquad \begin{array}{lll} \text{at} & t = 0 & T = T_\infty \\ \text{at} & y = 0 & q'' = q_0'' \cos \omega t \\ \text{at} & y = \tfrac{1}{4}\,\text{in} & q'' = \bar{h}(T - T_\infty)_{y=1/4} \end{array}$$

(d) Compare the solutions obtained in parts (b) and (c).

4-20^c Consider a lubricated bearing in a piece of idle machinery on the surface of the moon. The initial temperature of the equipment is $-100°F$. When solar radiation strikes the bearing, the temperature of the bearing and the oil will increase. Figure 4-20P shows a

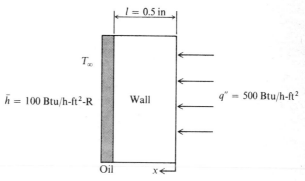

FIGURE 4-20P
Bearing cross section.

sketch of the bearing. The oil film is 0.1 in thick and the emittance of the outer wall surface is $\varepsilon = 0.8$. Neglect internal resistance in both the oil and wall. Derive and obtain solutions to the governing equations for various values of the governing non-dimensional parameters. How long will it take the oil temperature to increase by 100°F?

4-21   A vertical surface forms the wall of an open container filled with liquid. The temperature in the wall is initially uniform at 40°C. At time $t = 0$ the outer temperature T_o is suddenly changed to 200°C. The liquid is initially at a uniform temperature of 40°C. The liquid mass is 1.0 kg, and the specific heat is 1.0 kcal$_{IT}$/kg-K. While the temperature of the liquid changes from 40°C to 100°C, the convection heat transfer coefficient can be estimated by $\bar{h} = 50.0$ W/m²-K. After the temperature of the fluid reaches its saturation temperature, the convection heat transfer coefficient on the inside surface is given by $\bar{h} = 87.0\,(T_w - T_{sat})^{1/7}$, for $1.0 < T_w - T_{sat} < 5.0$°C. Because of a change in the boiling characteristics, the convection heat transfer coefficient is given by $\bar{h} = 0.240(T_w - T_{sat})^3$, for $T_w - T_{sat} \geq 5.0$°C.

(a)^c During the initial transient period while the liquid temperature is increasing, we can write for the liquid

$$\bar{h}A_1(T_w - T_f) = \frac{\rho c V\, dT_f}{dt}$$

and

$$\frac{\partial^2 T}{\partial x^2} = \frac{1}{\alpha}\frac{\partial T}{\partial t} \qquad \text{for the wall}$$

At $x = l$, one can write $T = T_w$, and $-(k_s\,\partial T/\partial x)_{x=l} = \bar{h}(T_w - T_f)$. For $l = 1.0$ cm, $\alpha = 0.3 \times 10^{-5}$ m²/s, and $k_s = 20.0$ W/m-K, obtain the transient temperature response in the wall until the liquid temperature reaches the saturation temperature.

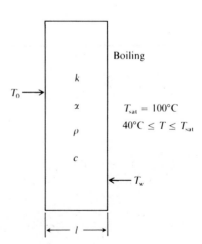

FIGURE 4-21P
Vertical surface with boiling.

(b)ᶜ Continue the analysis of the transient response in the wall, accounting for the change in the value of $\bar{h}$ due to the regions of boiling encountered.

(c) What is the final steady-state temperature profile within the wall?

4-22ᶜ A study is to be made of more effective ways of using fire-fighting equipment. Consider a small section of a wooden wall which receives radiation from an internal fire. When the outside surface of the wall is exposed to the air, the convection heat transfer coefficient is $\bar{h} = 5.0$ Btu/h-ft²-°R.

(a) Assume that the initial temperature distribution in the wall is uniform at 70°F and at $t = 0$ the inside surface of the wall is suddenly exposed to a surface heat transfer rate of 10^5 Btu/h-ft². Calculate the transient temperature distribution in the wall.

(b) A certain time later (depending upon when the fire truck arrives) a stream of water is directed over the surface under consideration for 5.0 s. Then it is directed to some other location for 10.0 s, returns for 5.0 s, and repeats this cyclic process. Obtain solutions for the transient temperature distribution in the wall for various initial conditions.

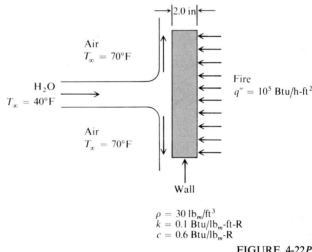

FIGURE 4-22P
Wooden wall.

4-23 A turbine blade is cooled at its base by a flowing coolant at a temperature of 200°F. A convection heat transfer coefficient of $\bar{h}_e = 500$ Btu/h-ft²-°R exists between the coolant and the base of the blade. The temperature of the hot gases flowing over the blade is $T = 1,000°F$, and a surface convection heat transfer coefficient of $\bar{h}_s = 50$ Btu/h-ft²-°R exists over the blade surface. The end of the blade is adiabatic. At steady-

state operating conditions the temperature distribution within the blade $\theta = (T - T_\infty)/$
$(T_w - T_\infty)$ is given by

$$\frac{d^2\theta}{dx^2} - m^2\theta = 0$$

where

$$m = \sqrt{\frac{\bar{h}_s p}{kA}}$$

with the boundary conditions

$$\text{at} \quad x = 0 \quad -k\frac{\partial\theta}{\partial x} = \bar{h}_e \theta$$

$$\text{at} \quad x = l \quad \frac{\partial\theta}{\partial x} = 0$$

(a) Obtain an analytical solution for the steady-state temperature distribution in the blade.

(b)^c To suddenly increase the power of the turbine during critical periods, it is contemplated that hot gases from an MHD generator can be directed through the turbine. When this move is done, the temperature of the gas flowing over the blade is suddenly increased to 2,000°F. Study the transient thermal response of the blade under these conditions. The initial condition is given by the steady-state solution in part (a). What effect would be realized if the temperature of the coolant were suddenly reduced to $T_B = 100$°F at the same time that the gas temperature is increased to 2,000°F?

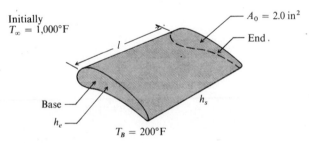

FIGURE 4-23P
Turbine blade.

4-24^c A school for retarded children plans to teach a course in cooking. It is extremely important to provide a safe environment for the class. It is anticipated that spoons and other implements may be left in boiling water for several minutes before they are removed by the students. You are asked to design implements which are safe. Because of other considerations it is desirable for the implements to be made from only one material. Analyze the transient thermal response for possible designs. Treat the implements as fins with a uniform initial temperature. At $t = 0$, the base is suddenly maintained at 100°C. The convection heat transfer coefficient over the exposed surface is 17 W/m²-K, and the environmental temperature around the implements may be taken as 30°C. The boundary condition at the end of the fin depends upon the fin design.

5

LAMINAR CONVECTION

5-1 Introduction

In the previous chapters heat transfer by radiation and conduction has been discussed. In these discussions the convective heat transfer at a solid-fluid interface was frequently of interest and was expressed in terms of a known convection coefficient. The present chapter is concerned with determining this convection coefficient for laminar convection.

The convective heat transfer at a solid-fluid interface is intimately connected with the characteristics of the thin layer of fluid near the surface. Under many flow conditions the effects of viscosity are confined to this thin layer of fluid which is called a *boundary layer*. The concept of the boundary layer was first proposed by L. Prandtl at the beginning of this century. (See Ref. 5-2.) Within the boundary layer the fluid velocity varies from that at the edge of the boundary layer to zero velocity at the solid surface. This retardation is due to viscous shearing stresses which act between adjacent fluid layers. The concept of a *thermal* boundary layer is similar to a *velocity* boundary layer. A thermal boundary layer is the thin region near the surface where the temperature varies from the free stream temperature to the surface temperature.

Heat transfer in boundary-layer flow may in general be classed as either forced or free convection. Forced-convection flow over a surface is caused by an external inertial driving force. Free convection is caused by body forces, such as gravity, acting on the fluid within the boundary layer. For example, when a surface is heated or cooled, the change in the fluid density due to heat transfer across the solid-fluid interface causes an unbalanced body force in the fluid. This force causes an acceleration of the fluid near the surface, and a boundary-layer flow may develop.

In fluid mechanics, the *Reynolds number* is used to indicate the nondimensional ratio of inertial forces to viscous forces. Mathematically, the Reynolds number is given by $Re = UL/v$, where U is a reference fluid velocity, v is the fluid kinematic viscosity, $v = \mu/\rho$, and L is a reference length. Experimental observation of external forced convection shows that laminar boundary-layer flow occurs when $Re < 5 \times 10^5$, and turbulent boundary-layer flow occurs for larger values of Re. In laminar flow there is no mixing between adjacent layers of the fluid, and heat transfer must occur by molecular conduction between the layers. In turbulent flow the molecular conduction is supplemented by energy transfer because of turbulent mixing between the fluid layers.

Free convection is characterized by the ratio of the body forces to viscous forces as given by the nondimensional *Grashof number*. Mathematically, the Grashof number is $Gr = g\bar{\beta}(T_w - T_\infty)L^3/v^2$, where g is the local acceleration due to gravity, and $\bar{\beta}$ is the isobaric compressibility of the fluid. For gases, external, laminar free convection is observed to occur for $Gr < 10^8$. Turbulent free convection occurs for larger values of Gr. A boundary-layer flow can occur for free convection in gases when $Gr > 10^4$. For smaller values of Gr, the body force is not sufficient to provide enough fluid acceleration to form a thin boundary layer.

Analysis of the complete boundary-layer equations in Ref. 5-3 shows that when the ratio of Gr/Re^2 is near unity, the effects due to both free and forced convection must be considered simultaneously. When $Gr/Re^2 >> 1.0$, the flow is pure free convection; whereas when $Gr/Re^2 << 1.0$, only forced convection is important.

In this chapter the velocity and temperature profiles that occur in steady, two-dimensional, laminar boundary layers in both forced and free convection are analyzed. This analysis leads to the determination of the local and average values of the convection heat transfer coefficients. The method of analysis is to choose a differential control volume within the laminar boundary layer and apply the basic conservation laws of mass, momentum, and energy to derive the governing equations. This method leads to nonlinear partial differential equations which govern the heat transfer and fluid flow behavior.

5-2 Continuity and Momentum Equations

The differential laminar boundary-layer control volume shown in Fig. 5-1 has height dy, width dx, and unit depth. The flow within the boundary layer is represented by a continuum model of a homogeneous fluid. This representation leads to a mathematical

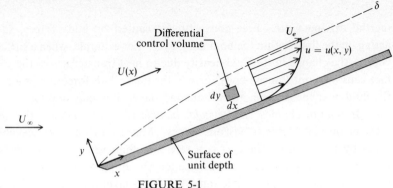

FIGURE 5-1
Differential control volume in velocity boundary layer.

model which relates velocity, pressure, and density as a function of the space coordinates.

For steady flow through the stationary differential control volume, the conservation of mass principle requires that the net mass flow rate equal zero. Since there are no sources or sinks within the control volume, the rate at which mass leaves the control volume must equal the rate at which mass enters the control volume. The fluid density and velocity in the x direction at the center of the control volume are indicated by ρ and u, respectively. Using a first-order Taylor series expansion about ρu, the mass entering the left face of the control volume per unit time at the location $x - dx/2$ through the surface area $dy(1)$ is

$$\dot{m}_{x-dx/2} = \left[\rho u + \frac{\partial(\rho u)}{\partial x} \frac{-dx}{2} \right] dy \qquad (5\text{-}1)$$

Similarly, for the mass flow rate leaving the control volume through the right face at $x + dx/2$,

$$\dot{m}_{x+dx/2} = \left[\rho u + \frac{\partial(\rho u)}{\partial x} \frac{dx}{2} \right] dy \qquad (5\text{-}2)$$

The mass entering the bottom of the control volume per unit time at $y - dy/2$ through the area $dx(1)$ is

$$\dot{m}_{y-dy/2} = \left[\rho v + \frac{\partial(\rho v)}{\partial y} \frac{-dy}{2} \right] dx \qquad (5\text{-}3)$$

where v is the velocity in the y direction.

The mass flow rate leaving through the top of the control volume at $y + dy/2$ is

$$\dot{m}_{y+dy/2} = \left[\rho v + \frac{\partial(\rho v)}{\partial y} \frac{dy}{2} \right] dx \qquad (5\text{-}4)$$

The conservation of mass principle requires that

$$\dot{m}_{x+dx/2} + \dot{m}_{y+dy/2} = \dot{m}_{x-dx/2} + \dot{m}_{y-dy/2}$$

Use of Eqs. (5-1) through (5-4) leads to

$$\frac{\partial(\rho u)}{\partial x} + \frac{\partial(\rho v)}{\partial y} = 0 \qquad (5\text{-}5a)$$

If we let $G_x = \rho u$ and $G_y = \rho v$ be the mass flow rates per unit area in the x and y directions, respectively, we can write

$$\frac{\partial G_x}{\partial x} + \frac{\partial G_y}{\partial y} = 0 \qquad (5\text{-}5b)$$

For incompressible flow, $\rho = $ constant, and the continuity equation reduces to

$$\frac{\partial u}{\partial x} + \frac{\partial v}{\partial y} = 0 \qquad (5\text{-}5c)$$

Equation (5-5c) is the continuity equation for steady, two-dimensional, incompressible flow of a continuous, homogeneous fluid.

Newton's second law of motion requires that the net forces acting on the control volume equal the time rate of change of momentum crossing the control volume surfaces. The forces can consist of body forces, such as gravity, and surface forces due to both normal stresses and shearing stresses acting on areas of the control volume surfaces.

The x components of the forces and momentum flux terms are indicated in Fig. 5-2 for a steady, two-dimensional, laminar, forced-convection boundary layer. The shearing stress per unit area is τ. In mathematical form, if one uses the notation indicated in Fig. 5-2, Newton's second law is written

$$- \left(\tau_{xx} + \frac{\partial \tau_{xx}}{\partial x} \frac{-dx}{2} \right) dy + \left(\tau_{xx} + \frac{\partial \tau_{xx}}{\partial x} \frac{dx}{2} \right) dy$$

$$- \left(\tau_{yx} + \frac{\partial \tau_{yx}}{\partial y} \frac{-dy}{2} \right) dx + \left(\tau_{yx} + \frac{\partial \tau_{yx}}{\partial y} \frac{dy}{2} \right) dx$$

$$= \left[\rho u^2 + \frac{\partial(\rho u^2)}{\partial x} \frac{dx}{2} \right] dy - \left[\rho u^2 + \frac{\partial(\rho u^2)}{\partial x} \frac{-dx}{2} \right] dy$$

$$+ \left[\rho uv + \frac{\partial(\rho uv)}{\partial y} \frac{dy}{2} \right] dx - \left[\rho uv + \frac{\partial(\rho uv)}{\partial y} \frac{-dy}{2} \right] dx \qquad (5\text{-}6)$$

This equation reduces to

$$\frac{\partial(\rho u^2)}{\partial x} + \frac{\partial(\rho uv)}{\partial y} = \frac{\partial \tau_{xx}}{\partial x} + \frac{\partial \tau_{yx}}{\partial y}$$

or

$$\frac{\partial(G_x u)}{\partial x} + \frac{\partial(G_y u)}{\partial y} = \frac{\partial \tau_{xx}}{\partial x} + \frac{\partial \tau_{yx}}{\partial y} \qquad (5\text{-}7)$$

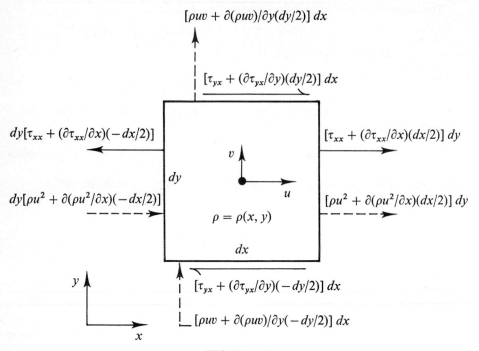

FIGURE 5-2
Enlarged control volume within velocity boundary layer.

Carrying out the differentiation in Eq. (5-7) gives

$$G_x \frac{\partial u}{\partial x} + u \frac{\partial G_x}{\partial x} + G_y \frac{\partial u}{\partial y} + u \frac{\partial G_y}{\partial y} = \frac{\partial \tau_{xx}}{\partial x} + \frac{\partial \tau_{yx}}{\partial y} \qquad (5\text{-}8)$$

The sum of the second and fourth terms is equal to zero as required by the continuity equation (5-5b).

The normal shearing stress may be replaced by the negative of the thermodynamic pressure (Ref. 5-3). Also, the shearing stress within the laminar boundary layer for a Newtonian fluid is given by

$$\tau_{yx} = \mu \frac{\partial u}{\partial y} \qquad (5\text{-}9)$$

Equation (5-8) may be written as

$$\rho u \frac{\partial u}{\partial x} + \rho v \frac{\partial u}{\partial y} = -\frac{\partial P}{\partial x} + \frac{\partial}{\partial y}\left(\mu \frac{\partial u}{\partial y}\right) \qquad (5\text{-}10a)$$

The dynamic viscosity of the fluid μ is a function of temperature. For gases, the temperature dependence is not great, and a constant value of μ may be used in the analysis when the temperature difference within the boundary layer is only a few hundred degrees or less. On the other hand, oils and organic liquids have strongly dependent dynamic viscosities. Convective heat transfer in fluids with variable dynamic viscosity is discussed in Chap. 7.

For constant dynamic viscosity, Eq. (5-10a) becomes

$$u \frac{\partial u}{\partial x} + v \frac{\partial u}{\partial y} = -\frac{1}{\rho} \frac{\partial P}{\partial x} + v \frac{\partial^2 u}{\partial y^2} \qquad (5\text{-}10b)$$

Equation (5-10b) is the x component of the momentum equation for a steady, two-dimensional, laminar, constant-property boundary-layer flow of a Newtonian fluid in forced convection. The two terms on the left side are the nonlinear convection terms. The two terms on the right side arise from inertial forces and viscous shearing forces, respectively.

Assuming $\partial v/\partial x << \partial u/\partial x$ and $\partial v/\partial y << \partial u/\partial y$, the y component of the momentum equation reduces to

$$\frac{\partial P}{\partial y} = 0 \qquad (5\text{-}11)$$

Physically, this expression means that the pressure in a laminar boundary layer does not vary normal to the surface. This characteristic is due to the thin boundary layer. Thus, the pressure is constant across the boundary layer and $P = P(x)$. When this is true, the pressure variation in the inviscid flow at the edge of the boundary layer is imposed directly on the surface. This fact implies that the pressure distribution may be calculated from inviscid flow solutions, such as discussed in Ref. 5-1. These are obtained without regard to the presence of the thin viscous boundary layer.

The mathematical boundary conditions associated with Eq. (5-10b) are

$$\begin{array}{lll} y = 0 & \quad u = 0 \\ y = 0 & \quad v = 0 & \qquad (5\text{-}12) \\ y \to \infty & \quad u \to U(x) \end{array}$$

The first equation specifies no fluid velocity in the x direction relative to the solid surface for the fluid physically in contact with the surface at $y = 0$. This is called the *no-slip* condition. The second equation specifies no fluid velocity normal to the surface at $y = 0$. Thus, we do not allow suction, blowing, or a phase change at the surface, i.e., no mass transfer. The third equation requires that the viscous velocity in the boundary layer approach the inviscid velocity at large (on the scale of the boundary layer) distances above the surface.

The flow outside the boundary layer is inviscid flow, and Euler's equation can be used to relate the velocity and pressure fields. Euler's equation is (Ref. 5-1)

$$\frac{P}{\rho} + \frac{U^2}{2} + gz = \text{constant} \qquad (5\text{-}13)$$

Differentiating with respect to x and neglecting potential energy changes due to changes in elevation ($dz = 0$) gives

$$\frac{1}{\rho}\frac{dP}{dx} + U\frac{dU}{dx} = 0 \qquad (5\text{-}14)$$

Combining Eqs. (5-14) and (5-10b) gives

$$u\frac{\partial u}{\partial x} + v\frac{\partial u}{\partial y} = U\frac{dU}{dx} + v\frac{\partial^2 u}{\partial y^2} \qquad (5\text{-}15)$$

The mathematical boundary conditions for a laminar boundary layer do not introduce a characteristic length. Thus, under certain conditions the velocity profiles at various distances from the leading edge are similar. Similarity implies that the velocity profiles $u(y)$ at various positions x can be made identical by selecting suitable scale factors for u and y. When velocity profiles are similar, the governing partial differential equations can be reduced to an ordinary differential equation by introducing a suitable mathematical transformation.

The governing equations for the type of boundary-layer flow being discussed are the momentum and continuity equations as given by Eqs. (5-15) and (5-5c). These two equations have two unknown dependent variables u and v. A stream function ψ which satisfies the continuity equation can be introduced, where $u = \partial\psi/\partial y$ and $v = -\partial\psi/\partial x$. By use of the stream function, Eq. (5-15) can be written

$$\frac{\partial\psi}{\partial y}\frac{\partial^2\psi}{\partial x\,\partial y} - \frac{\partial\psi}{\partial x}\frac{\partial^2\psi}{\partial y^2} = U\frac{dU}{dx} + v\frac{\partial^3\psi}{\partial y^3}$$

The boundary conditions are then

$$y = 0 \qquad \frac{\partial\psi}{\partial x} = \frac{\partial\psi}{\partial y} = 0$$

$$y \to \infty \qquad \frac{\partial\psi}{\partial y} \to U(x)$$

Only certain types of inviscid flow velocity will allow similar velocity profiles in the boundary layer. As shown in Ref. 5-3, this occurs when the inviscid flow velocity outside the boundary layer $U(x)$ is given by $U = Cx^m$, where C and m are constants.

Imposing this requirement in Eq. (5-15) gives

$$u \frac{\partial u}{\partial x} + v \frac{\partial u}{\partial y} = m \frac{U^2}{x} + v \frac{\partial^2 u}{\partial y^2} \qquad (5\text{-}16)$$

The required mathematical similarity transformation needed to transform Eq. (5-16) to an ordinary differential equation is derived in Ref. 5-4. This equation is given by

$$\eta = y \sqrt{\frac{m+1}{2} \frac{U}{vx}} = y \sqrt{\frac{m+1}{2} \frac{C}{v}} x^{(m-1)/2}$$

$$(5\text{-}17)$$

and

$$f(\eta) = \frac{\psi}{\sqrt{\dfrac{2}{m+1}} \, vxCx^m} = \frac{\psi}{\sqrt{\dfrac{2}{m+1}} \, Cv \, x^{(m+1)/2}}$$

In making the mathematical transformations, two partial derivatives continually arise. These derivatives are given by

$$\frac{\partial \eta}{\partial y} = \sqrt{\frac{m+1}{2} \frac{C}{v}} x^{(m-1)/2} \qquad (5\text{-}18)$$

and

$$\frac{\partial \eta}{\partial x} = y \sqrt{\frac{m+1}{2} \frac{C}{v}} \frac{m-1}{2} x^{(m-3)/2} = \frac{\eta(m-1)}{2x} \qquad (5\text{-}19)$$

Using the chain rules of differentiation gives

$$u = \frac{\partial \psi}{\partial y} = \frac{\partial \psi}{\partial \eta} \frac{\partial \eta}{\partial y}$$

$$= f' \sqrt{\frac{2}{m+1}} \, Cv \, x^{(m+1)/2} \sqrt{\frac{m+1}{2} \frac{C}{v}} x^{(m-1)/2} = f'U \qquad (5\text{-}20)$$

In a similar manner, using primes to indicate differentiation with respect to η gives

$$v = -\frac{\partial \psi}{\partial x} = -\frac{\partial \psi}{\partial \eta} \frac{\partial \eta}{\partial x}$$

$$= -\left(f \sqrt{\frac{2}{m+1}} \, Cv \, \frac{m+1}{2} x^{(m-1)/2} \frac{\partial x}{\partial \eta} + \sqrt{\frac{2}{m+1}} \, Cv \, x^{(m+1)/2} f' \right) \frac{\eta(m-1)}{2x}$$

$$= -\sqrt{\frac{m+1}{2}} \, Cv \, x^{(m-1)/2} \left(f + f'\eta \frac{m-1}{m+1} \right) \qquad (5\text{-}21)$$

Expressions for the velocity derivatives then follow as

$$\frac{\partial u}{\partial y} = \frac{\partial u}{\partial \eta}\frac{\partial \eta}{\partial y} = f''U\sqrt{\frac{m+1}{2}\frac{C}{v}}x^{(m-1)/2} \tag{5-22}$$

$$\frac{\partial^2 u}{\partial y^2} = \frac{\partial}{\partial \eta}\left(\frac{\partial u}{\partial y}\right)\frac{\partial \eta}{\partial y}$$

$$= f'''U\sqrt{\frac{m+1}{2}\frac{C}{v}}x^{(m-1)/2}\sqrt{\frac{m+1}{2}\frac{C}{v}}x^{(m-1)/2}$$

$$= f'''U\frac{m+1}{2}\frac{C}{v}x^{m-1} \tag{5-23}$$

and $$\frac{\partial u}{\partial x} = \frac{\partial u}{\partial \eta}\frac{\partial \eta}{\partial x}$$

$$= \left[Cx^m f'' + f'Cmx^{m-1}\frac{2x}{\eta(m-1)}\right]\frac{\eta(m-1)}{2x}$$

$$= f''U\frac{\eta(m-1)}{2x} + f'Cmx^{m-1} \tag{5-24}$$

Substituting these expressions for the velocities and their derivatives into Eq. (5-14) and reducing give

$$f''' + ff'' + \beta(1 - f'^2) = 0 \tag{5-25}$$

where $\beta = 2m/(m+1)$. The physical meaning of β is indicated in Fig. 5-3. The value of m specifies the inviscid flow field which produces a certain inviscid pressure gradient. For example, $\beta = 0$ for flow over a flat plate with $dP/dx = 0$. Thus $m = 0$ and $U = C$, a constant free-stream velocity. For stagnation flow, $\beta = 1.0$ and $m = 1.0$. This requires that the free-stream velocity be given by $U = Cx$. For flow over a 90° wedge, $\beta = \frac{1}{2}$, $m = \frac{1}{3}$, and $U = Cx^{1/3}$. The three transformed boundary conditions necessary for obtaining a solution to Eq. (5-25), a third-order nonlinear ordinary differential equation, are given by

$$\begin{aligned} f(0) &= f'(0) = 0 \\ f'(\eta &\to \infty) \to 1.0 \end{aligned} \tag{5-26}$$

Equation (5-25) and its associated boundary conditions is called the *Falkner-Skan equation*. Numerical solutions to this equation are obtained in the next section.

5-3   Numerical Solutions to the Momentum Equation

Equation (5-25) and the boundary conditions of Eqs. (5-26) form an asymptotic, two-point boundary-value problem. Recall that the two-point boundary-value problems in Chap. 2 had finite boundary conditions. In Chap. 2 the Newton-Raphson method was

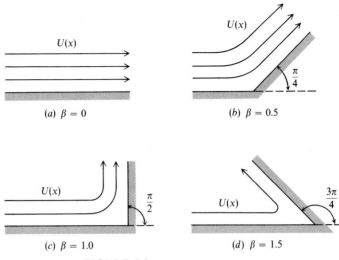

$(a)\ \beta = 0$

$(b)\ \beta = 0.5$

$(c)\ \beta = 1.0$

$(d)\ \beta = 1.5$

FIGURE 5-3
Potential flows associated with Falkner-Skan solutions.

used for estimating gradients needed in the iteration scheme. Although the Newton-Raphson iteration scheme assures convergence to the required outer boundary condition, it does not insure asymptotic convergence. Asymptotic convergence requires that $f''(\eta) \rightarrow 0$ at $\eta = \eta_{\max}$ as well as $f'(\eta) \rightarrow 1.0$ at $\eta = \eta_{\max}$. The required value of $\eta_{\max}$ must be determined by trial and error during the numerical integration process.

A Nachtsheim-Swigert iteration scheme is used to solve the asymptotic, two-point boundary-value problem. (See Ref. 5-5.) This technique is discussed in Appendix B. The procedure is to estimate the unknown value of $f''(0)$ and integrate the Falkner-Skan equation from $\eta = 0$ to $\eta = \eta_{\max}$ (a specified value). The value of $\eta_{\max}$ is the numerical equivalent of infinity as specified in the boundary conditions. Upon completion of the integration, the value of $f'(\eta_{\max})$ is compared with the required value, 1.0, a second estimate of $f''(0)$ is made, and a second integration performed. In the Nachtsheim-Swigert iteration scheme the successive estimates of $f''(0)$ are obtained such that the sum of the squares of the errors $\varepsilon_1{}^2 + \varepsilon_2{}^2$ is a minimum, where $f''(\eta_{\max}) = \varepsilon_2$ and $f'(\eta_{\max}) = 1 \pm \varepsilon_1$.

When the initial estimate of $f''(0)$ is not close to the correct value, the attempted solutions may diverge. A procedure which helps the divergence problem is to initially limit the value of $\eta_{\max}$ and use the Nachtsheim-Swigert iteration scheme to improve the estimation of $f''(0)$. Then when better estimates of $f''(0)$ are known, $\eta_{\max}$ is increased. This procedure is also discussed in Appendix B. In laminar forced convection, a maximum value of $\eta_{\max} = 6.0$ allows a correct solution to the Falkner-Skan equation.

The first step in the numerical solution is to write the third-order nonlinear ordinary differential equation given by Eq. (5-25) as three equivalent first-order equations. Following the procedure described in Appendix A, we write

$$
\begin{aligned}
Z &= f \\
Y &= f' \\
X &= f''
\end{aligned}
\qquad (5\text{-}27)
$$

and then let

$$
\begin{aligned}
F_1 &= Y \\
F_2 &= X \\
F_3 &= -ZX - \beta(1 - Y^2)
\end{aligned}
\qquad (5\text{-}28)
$$

These are the three first-order equations, since $F_1 = dZ/d\eta$, $F_2 = dY/d\eta$, and $F_3 = dX/d\eta$. The required boundary conditions needed at $\eta = 0$ are $Z(0)$, $Y(0)$, and $X(0)$.

A BASIC computer program written to solve the Falkner-Skan equation is listed below. (See Listing 5-1 and Flow Chart 5-1.) The variables are identified in Table 5-1.

The listing of the FSKAN program shows that it is divided into four parts: optional program explanation, Runge-Kutta integration scheme, Nachtsheim-Swigert iteration scheme, and various printing routines. In Statements 200, 210, and 220, initial values are assigned to the logical control variables. The desired value of beta, [B], the Falkner-Skan pressure gradient parameter, is interactively requested in Statements 910 and 920. An estimate for the unknown initial condition $f''(0)$, [XO], is required in Statements 950 and 960. The value of $\eta_{\max}$, [T], and the step size h, [H], are set in Statements 1020 and 1030, respectively. The step size does not change, but the value of $\eta_{\max}$ is subsequently changed in Statement 2265. The initial conditions are set in 1040 to 1090.

The fourth-order Runge-Kutta integration scheme is given in Statements 1140 to 1680. If at Statement 1670 $E \geq T$, then the program is switched to Statements 1720 and 1725 where asymptotic convergence to the correct outer boundary condition is tested. The Nachtsheim-Swigert iteration subroutine begins at Statement 2080. The initial guess of $f''(0)$, [XO], is arbitrarily increased by 0.001 in Statement 1780. The program is then returned to Statement 1040, and the second pass through the program is made. After the first integration, values of f', f'' at $\eta_{\max}$, and $f''(0)$ are saved as A5, C5, and B5 in Statements 2080 to 2100. After the second integration, values of f', f'' at $\eta_{\max}$, and $f''(0)$ are saved as A6, C6, and B6 in Statements 2120 to 2140. Statements 2150 and 2160 calculate the derivatives $\Delta f'(\eta_{\max})/\Delta f''(0)$ and $\Delta f''(\eta_{\max})/\Delta f''(0)$ as C1 and C2.

The Nachtsheim-Swigert correction to $f''(0)$, [A1], is calculated in Statement 2170 and added to the previous value of $f''(0)$, [XO], in Statement 2180. On each subsequent pass Statement 1740 transfers the program to Statement 2120 where the current values of f', f'' at η_{max}, and $f''(0)$ are assigned to A6, C6, and B6, respectively. Hence, new Nachtsheim-Swigert iteration derivatives are calculated for each subsequent integration. After three integrations the value of A in Statement 2264 is 8.0. Thus the value of η_{max}, [T], is increased by 2.0 and A is reset to 5.0. After three more integrations, η_{max} is again increased by 2.0 but additional increases when $\eta_{max} > 5.5$ are prohibited by Statement 2263. Thus the program performs three integrations with $\eta_{max} = 2.0$, three with $\eta_{max} = 4.0$, and all additional integrations with $\eta_{max} = 6.0$. Integration continues until convergence to the required outer boundary condition to within the specified accuracy ε_1 is achieved.

When convergence has been achieved, Statement 1725 switches the program to Statements 1830 to 1890 where a request for the full results can be made. Statements 1960 to 2040 ask if an additional solution is desired.

A solution to the Falkner-Skan equation for $\beta = 0$ corresponds to a flat plate with $dP/dx = 0$. The solution for the unknown value of $f''(0)$ gives a value of 0.4696 and can be verified by using the program FSKAN. This value was published in 1908 by H. Blasius (Ref. 5-6). This value is useful when estimating the unknown values of $f''(0)$ for solutions to the Falkner-Skan equation with $\beta > 0$.

Consider a solution with $\beta = 0.25$. This value corresponds to two-dimensional flow over a wedge with a half-angle of $\pi/8$. Flows with $\beta > 0$ have a favorable pressure gradient in the inviscid flow. That is, the pressure decreases in the direction of the flow. This favorable pressure gradient will produce a thinner boundary layer, and this causes a larger velocity gradient at the surface. Since the velocity gradient is directly proportional to f'', as shown in Eq. (5-22), the initial estimate for $f''(0)$ with $\beta = 0.25$ should be greater than 0.4696.

A numerical solution obtained from FSKAN is given below for $\beta = 0.25$. Based upon the above reasoning, the initial guess of $f''(0)$ was 0.6. Convergence to the correct value was rapidly achieved, and the solution gives a value $f''(0) = 0.7319$.

By use of Eq. (5-22) and the numerical solution for $f''(0)$, values of wall shearing stress as a function of distance along the wall can be calculated. For example, to calculate the local wall shearing stress as a function of x and the Reynolds number, the proper expression is

$$\tau(x, 0) = \mu\left(\frac{\partial u}{\partial y}\right)_{y=0} = \mu U \sqrt{\frac{m+1}{2}\frac{C}{v}}\, x^{(m-1)/2} f''(0)$$

For a flat plate, since $\beta = m = 0$ and $U = C$, this equation becomes

$$\tau(x, 0) = \mu U \sqrt{\frac{U}{2vx}}\, f''(0)$$

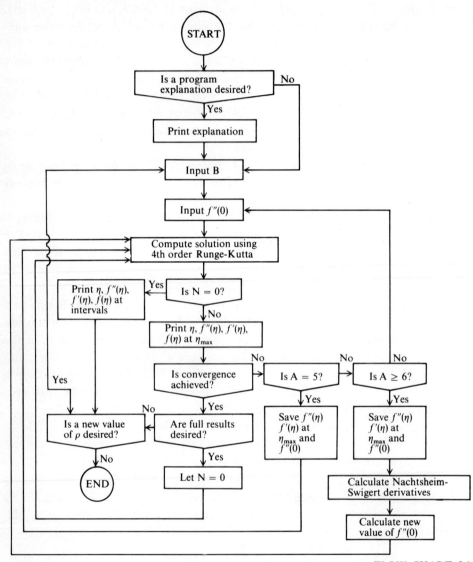

FLOW CHART 5-1

If one uses the value $f''(0) = 0.4696$, the dimensional wall shearing stress is given by

$$\tau(x, 0) = 0.332 \frac{\mu U}{x} \text{Re}_x^{1/2} \qquad (5\text{-}29)$$

where the local Reynolds number $\text{Re}_x = Ux/\nu$. Defining a local nondimensional skin friction as

$$C_{f_x} = \frac{\tau(x, 0)}{\rho U^2/2}$$

one obtains

$$C_{f_x} = \sqrt{2} f''(0) \text{Re}_x^{-1/2}$$

Table 5-1   VARIABLES IN THE FSKAN PROGRAM

Variable	Definition
A	Logical control variable
A1	Nachtsheim-Swigert correction to initial condition
A5	Initial value of y at η_{max}
A6	Subsequent value of y at η_{max}
B	Falkner-Skan pressure gradient parameter
B5	Initial estimate of X0
B6	Subsequent estimate of X0
C1	Derivative $\Delta f'(\eta = \eta_{max})/\Delta f''(0)$
C2	Derivative $\Delta f''(\eta = \eta_{max})/\Delta f''(0)$
C5	Initial value of X at η_{max}
C6	Subsequent value of X at η_{max}
E	Eta, the independent variable
F	Nondimensional stream function and variable F_3 in Runge-Kutta integration scheme
G	Variable F_2 in Runge-Kutta integration scheme
H	Runge-Kutta step size
J	Variable F_1 in Runge-Kutta integration scheme
K1, 2, 3, 4	Variables in Runge-Kutta integration scheme
L	Logical control variable
L1, 2, 3, 4	Variables in Runge-Kutta integration scheme
M	Logical control variable
M1, 2, 3, 4	Variables in Runge-Kutta integration scheme
N, N1	Logical control variables
R	Logical control variable
T	Maximum value of the independent variable η_{max}
T8, T9	Logical control variables
X	Corresponds to the second derivative of the nondimensional stream function—the shearing stress
X0	Initial value of X
Y	Corresponds to the first derivative of the nondimensional stream function—the velocity ratio
Y0	Initial value of Y
Z	Corresponds to the nondimensional stream function
Z0	Initial value of Z

Listing 5-1 FSKAN

```
200  LETN1=0
210  LETN=1
220  LETA=5
230  PRINT"DO YOU WISH AN EXPLANATION OF THE PROGRAM ?"
240  PRINT" IF YES TYPE 1.   IF NO TYPE 2.";
250  INPUT L
260  IF L=1 THEN 320
270  IF L=2 THEN 900
280  REM
290  REM
300  REM PROGRAM EXPLANATION
310  REM
320  PRINT
330  PRINT
340  PRINT
350  PRINT"THIS PROGRAM NUMERICALLY INTEGRATES THE FALKNER-SKAN EQUATION"
360  PRINT"USING A FOURTH ORDER RUNGE-KUTTA INTEGRATION SCHEME. THE FALKNER-"
370  PRINT"SKAN EQUATION IS A THIRD ORDER NON-LINEAR ORDINARY DIFFERENTIAL"
380  PRINT"EQUATION, FOR WHICH NO KNOWN ANALYTICAL SOLUTION EXISTS."
390  PRINT"IT IS USUALLY ASSOCIATED WITH A TWO-POINT ASYMPTOTIC"
400  PRINT"BOUNDARY VALUE PROBLEM.  THAT IS,"
410  PRINT
420  PRINT
430  PRINT"     F*** + FF** + B(1-F*F*)=0"
440  PRINT
450  PRINT"        F(0)=F*(0)=0"
460  PRINT
470  PRINT"        F*(INFINITY) = 1.0"
480  PRINT
490  PRINT
500  PRINT"WHERE HERE THE * REFERS TO DIFFERENTIATION WITH RESPECT TO"
510  PRINT"ETA AND F IS A NON-DIMENSIONAL STREAM FUNCTION."
520  PRINT
530  PRINT
540  PRINT"SINCE THE NUMERICAL INTEGRATION SCHEME REQUIRES AN INITIAL VALUE"
550  PRINT"PROBLEM IT IS NECESSARY TO ESTIMATE THE UNKNOWN INITIAL"
560  PRINT"CONDITION F**(0), IN ORDER TO START THE INTEGRATION.  THE"
570  PRINT"INTEGRATION IS THEN CARRIED OUT TO SOME SMALL VALUE (2.0) OF THE"
580  PRINT"INDEPENDENT VARIABLE AND THE OUTER BOUNDARY CONDITION"
590  PRINT"F*(INFINITY) CHECKED."
600  PRINT
610  PRINT"IF F*(INFINITY) = 1.0, A SOLUTION HAS BEEN OBTAINED"
620  PRINT
630  PRINT"IF F*(INFINITY) IS NOT EQUAL TO ONE THEN A NEW INITIAL"
640  PRINT"CONDITION MUST BE CALCULATED AND THE PROCESS REPEATED UNTIL"
650  PRINT"THE CORRECT OUTER BOUNDARY CONDITION IS ACHIEVED. THE"
660  PRINT"MAXIMUM VALUE OF THE INDEPENDENT VARIABLE MUST ALSO BE INCREASED."
665  PRINT
670  PRINT"THE CONTROLLING FACTOR ON F**(0) IS THE FALKNER-SKAN"
680  PRINT"PRESSURE GRADIENT PARAMETER B(BETA).  REASONABLE VALUES ARE"
690  PRINT"0<=B<2.0.  REASONABLE VALUES OF F**(0) ARE 0<F**(0)<1.5."
710  PRINT
720  PRINT"CHOOSING B=0 REDUCES THE FALKNER-SKAN EQUATION TO THE"
730  PRINT"FAMILIAR BLASIUS EQUATION FOR THE VISCOUS INCOMPRESSIBLE"
740  PRINT"BOUNDARY LAYER FLOW OVER A FLAT PLATE."
750  PRINT
760  PRINT
770  PRINT"A NACHTSCHEIM-SWIGERT ITERATION SCHEME IS INCORPORATED"
780  PRINT"INTO THE PROGRAM.  ONE INITIAL GUESS OF THE UNKNOWN INITIAL"
790  PRINT"CONDITION F**(0) IS REQUIRED BEFORE AUTOMATIC ITERATION BEGINS."
840  PRINT
900  REM
910  PRINT"WHAT VALUE OF THE FALKNER-SKAN PARAMETER,B, DO YOU WISH";
920  INPUT B
930  PRINT
940  PRINT
950  PRINT"WHAT VALUE OF THE INITIAL CONDITION F**(0) DO YOU CHOOSE";
960  INPUTXO
```

Listing 5-1 **FSKAN** (*continued*)

```
970   PRINT
980   PRINT
990   REM
1000  REM SET ETA MAX, STEP SIZE AND INITIAL CONDITIONS
1010  REM
1020  LETT=2
1030  LETH=.01
1040  LETE=0
1050  LETYO=0
1060  LETZO=0
1070  LETX=XO
1080  LETY=YO
1090  LETZ=ZO
1095  IF T<5.5 THEN 1130
1096  IF A<7 THEN 1130
1100  PRINT"ETA","F**","F*","F"
1110  PRINT
1120  PRINTE,X,Y,Z
1130  REM
1140  REM FOURTH ORDER RUNGE-KUTTA INTEGRATION SCHEME
1150  REM
1160  LETF=-(Z*X+B*(1-Y^2))
1170  LETG=X
1180  LETJ=Y
1190  IF N1=1 THEN 1300
1200  IF N1=2 THEN 1380
1210  IF N1=3 THEN 1460
1220  LETK1=H*F
1230  LETL1=H*G
1240  LETM1=H*J
1250  LETX=X+.5*K1
1260  LETY=Y+.5*L1
1270  LETZ=Z+.5*M1
1280  LETN1=1
1290  GOTO 1160
1300  LETK2=H*F
1310  LETL2=H*G
1320  LETM2=H*J
1330  LETX=X-.5*K1+.5*K2
1340  LETY=Y-.5*L1+.5*L2
1350  LETZ=Z-.5*M1+.5*M2
1360  LETN1=2
1370  GOTO 1160
1380  LETK3=H*F
1390  LETL3=H*G
1400  LETM3=H*J
1410  LETX=X-.5*K2+K3
1420  LETY=Y-.5*L2+L3
1430  LETZ=Z-.5*M2+M3
1440  LETN1=3
1450  GOTO 1160
1460  LETK4=H*F
1470  LETL4=H*G
1480  LETM4=H*J
1490  LETX=X-K3+(K1+2*K2+2*K3+K4)/6
1500  LETY=Y-L3+(L1+2*L2+2*L3+L4)/6
1510  LETZ=Z-M3+(M1+2*M2+2*M3+M4)/6
1520  LETN1=0
1530  LETE=E+H
1540  LETT8=T8+1
1550  REM
1560  REM PRINT SEQUENCE
1570  REM
1580  IF N>0 THEN 1650
1590  IF T9=T8 THEN 1610
1600  GOTO 1160
1610  PRINTE,X,Y,Z
1620  LETT9=T8+20
```

Listing 5-1 FSKAN (*continued*)

```
1630 IF E>=T THEN 1940
1640 GOTO 1160
1650 IF E<T THEN 1680
1655 IF T<5.5 THEN 1670
1656 IF A<7 THEN 1670
1660 PRINTE,X,Y,Z
1670 IF E>=T THEN 1720
1680 GOTO 1160
1690 REM
1700 REM CONVERGENCE TEST
1710 REM
1720 IF ABS(1-Y)>5.0E-7 THEN 1730
1725 IF ABS(X)<5E-6 THEN 1810
1730 IF A=5 THEN 2080
1740 IF A>=6 THEN 2120
1770 LETA=A+1
1780 LETX0=X0+0.001
1800 GO TO 1040
1810 PRINT
1820 PRINT
1830 PRINT"CONVERGENCE ACHIEVED ON F* WITHIN 5 X 10**-7."
1835 PRINT"AND ON F** WITHIN 5 X 10**-6"
1840 PRINT"DO YOU WISH THE FULL RESULTS? IF YES TYPE 1.  IF NO TYPE 2.";
1850 INPUT R
1860 PRINT
1870 PRINT
1880 IF R=1 THEN 1900
1890 IF R=2 THEN 1940
1900 LETN = 0
1910 LETT8=0
1920 LETT9=20
1930 GOTO 1040
1940 PRINT
1950 PRINT
1960 PRINT"DO YOU WISH TO CHOOSE A NEW VALUE OF THE FALKNER-SKAN"
1970 PRINT"PARAMETER B?  IF YES TYPE 1. IF NO TYPE 2.";
1980 INPUT M
1990 LETT8=0
2000 LETT9=20
2010 LETN=1
2020 LETA=5
2030 IF M=1 THEN 900
2040 IF M=2 THEN 2280
2050 REM
2060 REM NACHTSCHEIM-SWIGERT ITERATION SCHEME
2070 REM
2080 LETA5=Y
2090 LETB5=X0
2100 LETC5=X
2110 GO TO 1770
2120 LETA6=Y
2130 LETB6=X0
2140 LETC6=X
2150 LETC1=(A6-A5)/(B6-B5)
2160 LETC2=(C6-C5)/(B6-B5)
2170 LETA1=((1-Y)*C1+X*C2)/(C1*C1+C2*C2)
2180 LETX0=X0+A1
2190 LETA5=A6
2200 LETB5=B6
2210 LETC5=C6
2220 PRINT
2260 LETA=A+1
2263 IF T>5.5 THEN 1030
2264 IF A<8 THEN 1030
2265 LETT=T+2
2266 LETA=5
2270 GO TO 1030
2280 END
```

Computer Results 5-1 FSKAN

```
DO YOU WISH AN EXPLANATION OF THE PROGRAM ?
 IF YES TYPE 1.   IF NO TYPE 2.? 2
WHAT VALUE OF THE FALKNER-SKAN PARAMETER,B, DO YOU WISH? .25


WHAT VALUE OF THE INITIAL CONDITION F**(0) DO YOU CHOOSE? 0.6
```

ETA	F**	F*	F
0	0.731973	0	0
6.01	1.03218 E-5	1.0001	5.06502

ETA	F**	F*	F
0	0.731947	0	0
6.01	8.70426 E-7	1.	5.06468

ETA	F**	F*	F
0	0.731946	0	0
6.01	5.46437 E-7	1.	5.06467

```
CONVERGENCE ACHIEVED ON F* WITHIN 5 X 10**-7.
AND ON F** WITHIN 5 X 10**-6
DO YOU WISH THE FULL RESULTS? IF YES TYPE 1.   IF NO TYPE 2.? 1
```

ETA	F**	F*	F
0	0.731946	0	0
0.2	0.681619	0.141373	1.43049 E-2
0.4	0.629571	0.272531	5.58687 E-2
0.6	0.57475	0.393014	0.122606
0.8	0.516857	0.502224	0.212323
1.	0.456352	0.599581	0.322705
1.2	0.394388	0.684666	0.451336
1.4	0.33266	0.75735	0.595743
1.6	0.273156	0.817878	0.753464
1.8	0.217855	0.866893	0.922125
2.	0.168432	0.905412	1.09952
2.2	0.12603	0.934735	1.28368
2.4	9.11469 E-2	0.956326	1.4729
2.6	6.36442 E-2	0.971686	1.66579
2.8	4.28709 E-2	0.982233	1.86125
3.	2.78397 E-2	0.989217	2.05844
3.2	0.01742	0.993675	2.25677
3.4	1.04988 E-2	0.996417	2.4558
3.6	6.09265 E-3	0.998042	2.65526
3.8	3.4037 E-3	0.998968	2.85497
4.	1.83022 E-3	0.999476	3.05482
4.2	9.47163 E-4	0.999745	3.25474
4.4	4.71767 E-4	0.999881	3.4547
4.6	2.26211 E-4	0.999948	3.65469
4.8	1.04488 E-4	0.999979	3.85468
5.	4.65661 E-5	0.999993	4.05468
5.2	2.00957 E-5	0.999999	4.25467
5.4	8.45372 E-6	1.	4.45467
5.6	3.49447 E-6	1.	4.65467
5.8	1.43057 E-6	1.	4.85467
6.	5.72771 E-7	1.	5.05467
6.19999	2.09419 E-7	1.	5.25467

```
DO YOU WISH TO CHOOSE A NEW VALUE OF THE FALKNER-SKAN
PARAMETER B?   IF YES TYPE 1. IF NO TYPE 2.? 2
```

Thus, the final expression is given by

$$C_{f_x} = 0.664 \mathrm{Re}_x^{-1/2} \qquad (5\text{-}30)$$

The skin friction coefficient varies inversely as the square root of the local Reynolds number.

By obtaining solutions for various values of β, for $0 \le \beta < 2.0$, the reader can study the effect of various pressure gradients on the local wall shear stress. Some solutions are given in Table 5-2. Values of $\beta < 0$ correspond to flow with adverse pressure gradients.

When the initial estimate of $f''(0)$ is too far from the actual value, the computer solution may not be convergent. Several false starts may be necessary to find a more accurate estimate of the unknown value of $f''(0)$. This is especially true for values of β near 2.0.

In addition to the wall shearing stress, the solution obtained by using FSKAN gives the complete profiles of f, f', and f'' as a function of η through the boundary layer. A dimensional velocity profile $u = u(x, y)$ and $v = v(x, y)$ can be obtained by use of Eqs. (5-20) and (5-21).

The displacement and momentum thicknesses are also used in fluid mechanics to describe the behavior of boundary layers. The displacement thickness may be interpreted as the decrease due to viscous effects, with respect to the equivalent inviscid flow, in the mass flow rate between the surface and a streamline at a large distance from the surface. In terms of the defined similarity parameters the displacement thickness δ^* is defined as follows for incompressible flow:

$$\delta^* = \int_0^\infty \left(1 - \frac{u}{U}\right) dy = \sqrt{\frac{2}{m+1} \frac{vx}{U}} \int_0^\infty (1 - f') \, d\eta \qquad (5\text{-}31)$$

Table 5-2   RESULTS FOR THE FALKNER-SKAN EQUATION

β	$f''(0)$	I1	I2	η_{max}	$\eta_{convergence}$
2.00	1.687218	0.4974331	0.2307831	6.0	4.6
1.60	1.521514	0.5440214	0.2504147	6.0	4.8
1.20	1.335721	0.6068977	0.2761104	6.0	5.0
1.00	1.232588	0.6479002	0.2923434	6.0	5.2
0.80	1.120268	0.698680	0.3118461	6.0	5.2
0.60	0.9958365	0.7639711	0.3359076	6.0	5.4
0.50	0.9276801	0.804584	0.3502693	6.0	5.4
0.40	0.8544213	0.8526334	0.3666903	6.0	5.6
0.30	0.7747546	0.9109929	0.3857349	6.0	5.6
0.20	0.6867083	0.9841576	0.4082296	6.0	5.8
0.10	0.5870354	1.080319	0.4354562	6.0	5.8
0.05	0.5311299	1.141735	0.4514675	6.0	6.0
0	0.4696005	1.216778	0.469598	6.0	6.0

The momentum thickness is a measure of the decrease in momentum of the mass flow in the boundary layer with respect to the value it would have in the equivalent inviscid flow field. Mathematically, the momentum thickness Θ for incompressible flow is given by

$$\Theta = \int_0^\infty \frac{u}{U}\left(1 - \frac{u}{U}\right) dy = \sqrt{\frac{2}{m+1}\frac{vx}{U}} \int_0^\infty f'(1-f')\, d\eta \qquad (5\text{-}32)$$

Let

$$I_1 = \int_0^\infty (1-f')\, d\eta \qquad \text{and} \qquad I_2 = \int_{f_0}^\infty f'(1-f')\, d\eta$$

The following program modifications to FSKAN produce values for these integrals.

```
1052   LET  U0 = 0
1054   LET  V0 = 0
1092   LET  U = U0
1094   LET  V = V0
1182   LET  I1 = 1 − Y
1184   LET  I2 = Y*(1 − Y)
1242   LET  S1 = H*I1
1244   LET  R1 = H*I2
1272   LET  U = U + 0.5*S1
1274   LET  V = V + 0.5*R1
1322   LET  S2 = H*I1
1324   LET  R2 = H*I2
1352   LET  U = U − 0.5*S1 + 0.5*S2
1354   LET  V = V − 0.5*R1 + 0.5*R2
1402   LET  S3 = H*I1
1404   LET  R3 = H*I2
1432   LET  U = U − 0.5*S2 + S3
1434   LET  V = V − 0.5*R2 + R3
1482   LET  S4 = H*I1
1484   LET  R4 = H*I2
1512   LET  U = U − S3 + (S1 + 2*S2 + 2*S3 + S4)/6
1514   LET  V = V − R3 + (R1 + 2*R2 + 2*R3 + R4)/6
1825   PRINT  "I1 = "U,  "I2 = "V
1945   PRINT  "I1 = "U,  "I2 = "V
```

Some results for I_1 and I_2 are given in Table 5-2. The displacement thickness and the momentum thickness both increase with decreasing values of β when $\beta > 0$.

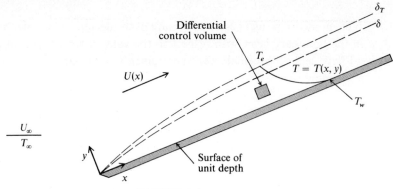

FIGURE 5-4
Differential control volume in thermal boundary layer.

Other problems in laminar boundary-layer flow are analyzed in Ref. 5-3. These include axisymmetric flow, convection with mass transfer at the wall, and nonsimilar boundary layers. Solutions to these problems can be obtained by making minor modifications to the FSKAN program once the formal mathematical transformations and boundary conditions are available.

5-4   Energy Equation

The conservation of energy principle is now applied to a thermal boundary-layer control volume as shown in Fig. 5-4. An enlarged control volume which shows the energy crossing the control volume surfaces is shown in Fig. 5-5. Within the thermal boundary layer the temperature changes continuously from the value T_e at the edge of the thermal boundary layer to the value T_w at the surface. The thickness of the thermal boundary layer δ_T can be greater or less than the thickness of the velocity boundary layer δ, depending upon the fluid properties.

The nondimensional fluid property parameter that arises in the analysis of convective heat transfer is the *Prandtl number*. The Prandtl number is given by $\mathrm{Pr} = \mu c_p/k = \nu/\alpha$. It is a ratio of kinematic viscosity to thermal diffusivity. Physically, it relates the viscous effects to the thermal effects. When $\mathrm{Pr} > 1.0$ then $\nu > \alpha$ and a momentum disturbance propagates farther into the free stream than a thermal disturbance. This fact infers $\delta > \delta_T$. The reverse is true for $\mathrm{Pr} < 1.0$. An order-of-magnitude analysis for laminar boundary-layer flow given in Ref. 5-3 results in $\delta_T = \delta/\mathrm{Pr}^{1/2}$. Values of the Prandtl number range from 10^{-3} for liquid metals to 10^3 for oils and organic fluids. The value for most gases is $0.6 < \mathrm{Pr} < 1.0$, and the Prandtl number for water varies from 2.0 to 14, for $0°C < T < 100°C$.

Consider steady, two-dimensional, constant-property, laminar boundary-layer flow in the presence of heat transfer at the solid surface. For boundary-layer flow it is

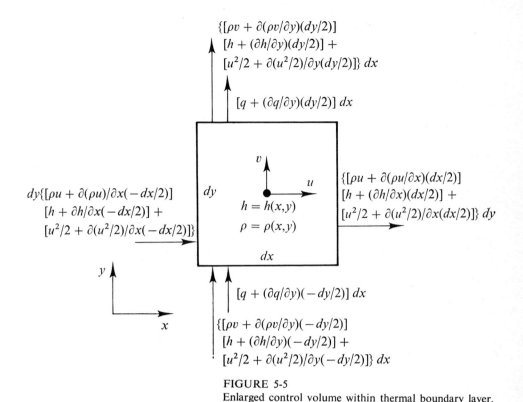

$\{[\rho v + \partial(\rho v/\partial y)(dy/2)]$
$[h + (\partial h/\partial y)(dy/2)] +$
$[u^2/2 + \partial(u^2/2)/\partial y(dy/2)]\}\ dx$

$[q + (\partial q/\partial y)(dy/2)]\ dx$

$dy\{[\rho u + \partial(\rho u)/\partial x(-dx/2)]$
$[h + \partial h/\partial x(-dx/2)] +$
$[u^2/2 + \partial(u^2/2)/\partial x(-dx/2)]\}$

dy

v

u

$h = h(x,y)$

$\rho = \rho(x,y)$

dx

$\{[\rho u + \partial(\rho u/\partial x)(dx/2)]$
$[h + (\partial h/\partial x)(dx/2)] +$
$[u^2/2 + \partial(u^2/2)/\partial x(dx/2)]\}\ dy$

y

x

$[q + (\partial q/\partial y)(-dy/2)]\ dx$

$\{[\rho v + \partial(\rho v/\partial y)(-dy/2)]$
$[h + (\partial h/\partial y)(-dy/2)] +$
$[u^2/2 + \partial(u^2/2)/\partial y(-dy/2)]\}\ dx$

FIGURE 5-5
Enlarged control volume within thermal boundary layer.

appropriate to assume $\partial T/\partial x << \partial T/\partial y$. This assumption is based upon the experimental observation that the heat transfer by conduction within the thermal boundary layer in the x direction is much less than the heat conduction across the fluid layers in the y direction. Physically, this effect is due to the large temperature gradient $\partial T/\partial y$ that occurs because of the thin thermal boundary layer.

The conservation of energy principle for the differential control volume requires that the sum of the net rate of energy crossing the boundary in the form of work and heat plus the net energy change due to convection across the boundaries be equal to zero. Energy is carried by convection across all four surfaces of the control volume. Energy transfer by conduction occurs only at the top and bottom surfaces because of the assumption that $\partial T/\partial x << \partial T/\partial y$. Work is done at the top and bottom surfaces of the boundary-layer control volume. This work is the product of the force due to a shearing stress acting between the horizontal layers of the fluid and the distance over which this force acts. The force is the product of the shearing stress and the differential area over which it acts. The time rate of work is the product of the force and the fluid velocity at the control volume surface. The net rate of work by viscous shearing stresses acting in the x direction is given by

$$\left(\tau_{yx} + \frac{\partial \tau_{yx}}{\partial y}\frac{dy}{2}\right)\left(u + \frac{\partial u}{\partial y}\frac{dy}{2}\right)dx - \left(\tau_{yx} + \frac{\partial \tau_{yx}}{\partial y}\frac{-dy}{2}\right)\left(u + \frac{\partial u}{\partial y}\frac{-dy}{2}\right)dx$$

$$\approx u\frac{\partial \tau_{yx}}{\partial y}\,dy\,dx + \tau_{yx}\frac{\partial u}{\partial y}\,dy\,dx \qquad (5\text{-}33)$$

where the higher-order terms have been neglected.

For a Newtonian fluid in a laminar boundary layer,

$$\tau_{yx} = \mu\frac{\partial u}{\partial y} \qquad (5\text{-}34)$$

and the net rate of work per unit time is given by

$$u\frac{\partial}{\partial y}\left(\mu\frac{\partial u}{\partial y}\right)dy\,dx + \mu\left(\frac{\partial u}{\partial y}\right)^2 dy\,dx$$

If each term in the momentum equation (5-10a) is multiplied by u, the equality

$$u\frac{\partial}{\partial y}\left(\mu\frac{\partial u}{\partial y}\right) = \rho u^2\frac{\partial u}{\partial x} + \rho uv\frac{\partial u}{\partial y} + u\frac{\partial P}{\partial x}$$

$$= G_x u\frac{\partial u}{\partial x} + G_y u\frac{\partial u}{\partial y} + u\frac{\partial P}{\partial x} \qquad (5\text{-}35)$$

can be written. Finally, the net rate of work done by the shearing stresses on the control volume is

$$\left[G_x u\frac{\partial u}{\partial x} + G_y u\frac{\partial u}{\partial y} + u\frac{\partial P}{\partial x} + \mu\left(\frac{\partial u}{\partial y}\right)^2\right]dy\,dx$$

The net rate of energy transfer in the form of heat conduction is obtained by using a first-order Taylor series expansion to express $q_{y\pm dy/2}$ in terms of q_y. This process gives

$$q = \left[-k\frac{\partial T}{\partial y} + \frac{\partial}{\partial y}\left(-k\frac{\partial T}{\partial y}\right)\frac{dy}{2}\right]dx - \left[-k\frac{\partial T}{\partial y} + \frac{\partial}{\partial y}\left(-k\frac{\partial T}{\partial y}\right)\left(\frac{-dy}{2}\right)\right]dx$$

or

$$q = -\frac{\partial}{\partial y}\left(k\frac{\partial T}{\partial y}\right)dy\,dx \qquad (5\text{-}36)$$

This equation is based upon the assumption that Fourier's law of heat conduction applies for the conductive heat transfer due to the temperature gradient between the unmixed laminar fluid layers in the boundary layer.

The energy convected across the control volume surfaces can be in the form of enthalpy, kinetic energy, and potential energy. The time rate of energy transfer due to the convective motion of the fluid is the product of the mass flow rate and the sum of these three energy terms.

The net energy transferred across the control volume surfaces because of convection can also be expressed by use of a first-order Taylor series expansion. The net change of convected energy per unit time, neglecting changes in potential energy, is written

$$\left[\rho u + \frac{\partial(\rho u)}{\partial x}\frac{dx}{2}\right]\left[h + \frac{\partial h}{\partial x}\frac{dx}{2} + \frac{u^2}{2} + \frac{\partial}{\partial x}\left(\frac{u^2}{2}\right)\frac{dx}{2}\right]dy$$

$$-\left[\rho u + \frac{\partial(\rho u)}{\partial x}\frac{-dx}{2}\right]\left[h + \frac{\partial h}{\partial x}\frac{-dx}{2} + \frac{u^2}{2} + \frac{\partial}{\partial x}\left(\frac{u^2}{2}\right)\frac{-dx}{2}\right]dy$$

$$\approx \frac{\partial}{\partial x}\left[\rho u \left(h + \frac{u^2}{2}\right)dx\,dy\right]$$

$$= \frac{\partial}{\partial x}\left[G_x\left(h + \frac{u^2}{2}\right)dx\,dy\right] \qquad (5\text{-}37)$$

where the higher-order terms have been neglected. The static enthalpy of the fluid is indicated by the symbol h. Similarly, the net change of convected energy per unit time in the y direction is

$$\frac{\partial}{\partial y}\left[G_y\left(h + \frac{u^2}{2}\right)dy\,dx\right]$$

The conservation of energy principle requires that the sum of the net energy transfer rate due to heat transfer, work per unit time, and convected energy equal zero. Thus,

$$-\frac{\partial}{\partial y}\left(k\frac{\partial T}{\partial y}\right) - \left[G_x u\frac{\partial u}{\partial x} + G_y u\frac{\partial u}{\partial y} + u\frac{\partial P}{\partial x} + \mu\left(\frac{\partial u}{\partial y}\right)^2\right]$$

$$+ \frac{\partial}{\partial x}\left[G_x\left(h + \frac{u^2}{2}\right)\right] + \frac{\partial}{\partial y}\left[G_y\left(h + \frac{u^2}{2}\right)\right] = 0 \qquad (5\text{-}38)$$

The negative sign on the work term is required since the work due to the shearing stresses is done *on* the control volume. Carrying out the indicated derivatives, using the continuity equation (5-5b), and reducing gives

$$G_x\frac{\partial h}{\partial x} + G_y\frac{\partial h}{\partial y} - \frac{\partial}{\partial y}\left(k\frac{\partial T}{\partial y}\right) - \mu\left(\frac{\partial u}{\partial y}\right)^2 - u\frac{\partial P}{\partial x} = 0 \qquad (5\text{-}39)$$

or

$$\rho u\frac{\partial h}{\partial x} + \rho v\frac{\partial h}{\partial y} - \frac{\partial}{\partial y}\left(k\frac{\partial T}{\partial y}\right) - \mu\left(\frac{\partial u}{\partial y}\right)^2 - u\frac{\partial P}{\partial x} = 0 \qquad (5\text{-}40)$$

For incompressible flow the density is constant and the work due to compression $u(\partial P/\partial x)$ can be neglected in the energy equation (5-40). In addition, the thermal conductivity and dynamic viscosity are treated as constants and the thermodynamic relation $dh = c_p \, dT$ applies, where c_p is the specific heat at constant pressure. This procedure gives

$$u \frac{\partial T}{\partial x} + v \frac{\partial T}{\partial y} = \alpha \frac{\partial^2 T}{\partial y^2} + \frac{\mu}{\rho c_p} \left(\frac{\partial u}{\partial y} \right)^2 \tag{5-41}$$

The boundary conditions for an isothermal wall are

$$
\begin{array}{ll}
y = 0 & T = T_w \\
y \to \infty & T \to T_e
\end{array}
$$

For gases in forced convection, the incompressible flow assumption is valid for velocities up to 60 percent of the local sonic velocity.

The same similarity transformation used in Sec. 5-2 can be used for Eq. (5-41). In addition, a nondimensional temperature $\theta = (T - T_w)/(T_e - T_w)$ is introduced. For constant T_w and T_e, it follows that

$$\frac{\partial T}{\partial x} = \frac{\partial \theta}{\partial \eta} \frac{\partial T}{\partial \theta} \frac{\partial \eta}{\partial x} = (T_e - T_w) \frac{\eta(m-1)}{2x} \theta' \tag{5-42}$$

and

$$\frac{\partial T}{\partial y} = \frac{\partial \theta}{\partial \eta} \frac{\partial T}{\partial \theta} \frac{\partial \eta}{\partial y} = (T_e - T_w) \sqrt{\frac{m+1}{2} \frac{U}{vx}} \theta' \tag{5-43}$$

where the primes indicate differentiation with respect to η. When the existence of a thermal boundary layer at the solid-fluid interface does not strongly affect the temperature field in the external inviscid flow, one can assume $T_e = T_\infty = $ constant. This assumption is normally valid for laminar, external forced convection. A constant wall temperature T_w is also a useful approximation. It is generally valid for small Biot numbers. (See Chap. 4.) The second derivative of temperature is expressed as

$$\frac{\partial^2 T}{\partial y^2} = \frac{\partial}{\partial \eta} \left(\frac{\partial T}{\partial \eta} \right) \frac{\partial \eta}{\partial y} = (T_e - T_w) \frac{m+1}{2} \frac{U}{vx} \theta'' \tag{5-44}$$

Using these results along with the previous expressions for u, v, and $\partial u/\partial y$ in Eq. (5-41) leads to the following ordinary differential equation:

$$\theta'' + \mathrm{Pr} \, f\theta' = \mathrm{PrE} \, x^{2m} f''^2 \tag{5-45}$$

Two nondimensional parameters Pr and E appear in this equation. The physical meaning of the Prandtl number was discussed earlier. The second nondimensional parameter is called the *Eckert number* and is defined as $\mathrm{E} = U^2/c_p(T_w - T_e)$. The Eckert number is a measure of the heating effect due to the work done on the fluid by the shearing stresses. This generation of thermal energy is called *viscous dissipation*.

The boundary conditions for the second-order, transformed energy equation are now expressed in terms of the independent variable η. These conditions are given by

$$\begin{array}{ll} \eta = 0 & \theta = 1.0 \\ \eta \to \infty & \theta \to 0 \end{array} \tag{5-46}$$

Note that the independent variable x appears in Eq. (5-45). This x dependence must be eliminated if similar solutions are to be obtained since the transformed ordinary differential equation can have only one independent variable (η). This will occur for flow over a flat plate ($m = 0$) or for flow with negligible viscous dissipation ($E = 0$). Similar solutions can also occur for certain conditions of a nonisothermal surface with a nonzero pressure gradient ($\beta \neq 0$) in the external inviscid flow.

5-5 Numerical Solutions for Forced Convection

The governing set of equations to be analyzed is given by

$$\text{momentum:} \quad f''' + ff'' + \beta(1 - f'^2) = 0 \quad \text{(5-25)}$$

$$\text{energy:} \quad \theta'' + \Pr f\theta' = \Pr E\, x^{2m} f''^2 \quad \text{(5-45)}$$

with boundary conditions

$$\begin{aligned} f(0) &= f'(0) = 0 \\ f'(\eta \to \infty) &= 1.0 \\ \theta(0) &= 0 \\ \theta(\eta \to \infty) &= 1.0 \end{aligned} \tag{5-46}$$

Since the momentum equation is independent of the energy equation (because of the assumption of constant properties), it can be solved independently of the energy equation as was done in Sec. 5-3. Then the results for $f(\eta)$ and $f''(\eta)$ are used in the energy equation to obtain solutions for $\theta(\eta)$ and $\theta'(\eta)$. The correct value of $\theta'(0)$ must be obtained in an iterative manner similar to the way $f''(0)$ was found for a solution to the momentum equation. The heat transfer rate at the surface due to convection can be determined from a knowledge of $\theta'(0)$ as shown below.

The first special case of interest is when $dP/dx \leq 0$ and the Prandtl-Eckert product is small enough that the viscous dissipation term in the energy equation can be neglected. This has been observed to be true for convection of air at subsonic velocities with Mach numbers less than 0.6. The second special case is the flow over a flat plate with $dP/dx = 0$, but $\Pr E >> 0$. Solutions to these two problems are obtained below.

The computer program FORCBL can be used to solve the set of equations given by Eqs. (5-25) and (5-45), with the boundary conditions given by Eqs. (5-46). The energy equation is first transformed to two equivalent first-order ordinary differential equations by writing

$$W = \theta'$$

$$R = \frac{dW}{d\eta} = -\text{Pr}\, f\theta' + \text{PrE}\, f''^2 \qquad (5\text{-}47)$$

These two equations are used along with the three first-order equations obtained in Sec. 5-3 for the momentum equation (5-28).

The variables used in the program FORCBL are identified in Table 5-3. (See Flow Chart 5-2, Listing 5-2, and Computer Results 5-2.) The FORCBL program works well for fluids with Prandtl numbers between 0.6 and 6.0, e.g., gases and subcooled water. Heat transfer to low Prandtl number fluids, such as liquid metals, and to high Prandtl number fluids, such as oils, is discussed in Chaps. 6 and 7.

Since the velocity field is uncoupled from the temperature field, the governing boundary-value problem for the velocity field is solved first in the FORCBL program. A numerical solution using the program FSKAN was previously discussed in Sec. 5-3. The FORCBL program uses the same logic as the FSKAN program to obtain the required value of $f''(0, \beta)$. This value is then used to generate the required values of $f(\eta), f'(\eta)$, and $f''(\eta)$, and then a solution to the energy equation is obtained.

A fixed step-size fourth-order Runge-Kutta integration scheme (Statements 1890 to 2450) is used with separate Nachtsheim-Swigert iteration schemes (Statements 2680 to 2930 and 3660 to 3860) for the solution of the Falkner-Skan equation and the subsequent solution of the energy equation. Statements 1910 to 1950 reveal how the functions $f(\eta), f'(\eta)$, and $f''(\eta)$ are again obtained simultaneously with the subsequent solution of the energy equation. This procedure is more efficient and more flexible than storing $f(\eta)$ and $f''(\eta)$ for specific values of eta and using these stored values in obtaining subsequent solutions of the energy equation.

The FORCBL program is divided into five parts: an optional program explanation, Runge-Kutta integration scheme, Nachtsheim-Swigert iteration schemes, a section used to define the form of the energy equation, and various printing routines. The program is written so that the interaction of the investigator and the mathematical model represented by the computer is stressed.

Integration of the Falkner-Skan equation begins in Statement 1610 with a request for the value of β. The initial estimate of the surface gradient $f''(0)$ is requested in Statement 1640, and the value of $\eta_{\max}$ is set in Statements 1700 to 1720. The fixed step-size for the Runge-Kutta integration scheme is set equal to 0.01 in Statement 1730. The initial value of the independent variable eta is assigned in Statement 1740 along with f'', f', f, θ', and θ in Statements 1750 to 1810.

Examination of the Runge-Kutta integration scheme in Statements 1890 to 2450 shows that the momentum and energy equations are being integrated simultaneously.

Table 5-3   VARIABLES IN THE FORCBL PROGRAM

Variable	Definition
A	Logical control variable
A1	Nachtsheim-Swigert correction to initial condition
A5	Value of Y at η_{max} for the second estimate of X0
A6	Value of Y at η_{max} for the third estimate of X0
A7	Value of P at η_{max} for the second estimate of S0
A8	Value of P at η_{max} for the third estimate of S0
B	Falkner-Skan pressure gradient parameter
B5, B6	Estimates of X0
B7, B8	Estimates of S0
C	Logical control variable
C1	Derivative $\Delta f'(\eta = \eta_{max})/\Delta f'(0)$
C2	Derivative $\Delta f''(\eta = \eta_{max})/\Delta f''(0)$
C5, C6	Values of X at η_{max}
E	The independent variable—eta
E1	Eckert number
F	Nondimensional stream function and variable in Runge-Kutta integration scheme
G	Variable in Runge-Kutta integration scheme
H	Runge-Kutta step size
J	Variable in Runge-Kutta integration scheme
K1, 2, 3	Variables in Runge-Kutta integration scheme
L	Logical control variable
L1, 2, 3, 4	Variables in Runge-Kutta integration scheme
M	Logical control variable
M1, 2, 3, 4	Variables in Runge-Kutta integration scheme
N, N1	Logical control variables
P	Nondimensional temperature function and variable in Runge-Kutta integration scheme
P0	Initial value of P
P1	Prandtl number
R	Variable in Runge-Kutta integration scheme
S	Nondimensional temperature gradient and variable in Runge-Kutta integration scheme
S0	Initial value of S
S9	Logical control variable
T	Maximum value of the independent variable η_{max}
T8, T9	Logical control variables
U1, 2, 3, 4	Variables in Runge-Kutta integration scheme
V1, 2, 3, 4	Variables in Runge-Kutta integration scheme
W	Variable in Runge-Kutta integration scheme
X	Corresponds to the second derivative of the nondimensional stream function
X0	Initial value of X
Y	Corresponds to the first derivative of the nondimensional stream function
Y0	Initial value of Y
Z	Corresponds to the nondimensional stream function
Z0	Initial value of Z

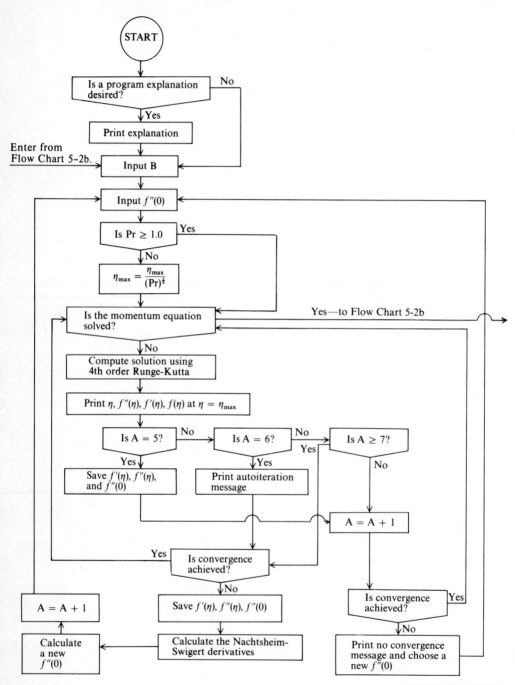

FLOW CHART 5-2a

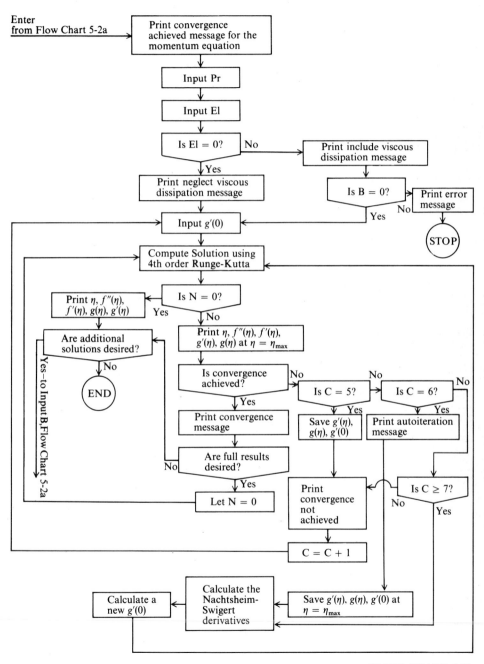

Enter
from Flow Chart 5-2a

Print convergence achieved message for the momentum equation

Input Pr

Input El

Is El = 0? — No → Print include viscous dissipation message

Yes

Print neglect viscous dissipation message

Is B = 0? — No → Print error message

Yes

(STOP)

Input g'(0)

Compute Solution using 4th order Runge-Kutta

Is N = 0? — Yes → Print η, $f''(\eta)$, $f'(\eta)$, $g(\eta)$, $g'(\eta)$

No

Print η, $f''(\eta)$, $f'(\eta)$, $g'(\eta)$, $g(\eta)$ at $\eta = \eta_{max}$

Are additional solutions desired? — No → END

Is convergence achieved? — No → Is C = 5? — No → Is C = 6? — No

Yes   Yes   Yes

Print convergence message

Save g'(\eta), g(\eta), g'(0)

Print autoiteration message

Are full results desired? — No

Yes

Print convergence not achieved

Is C ≥ 7? — No

Yes

Let N = 0

C = C + 1

Yes–to Input B, Flow Chart 5-2a

Calculate a new g'(0)

Calculate the Nachtsheim-Swigert derivatives

Save g'(\eta), g(\eta), g'(0) at $\eta = \eta_{max}$

FLOW CHART 5-2*b*

Listing 5-2 FORCBL

```
1000 LETN9=0
1010 LETN=1
1020 LETP1=1
1030 LETS0=0
1040 LETP0=0
1050 LETE1=0
1060 LETA=4
1070 LETC=2
1080 PRINT"DO YOU WISH AN EXPLANATION OF THE PROGRAM ?";
1090 PRINT" IF YES TYPE 1.  IF NO TYPE 2.";
1100 INPUTL
1110 IF L=1 THEN 1160
1120 IF L=2 THEN 1600
1130 REM
1140 REM PROGRAM EXPLANATION
1150 REM
1160 PRINT
1170 PRINT"THIS PROGRAM COMPUTES THE LAMINAR INCOMPRESSIBLE BOUNDARY"
1180 PRINT"LAYER WITH FORCED CONVECTION.  FOR THIS CASE THE MOMENTUM"
1190 PRINT"(FALKNER-SKAN)EQUATION AND THE ENERGY EQUATION ARE UNCOUPLED"
1200 PRINT"AND HENCE CAN BE INTEGRATED SUCCESSIVELY.  THE BOUNDARY-"
1210 PRINT"VALUE PROBLEM IS:"
1220 PRINT
1230 PRINT"     F*** + FF** + B(1-F*F*) = 0"
1240 PRINT
1250 PRINT"  G**+(PR)(FG*) = (PR)(E)(X^2M)(F**^2)"
1260 PRINT
1270 PRINT"      F(0)=F*(0)=G(0)=0"
1280 PRINT
1290 PRINT"      F(INFINITY)=G(INFINITY)=1.0"
1300 PRINT
1310 PRINT"      WHERE"
1320 PRINT
1330 PRINT"      G=(T-TW)/(T(INF.)-TW)"
1340 PRINT
1350 PRINT"      AND M=2B/(2-B)"
1360 PRINT
1370 PRINT"AND HERE THE * REFERS TO DIFFERENTIATION WITH RESPECT TO"
1380 PRINT"ETA, F IS A TRANSFORMED STREAM FUNCTION AND G IS A"
1390 PRINT"TRANSFORMED TEMPERATURE FUNCTION CORRESPONDING TO THETA"
1400 PRINT
1410 PRINT"THIS FORM OF THE BOUNDARY VALUE PROBLEM IS ASSOCIATED WITH"
1420 PRINT"INCOMPRESSIBLE VISCOUS BOUNDARY LAYER FLOW.  THE VELOCITY"
1430 PRINT"AT THE EDGE OF THE BOUNDARY LAYER IS ASSUMED TO HAVE"
1440 PRINT"A POWER LAW VARIATION."
1530 PRINT
1540 PRINT"A NACHTSHEIM-SWIGERT ITERATION TECHNIQUE IS INCORPORATED IN"
1550 PRINT"THE PROGRAM.  IMPLIMENTATION OF THIS ITERATION SCHEME REQUIRES"
1560 PRINT"ONE INITIAL GUESS OF THE UNKNOWN SURFACE GRADIENT F**(0)."
1590 PRINT
1600 PRINT
1610 PRINT"WHAT VALUE OF THE FALKNER-SKAN PARAMETER,B, DO YOU WISH";
1620 INPUTB
1630 PRINT
1640 PRINT"WHAT VALUE OF THE SURFACE GRADIENT F**(0) DO YOU CHOOSE";
1650 INPUTXO
1660 PRINT
1670 REM
1680 REM SET ETA MAX, STEP SIZE AND INITIAL CONDITIONS
1690 REM
1700 LET T=6
1710 IF P1>=1 THEN 1730
1720 LET T=T/(SQR(P1))
1730 LETH=.01
```

Listing 5-2   **FORCBL** (*continued*)

```
1740 LETE=0
1750 LETY0=0
1760 LETZ0=0
1770 LETX=X0
1780 LETY=Y0
1790 LETZ=Z0
1800 LETS=S0
1810 LETP=P0
1820 IF C>2 THEN 1870
1830 PRINT"ETA","F**","F*","F"
1840 PRINT
1850 PRINTE,X,Y,Z
1860 GOTO 1910
1870 PRINTE,X,Y,S,P
1880 REM
1890 REM FOURTH ORDER RUNGE-KUTTA INTEGRATION SCHEME
1900 REM
1910 LETF=-(Z*X+B*(1-Y^2))
1920 LETG=X
1930 LETJ=Y
1940 LETW=S
1950 LETR=-P1*Z*S +P1*E1*X*X
1960 IF N1=2 THEN 2110
1970 IF N1=3 THEN 2230
1980 IF N1=4 THEN 2350
1990 LETK1=H*F
2000 LETL1=H*G
2010 LETM1=H*J
2020 LETU1=H*R
2030 LETV1=H*W
2040 LETX=X+.5*K1
2050 LETY=Y+.5*L1
2060 LETZ=Z+.5*M1
2070 LETS=S+.5*U1
2080 LETP=P+.5*V1
2090 LETN1=2
2100 GOTO 1910
2110 LETK2=H*F
2120 LETL2=H*G
2130 LETM2=H*J
2140 LETU2=H*R
2150 LETV2=H*W
2160 LETX=X-.5*K1+.5*K2
2170 LETY=Y-.5*L1+.5*L2
2180 LETZ=Z-.5*M1+.5*M2
2190 LETS=S-.5*U1+.5*U2
2200 LETP=P-.5*V1+.5*V2
2210 LETN1=3
2220 GOTO 1910
2230 LETK3=H*F
2240 LETL3=H*G
2250 LETM3=H*J
2260 LETU3=H*R
2270 LETV3=H*W
2280 LETX=X-.5*K2+K3
2290 LETY=Y-.5*L2+L3
2300 LETZ=Z-.5*M2+M3
2310 LETS=S-.5*U2+U3
2320 LETP=P-.5*V2+V3
2330 LETN1=4
2340 GOTO 1910
2350 LETK4=H*F
2360 LETL4=H*G
2370 LETM4=H*J
```

Listing 5-2 FORCBL (*continued*)

```
2380 LETU4=H*R
2390 LETV4=H*W
2400 LETX=X-K3+(K1+2*K2+2*K3+K4)/6
2410 LETY=Y-L3+(L1+2*L2+2*L3+L4)/6
2420 LETZ=Z-M3+(M1+2*M2+2*M3+M4)/6
2430 LETS=S-U3+(U1+2*U2+2*U3+U4)/6
2440 LETP=P-V3+(V1+2*V2+2*V3+V4)/6
2450 LETE=E+H
2460 LETT8=T8+1
2470 LETN1=0
2480 IF C>2 THEN 3370
2490 IF E<T THEN 2530
2500 PRINTE,X,Y,Z
2510 LETN=1
2520 IF E>T THEN 2540
2530 GOTO 1910
2540 IF A=5 THEN 2700
2550 IF A=6 THEN 2740
2560 IF A >=7 THEN 2770
2570 REM
2580 REM CONVERGENCE TEST
2590 REM
2600 LETA=A+1
2610 IF ABS(1-Y)>5E-7 THEN 2630
2620 IF ABS(X)<5.0E-6 THEN 2970
2630 PRINT
2640 PRINT"CONVERGENCE NOT ACHIEVED.  INPUT NEW VALUE FOR F**(O)."?
2650 GOTO 1650
2660 PRINT
2670 REM
2680 REM NACHTSCHEIM-SWIGERT ITERATION SCHEME-F**(O)
2690 REM
2700 LETA5=Y
2710 LETB5=X0
2720 LETC5=X
2730 GOTO 2600
2740 PRINT
2750 PRINT"AUTOMATIC ITERATION IS NOW TAKING PLACE."
2760 PRINT
2770 IF ABS(1-Y)>5.0E-7 THEN 2790
2780 IF ABS(X)<5.0E-6 THEN 2970
2790 LETA6=Y
2800 LETB6=X0
2810 LETC6=X
2820 LETC1=(A6-A5)/(B6-B5)
2830 LETC2=(C6-C5)/(B6-B5)
2840 LETA1=((1-Y)*C1+X*C2)/(C1*C1+C2*C2)
2850 LETX0=X0+A1
2860 LETA5=A6
2870 LETB5=B6
2880 LETC5=C6
2890 PRINT
2900 PRINT"F**(O)=";X0
2910 PRINT
2920 LETA=A+1
2930 GOTO 1660
2940 REM
2950 REM PROGRAM EXPLANATION AND SELECTION OF FORM OF ENERGY EQUATION
2960 REM
2970 PRINT
2980 PRINT"CONVERGENCE ACHIEVED WITHIN 5E-7 ON F* AND 5E-6 ON F** FOR THE"
2990 PRINT"FALKNER-SKAN EQUATION.  THE ENERGY EQUATION IS NOW SOLVED."
3000 PRINT
3010 PRINT"YOU ARE REQUIRED TO GUESS ONE VALUE OF G*(O) TO INITIATE A"
```

Listing 5-2 FORCBL (*continued*)

```
3020 PRINT"NACHTSHEIM-SWIGERT ITERATION SCHEME."
3030 PRINT
3040 PRINT"CHOOSE A VALUE OF THE PRANDTL NUMBER. PR=";
3050 INPUTPI
3070 PRINT"CHOOSE A VALUE OF THE ECKERT NUMBER. E=";
3080 INPUTEI
3110 IF EI= 0 THEN 3180
3120 PRINT"VISCOUS DISSIPATION IS INCLUDED IN YOUR ANALYSIS."
3130 PRINT"THUS, FOR AN ISOTHERMAL SURFACE SIMILARITY REQUIRES A"
3140 PRINT"FLAT PLATE ANALYSIS WITH B=0."
3150 IF B<>0 THEN 4225
3160 GOTO 3270
3170 PRINT
3180 PRINT"YOU HAVE CHOSEN TO NEGLECT VISCOUS DISSIPATION."
3270 PRINT
3280 PRINT"WHAT VALUE OF G*(0) DO YOU CHOOSE";
3290 INPUTSO
3300 PRINT
3310 LET C=4
3320 REM
3330 REM PRINT ROUTINE
3340 REM
3350 PRINT"ETA","F**","F*","G*","G"
3360 GOTO 1660
3370 IF N>0 THEN 3420
3380 IF T9=T8 THEN 3400
3390 GOTO 1910
3400 PRINTE,X,Y,S,P
3410 LETT9=T8+20
3420 IF E<T THEN 1910
3430 IF N=0 THEN 3450
3440 PRINTE,X,Y,S,P
3450 IF E >=T THEN 3500
3460 GOTO 1910
3470 REM
3480 REM CONVERGENCE TEST
3490 REM
3500 IF ABS(1-P)>5.0E-7 THEN 3520
3510 IF ABS(S)<5.0E-6 THEN 3960
3520 PRINT
3530 IF C=5 THEN 3680
3540 IF C=6 THEN 3730
3550 IF C >=7 THEN 3750
3560 GOTO 3580
3570 PRINT
3580 PRINT
3590 PRINT"CONVERGENCE NOT ACHIEVED. INPUT NEW VALUE FOR G*(0).";
3600 INPUTSO
3610 PRINT
3620 PRINT"ETA","F**","F*","G*","G"
3630 LETC=C+1
3640 GOTO 1660
3650 REM
3660 REM NACHTSHEIM-SWIGERT ITERATION SCHEME-G*(0)
3670 REM
3680 LETA7=P
3690 LETB7=SO
3700 LETC7=S
3710 PRINT
3720 GOTO 3590
3730 PRINT"AUTOMATIC ITERATION IS NOW TAKING PLACE."
3740 LETC=C+1
3750 LETA8=P
3760 LETB8=SO
```

Listing 5-2 FORCBL (*continued*)

```
3770 LETC8=S
3780 IF ABS(1-P)>5.0E-7 THEN 3800
3790 IF ABS(S)<5.0E-6 THEN 3960
3800 LETC1=(A8-A7)/(B8-B7)
3810 LETC2=(C8-C7)/(B8-B7)
3820 LETA1=((1-P)*C1+S*C2)/(C1*C1+C2*C2)
3830 LETSO=S0+A1
3840 PRINT
3850 PRINT
3860 IF N>0 THEN 3920
3870 PRINT"BETA=";B
3880 PRINT"PRANDTL NUMBER=";P1
3900 PRINT"ECKERT NUMBER =";E1
3910 PRINT"F**(0)=";XO
3920 PRINT"G*(0)=";SO
3930 PRINT
3935 PRINT"ETA","F**","F*","G*","G"
3936 PRINT
3940 IF N=0 THEN 4080
3950 GOTO 1660
3960 PRINT
3970 IF N=0 THEN 4120
3980 PRINT"CONVERGENCE ACHIEVED WITHIN 5E-7 FOR F* AND G AND"
3990 PRINT"WITHIN 5E-6 FOR F** AND G*."
4000 PRINT
4010 PRINT"DO YOU WISH THE FULL RESULTS? YES TYPE 1. NO TYPE 2.";
4020 INPUTS8
4030 IF S8=2 THEN 4110
4040 LETN=0
4050 LETT8=0
4060 LETT9=20
4070 GOTO 3840
4080 PRINT"ETA","F**","F*","G*","G"
4090 PRINT
4100 GOTO 1700
4110 PRINT
4120 PRINT"DO YOU WISH ADDITIONAL SOLUTIONS? YES TYPE 1. NO TYPE 2.";
4130 INPUTS9
4140 PRINT
4150 IF S9=2 THEN 4230
4160 LETSO=0
4170 LETPO=0
4180 LETA=3
4190 LETC=2
4200 LETN=1
4210 LETP1=1
4220 GOTO 1610
4225 PRINT"SIMILARITY CONSIDERATIONS LIMITS THE ANALYSIS TO B=0 (FLAT PLATE)"
4226 PRINT"IF THE ECKERT NUMBER >0.  SEE DISCUSSION IN TEXT."
4230 END
```

Computer Results 5-2   FORCBL

DO YOU WISH AN EXPLANATION OF THE PROGRAM ? IF YES TYPE 1.   IF NO TYPE 2.? 2

WHAT VALUE OF THE FALKNER-SKAN PARAMETER,B, DO YOU WISH? .5

WHAT VALUE OF THE SURFACE GRADIENT F**(0) DO YOU CHOOSE? .7

ETA	F**	F*	F
0	0.7	0	0
6.01	-0.433155	-1.023	-7.39996 E-2

CONVERGENCE NOT ACHIEVED.   INPUT NEW VALUE FOR F**(0).? .8

ETA	F**	F*	F
0	0.8	0	0
6.01	-0.213447	-7.10299 E-2	2.33015

CONVERGENCE NOT ACHIEVED.   INPUT NEW VALUE FOR F**(0).? .9

ETA	F**	F*	F
0	0.9	0	0
6.01	-4.32606 E-2	0.776175	4.59482

AUTOMATIC ITERATION IS NOW TAKING PLACE.

F**(0)= 0.924409

ETA	F**	F*	F
0	0.924409	0	0
6.01	-5.04848 E-3	0.973738	5.13359

F**(0)= 0.92742

ETA	F**	F*	F
0	0.92742	0	0
6.01	⌣-4.00778 E-4	0.997913	5.19974

F**(0)= 0.927661

ETA	F**	F*	F
0	0.927661	0	0
6.01	-2.88046 E-5	0.99985	5.20504

F**(0)= 0.927679

ETA	F**	F*	F
0	0.927679	0	0
6.01	-1.99825 E-6	0.999989	5.20542

F**(0)= 0.92768

Computer Results 5-2   FORCBL (*continued*)

ETA	F**	F*	F
0	0.92768	0	0
6.01	-1.96416 E-10	0.999999	5.20545

F**(0)= 0.92768

ETA	F**	F*	F
0	0.92768	0	0
6.01	7.17656 E-8	1.	5.20545

CONVERGENCE ACHIEVED WITHIN 5E-7 ON F* AND 5E-6 ON F** FOR THE
FALKNER-SKAN EQUATION. THE ENERGY EQUATION IS NOW SOLVED.

YOU ARE REQUIRED TO GUESS ONE VALUE OF G*(0) TO INITIATE A
NACHTSHEIM-SWIGERT ITERATION SCHEME.

CHOOSE A VALUE OF THE PRANDTL NUMBER. PR=? .72
CHOOSE A VALUE OF THE ECKERT NUMBER. E=? 0
YOU HAVE CHOSEN TO NEGLECT VISCOUS DISSIPATION.

WHAT VALUE OF G*(0) DO YOU CHOOSE? .5

ETA	F**	F*	G*	G
0	0.92768	0	0.5	0
7.08	-3.02671 E-8	1.	2.99818 E-7	1.05127

CONVERGENCE NOT ACHIEVED. INPUT NEW VALUE FOR G*(0).? .4

ETA	F**	F*	G*	G
0	0.92768	0	0.4	0
7.08	-3.02671 E-8	1.	2.39855 E-7	0.841018

CONVERGENCE NOT ACHIEVED. INPUT NEW VALUE FOR G*(0).? .45

ETA	F**	F*	G*	G
0	0.92768	0	0.45	0
7.08	-3.02671 E-8	1.	2.69836 E-7	0.946146

AUTOMATIC ITERATION IS NOW TAKING PLACE.

G*(0)= 0.475614

ETA	F**	F*	G*	G
0	0.92768	0	0.475614	0
7.08	-3.02671 E-8	1.	2.85195 E-7	1

CONVERGENCE ACHIEVED WITHIN 5E-7 FOR F* AND G AND
WITHIN 5E-6 FOR F** AND G*.

DO YOU WISH THE FULL RESULTS? YES TYPE 1. NO TYPE 2.? 2

DO YOU WISH ADDITIONAL SOLUTIONS? YES TYPE 1. NO TYPE 2.? 2

However, since the initial values of S and E1 equal zero, $W = R = U1 = V1 = U2 = V2 = U3 = V3 = U4 = V4 = 0$, and hence the trivial solution $\theta(\eta) = 0$ throughout the boundary layer is obtained.

If $E \geq T$ at Statement 2490, the values of E, X, Y, and Z are printed for the last value of eta. This procedure has the effect of eliminating the time that would be consumed in printing values for each increment in eta for unsuccessful estimates of $f''(0)$. If at Statement 2520, $E > T$, then when $A < 5$, the program passes through Statements 2540 to 2590 to Statement 2600 where A is incremented. Asymptotic convergence to the correct boundary conditions is checked in Statements 2610 and 2620. If convergence is achieved, the program is switched to Statement 2970 where the solution of the energy equation begins.

During the second and third passes through the program, the value of A is 5 and 6, respectively, since A is initialized at 4.0 in Statement 1060. Values of f' and f'' at $\eta = \eta_{max}$ as well as $f''(0)$ are saved as A5, B5, and C5 in Statements 2700 to 2720 and as A6, B6, and C6 in Statements 2790 to 2810. These values are subsequently used in the Nachtsheim-Swigert iteration scheme. These values are saved for the second and third estimates of $f''(0)$ to allow the investigator to improve his initial guesses for $f''(0)$. This procedure gives more rapid convergence to the correct value of $f''(0)$ after automatic iteration takes place. Convergence is checked in Statements 2770 and 2780 and the correction to $f''(0)$ is determined in Statements 2820 to 2850.

When convergence is achieved, the program passes to Statement 2970 where integration of the energy equation begins. The value of $f''(0)$ which resulted in asymptotic convergence is saved and used in the subsequent solution of the energy equation. The Prandtl number is requested in Statement 3040, and the Eckert number is requested in Statement 3070. Then the initial estimate of $\theta'(0)$ is requested in Statement 3280. Statements 3300 to 3360 set $C = 4$ and print a new table heading appropriate to the beginning of the Runge-Kutta integration scheme. However, now S and E1 are not necessarily zero. Hence, W, R, U1, V1, U2, V2, U3, V3, U4, and V4 are not zero and integration of both the energy equation and momentum equation takes place. If a Prandtl number less than unity is specified, Statement 1720 increases the value of η_{max} by a factor of $1/Pr^{1/2}$. Since the thermal boundary layer is thicker than the velocity layer by a factor $1/Pr^{1/2}$ when $Pr < 1$, a larger integration range is required for convergence of the energy equation.

When convergence to the required asymptotic boundary conditions has been achieved for the energy equation, Statement 3790 causes the convergence message in Statements 3980 and 3990 to be printed. Statement 4010 then asks if the full results are required. If they are requested, Statements 4040 to 4060 set $N = 0$, $T8 = 0$, and $T9 = 20$, respectively. Then in Statement 3840 the values of β, Pr, E, $f''(0)$, and $\theta'(0)$ are printed. Finally, the program is returned for the final pass through the integration scheme in Statements 3940 and 4070 to 4090.

A sample solution for $\beta = 0.5$, $E = 0$, and $Pr = 0.72$ is shown following the program listing in Computer Results 5-2. The heat transfer can be calculated using the value $\theta'(0) = 0.4756$ obtained from the numerical solution. Using Eq. (5-43) gives

$$\frac{q}{A} = -k\left(\frac{\partial T}{\partial y}\right)_{y=0} = -k(T_\infty - T_w)\sqrt{\frac{m+1}{2}\frac{U}{vx}}\,\theta'(0)$$

Also, from Newton's law of cooling

$$\frac{q}{A} = h_x(T_w - T_\infty)$$

where h_x is a local convection coefficient.

Equating the above two expressions for q/A and realizing that $m = \frac{1}{3}$ when $\beta = \frac{1}{2}$ gives

$$\frac{h_x x}{k} = \text{Re}_x^{1/2}\theta'(0)\sqrt{\frac{m+1}{2}} = 0.388\,\text{Re}_x^{1/2} \qquad (5\text{-}48)$$

where $\text{Re}_x = Ux/v$. The ratio $h_x x/k$ is called the *local Nusselt number*. In general, the Nusselt number is proportional to the Reynolds number, Eckert number, and the nondimensional temperature gradient at the surface. The nondimensional temperature gradient is in turn a function of Pr and β. Thus, for external forced convection of a constant property fluid, $\text{Nu}_x = f(\text{Re}_x, Pr, E, \beta)$. Note that the Nusselt number is similar to the Biot number used in Chap. 4 except that the thermal conductivity of the fluid rather than the solid appears in the denominator.

For a flat plate with $\beta = 0$, $Pr = 1.0$, and $E = 0$, a solution obtained by using the program FORCBL gives $\theta'(0) = 0.4696$. This value is exactly the same as the value for $f''(0)$. In fact, the complete solution shows that the nondimensional velocity and temperature profiles are the same when $\beta = 0$, $E = 0$, and $Pr = 1.0$. For these values of the nondimensional parameters, the equality is expected since Eqs. (5-25) and (5-45) have the same form and the same boundary conditions. This analogy between the temperature and velocity profiles is the basis for Reynolds' analogy, which is discussed in Chap. 6.

Using the values $\theta'(0) = 0.4696$ and $m = 0$ in Eq. (5-48) gives a local Nusselt number

$$\text{Nu}_x = 0.332\,\text{Re}_x^{1/2} \qquad (5\text{-}49)$$

for flow over a flat plate. If one uses the definition for the average convection coefficient over a surface of unit depth and length L, the average convection coefficient is given by

$$\bar{h} = \frac{\int_0^L h_x\,dx}{L(1)} = 0.332\,\frac{k}{L}\left(\frac{U}{v}\right)^{1/2}\int_0^L x^{-1/2}\,dx \qquad (5\text{-}50)$$

Integration gives

$$\bar{h} = 0.664\,\frac{k}{L}\,\text{Re}_x^{1/2} \qquad (5\text{-}51)$$

Forming an average Nusselt number, $\overline{\mathrm{Nu}} = \bar{h}L/k$, gives

$$\overline{\mathrm{Nu}} = \frac{\bar{h}L}{k} = 0.664 \ \mathrm{Re}^{1/2} \qquad (5\text{-}52)$$

where $\mathrm{Re} = UL/v$. Thus, the average Nusselt number is twice the local value at $x = L$ for laminar forced convection over a flat plate.

Knowledge of the Reynolds number and the thermal conductivity of the fluid, along with the proper value of $\theta'(0)$ for the specific values of β, Pr, and E, leads to the value of the convection heat transfer coefficient as illustrated above. Values of the average convection heat transfer coefficient were used as known boundary conditions in the first four chapters. The theory and numerical techniques presented in this chapter allow the calculation of $\bar{h}$ for many flow situations.

For a limited Prandtl number range, $0.6 < \mathrm{Pr} < 15$, numerical solutions indicate that the Prandtl number effect can be accounted for by introducing $\mathrm{Pr}^{1/3}$ into Eqs. (5-49) and (5-52). Thus, for laminar flow over a flat plate, subject to the limiting assumptions imposed in this section, the local Nusselt number may be expressed as

$$\mathrm{Nu}_x = 0.332 \ \mathrm{Re}_x^{1/2} \mathrm{Pr}^{1/3} \qquad 0.6 < \mathrm{Pr} < 15 \qquad (5\text{-}53)$$

and the average Nusselt number is

$$\overline{\mathrm{Nu}} = 0.664 \ \mathrm{Re}^{1/2} \mathrm{Pr}^{1/3} \qquad 0.6 < \mathrm{Pr} < 15 \qquad (5\text{-}54)$$

The numerical solutions can be used to observe the effects of changing the non-dimensional parameters. For example, when β is increased from 0.0 to 0.5 with $\mathrm{Pr} = 1.0$ and $\mathrm{E} = 0$, $f''(0)$ increases from 0.4696 to 0.9277, and $\theta'(0)$ increases from 0.4696 to 0.5390. Thus, a favorable pressure gradient $dP/dx < 0$ has a much greater effect on the shearing stress at the surface than on the convective heat transfer at the surface. The Eckert number increases as the square of the inviscid velocity. As the velocity increases, the effect of viscous dissipation can become important. If the Eckert number is increased from 0 to 1.0 for $\beta = 0$ and $\mathrm{Pr} = 1.0$, numerical solutions obtained by using FORCBL show that $\theta'(0)$ decreases from 0.4696 to 0.2348. This reduction in temperature gradient at the surface is due to viscous heating within the boundary layer.

The various types of temperature profiles that can occur within a thermal boundary layer are illustrated in Fig. 5-6. Profile (a) represents the temperature profile without viscous dissipation effects and with $T_w < T_e$. Profile (b) shows the effect of viscous dissipation. The average temperature within the boundary layer has increased, and the temperature gradient at the surface has decreased. As the PrE product is increased, the value of $\theta'(0)$ decreases until a value of zero is reached. This result is illustrated by profile (c). When this condition is reached, no heat transfer occurs at the

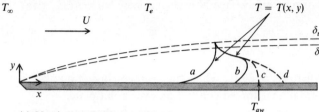

(a) No viscous dissipation: $T_w < T_e$
(b) Viscous dissipation: $T_w < T_{aw}$
(c) Viscous dissipation: $T_w = T_{aw}$
(d) Viscous dissipation: $T_w > T_{aw}$

FIGURE 5-6
Boundary-layer profiles with viscous dissipation.

surface, and the surface temperature is called the *adiabatic wall temperature* T_{aw}. Profile (d) can occur if T_w is independently maintained at a value greater than T_{aw}.

Since the heat transfer rate by convection is a function of $T_w - T_{aw}$ when viscous dissipation is important, the convection heat transfer coefficient is defined by writing Newton's law of cooling as

$$\frac{q}{A} = h_x(T_w - T_{aw}) \qquad (5\text{-}55)$$

Thus, $q = 0$ when $T_w = T_{aw}$. Also, when $T_w < T_{aw}$, heat transfer occurs from the fluid to the surface, and when $T_w > T_{aw}$, heat transfer occurs from the surface to the fluid.

The adiabatic wall temperature is normally different from the total temperature of the free-stream inviscid flow. The total temperature is reached when the high-speed flow is isentropically reduced to zero. For steady, reversible, adiabatic flow, the first law of thermodynamics is written

$$h_1 + \frac{U_1^2}{2} = h_\infty + \frac{U_\infty^2}{2} \qquad (5\text{-}56)$$

If the velocity U_∞ is isentropically reduced to zero, then $U_1 = 0$ and $h_1 = h_0$, the total enthalpy. Thus

$$h_0 = h_\infty + \frac{U_\infty^2}{2} = c_p T_\infty + \frac{U_\infty^2}{2}$$

or

$$T_0 = T_\infty + \frac{U_\infty^2}{2c_p} \qquad (5\text{-}57)$$

The total temperature T_0, the static temperature T_∞, and the dynamic temperature $U_\infty^2/2c_p$ appear in the above equation.

Since the retardation of the flow within a boundary layer is not isentropic, the total temperature does not equal the adiabatic wall temperature. The *recovery factor* indicates the actual dynamic temperature rise at the surface due to viscous dissipation. It is given by

$$R_T = \frac{T_{aw} - T_\infty}{T_0 - T_\infty} = \frac{T_{aw} - T_\infty}{U_\infty^2/2c_p} \qquad (5\text{-}58)$$

For gases, it has been shown experimentally that the recovery factor can be approximated by using

$$\text{laminar flow:} \qquad R_T = \text{Pr}^{1/2} \qquad (5\text{-}59)$$

$$\text{turbulent flow:} \qquad R_T = \text{Pr}^{1/3} \qquad (5\text{-}60)$$

These relations predict that $R_T = 1.0$ for an idealized gas with $\text{Pr} = 1.0$. All real gases have Prandtl numbers less than unity. Since $\text{Pr} < 1.0$ for gases, it is interesting to note that the recovery factor is greater in turbulent flow than in laminar flow. This effect is due to the added energy exchange between the fluid layers due to turbulent mixing. Characteristics of turbulent flow are discussed in the next chapter.

When large temperature gradients occur within the boundary layer, the assumption of constant fluid properties is no longer valid. Also, when $\partial P/\partial x \neq 0$, the term $u(\partial P/\partial x)$ must remain in the energy equation if the flow is compressible.

5-6 Free Convection

The governing equations for laminar free convection on a vertical, isothermal surface are derived by defining a differential control volume within the boundary layer as shown in Fig. 5-7. The continuity and energy equations (5-5a) and (5-41) are identical in both free and forced convection. However, a body force caused by a density variation within the boundary layer rather than a driving inertial force appears in the momentum equation.

Since free convection is caused by density changes in the boundary-layer fluid, a complete analysis of laminar free convection must be based upon treating the boundary-layer flow of a compressible fluid. This analysis is presented in Ref. 5-10. When the temperature changes across the boundary layer are not severe, a mathematical model based upon a modification to the incompressible, constant-property momentum equation for forced convection can be used. This modification is based upon replacing the inertial force with a body force, and leaving the term due to the shearing forces and the nonlinear convective terms unchanged.

The body force in the boundary layer due to a density change is caused by a temperature gradient within the boundary layer. The isobaric volumetric coefficient

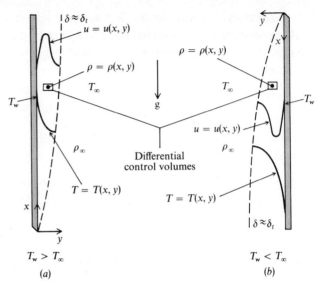

FIGURE 5-7
Free convection on a vertical surface for Pr ≈ 1.0.

of thermal expansion $\bar{\beta}$ is a thermodynamic property of the fluid which relates temperature changes to density changes and is given by

$$\bar{\beta} = -\frac{1}{\rho}\left(\frac{\partial \rho}{\partial T}\right)_p \qquad (5\text{-}61)$$

If we use a finite difference approximation to the derivative in Eq. (5-61), the density
change within a free-convection boundary layer can be written

$$\Delta\rho = -\rho\bar{\beta}\,\Delta T \qquad (5\text{-}62)$$

where ρ is the initial density before the change occurs. If the initial density of the
differential control volume element within the boundary layer is equal to ρ_∞, the
undisturbed density in the inviscid region, it follows that

$$\rho_\infty - \rho = \rho_\infty \bar{\beta}(T - T_\infty) \qquad (5\text{-}63)$$

Relative to the coordinate systems shown in Fig. 5-7, the body force per unit
volume which causes the flow in the boundary layer relative to the undisturbed
inviscid region is written as

$$g_x(\rho_\infty - \rho) = g_x\rho_\infty\bar{\beta}(T - T_\infty) \qquad (5\text{-}64)$$

where g_x is the local acceleration in the x direction due to gravity. When this body force is added to the momentum equation (5-10a) and the pressure gradient $\partial P/\partial x$ is neglected, the resulting equation is

$$\rho u \frac{\partial u}{\partial x} + \rho v \frac{\partial u}{\partial y} = g_x \rho_\infty \bar{\beta}(T - T_\infty) + \frac{\partial}{\partial y}\left(\mu \frac{\partial u}{\partial y}\right) \qquad (5\text{-}65)$$

Treating $\mu = $ constant in the shearing force term, and $\rho = \rho_\infty = $ constant in the non-linear convection terms yields

$$u \frac{\partial u}{\partial x} + v \frac{\partial u}{\partial y} = g_x \bar{\beta}(T - T_\infty) + v \frac{\partial^2 u}{\partial y^2} \qquad (5\text{-}66)$$

Equation (5-66) applies to both free-convection situations shown in Fig. 5-7. In Fig. 5-7a, g_x acts in the negative x direction and the product $g_x(T - T_\infty)$ is negative since $T > T_\infty$. Thus, the body force opposes the shearing forces acting within the boundary layer as required. The boundary layer shown in Fig. 5-7a is for a fluid in which the density decreases with increasing temperature. In Fig. 5-7b, g_x acts in the positive x direction, and the product $g_x(T - T_\infty)$ is again negative since $T < T_\infty$ and the body force is in opposition to the shearing forces.

Viscous dissipation is usually negligible in free convection because of the low velocities that normally occur. The energy equation follows directly from Eq. (5-41) and is given by

$$u \frac{\partial T}{\partial x} + v \frac{\partial T}{\partial y} = \alpha \frac{\partial^2 T}{\partial y^2} \qquad (5\text{-}67)$$

Equations (5-66) and (5-67) are coupled because of the appearance of temperature in the momentum equation, and they must be solved simultaneously. The boundary conditions are

$$\begin{array}{ll} y = 0 \quad u = v = 0 & T = T_w \\ y \to \infty \quad u \to 0 & T \to T_\infty \end{array} \qquad (5\text{-}68)$$

The required similarity transformation necessary to obtain a set of ordinary non-linear differential equations is given in Ref. 5-3. The results for constant T_w and T_∞ are given as

$$\eta = Cyx^{-1/4} \qquad (5\text{-}69)$$

and

$$f(\eta) = \frac{\psi}{4vCx^{3/4}} \qquad (5\text{-}70)$$

where

$$C = \left[\frac{g_x \bar{\beta}(T_w - T_\infty)}{4v^2}\right]^{1/4} \qquad (5\text{-}71)$$

We now make the required transformations

$$u = \frac{\partial \psi}{\partial y} = \frac{\partial \psi}{\partial \eta} \frac{\partial \eta}{\partial y}$$

$$= (4vCx^{3/4}f')(Cx^{-1/4}) = 4vC^2f'x^{1/2} \qquad (5\text{-}72)$$

and

$$v = -\frac{\partial \psi}{\partial x} = -\frac{\partial \psi}{\partial \eta} \frac{\partial \eta}{\partial x}$$

$$= -\left[4vCx^{3/4}f' + 4fvC(\tfrac{3}{4}x^{-1/4})\frac{\partial x}{\partial \eta}\right]\frac{\partial \eta}{\partial x} \qquad (5\text{-}73)$$

where

$$\frac{\partial \eta}{\partial x} = -(\tfrac{1}{4})Cyx^{-5/4} \qquad (5\text{-}74)$$

These equations lead to

$$v = vC^2yx^{-1/2}f' - 3fvCx^{-1/4} \qquad (5\text{-}75)$$

In a similar manner

$$\frac{\partial u}{\partial y} = \frac{\partial u}{\partial \eta} \frac{\partial \eta}{\partial y}$$

$$= (4vC^2f''x^{1/2})(Cx^{-1/4}) = 4vC^3f''x^{1/4} \qquad (5\text{-}76)$$

$$\frac{\partial u}{\partial x} = \frac{\partial u}{\partial \eta} \frac{\partial \eta}{\partial x}$$

$$= \left[4vC^2f''x^{1/2} + 4vC^2f'(\tfrac{1}{2}x^{-1/2})\frac{\partial x}{\partial \eta}\right]\frac{\partial \eta}{\partial x}$$

$$= -vC^3f''yx^{-3/4} + 2vC^2f'x^{-1/2} \qquad (5\text{-}77)$$

and

$$\frac{\partial^2 u}{\partial y^2} = \frac{\partial}{\partial \eta}\left(\frac{\partial u}{\partial y}\right)\frac{\partial \eta}{\partial y} = 4vC^4f'' \qquad (5\text{-}78)$$

Substitution of these expressions into the momentum equation (5-66) gives

$$f''' + 3ff'' - 2f'^2 + \theta = 0 \qquad (5\text{-}79)$$

where

$$\theta = \frac{T - T_\infty}{T_w - T_\infty} \qquad (5\text{-}80)$$

For the energy equation we also need

$$\frac{\partial T}{\partial x} = \frac{\partial \theta}{\partial \eta} \frac{\partial T}{\partial \theta} \frac{\partial \eta}{\partial x}$$

$$= (T_w - T_\infty)(-\tfrac{1}{4}Cyx^{-5/4})\theta' \qquad (5\text{-}81)$$

$$\frac{\partial T}{\partial y} = \frac{\partial \theta}{\partial \eta} \frac{\partial T}{\partial \theta} \frac{\partial \eta}{\partial y}$$

$$= (T_w - T_\infty)Cx^{-1/4}\theta' \qquad (5\text{-}82)$$

and

$$\frac{\partial^2 T}{\partial y^2} = \frac{\partial}{\partial \eta}\left(\frac{\partial T}{\partial y}\right)\frac{\partial \eta}{\partial y}$$

$$= (T_w - T_\infty)Cx^{-1/4}\theta''Cx^{-1/4} \qquad (5\text{-}83)$$

Substitution into the energy equation (5-67) gives

$$\theta'' + 3 \, \text{Pr} \, f\theta' = 0 \qquad \text{(5-84)}$$

The transformed boundary conditions are

$$
\begin{array}{cccc}
\eta = 0 & f = 0 & f' = 0 & \theta = 1 \\
\eta \rightarrow \infty & & f' \rightarrow 0 & \theta \rightarrow 0
\end{array}
\qquad \text{(5-85)}
$$

We are now in a position to solve the two-point, asymptotic boundary-value problem governing laminar free convection on a solid, isothermal, vertical surface.

5-7   Numerical Solutions for Free Convection

The set of equations governing the velocity and temperature profiles in free convection is given by

$$\text{momentum:} \qquad f''' + 3ff'' - 2f'^2 + \theta = 0 \qquad \text{(5-79)}$$

$$\text{energy:} \qquad \theta'' + 3 \, \text{Pr} \, f\theta' = 0 \qquad \text{(5-84)}$$

We first write the two coupled equations in the form of five equivalent first-order differential equations. Let

$$
\begin{aligned}
Z &= f \\
Y &= f' \\
X &= f'' \qquad \text{(5-86)} \\
P &= \theta \\
S &= \theta'
\end{aligned}
$$

and then write

$$
\begin{aligned}
F_1 &= Y \\
F_2 &= X \\
F_3 &= -3ZX + 2Y^2 - P \qquad \text{(5-87)} \\
F_4 &= S \\
F_5 &= -3 \, \text{Pr} \, ZS
\end{aligned}
$$

where $F_1 = dZ/d\eta$, $F_2 = dY/d\eta$, $F_3 = dX/d\eta$, $F_4 = dP/d\eta$, and $F_5 = dS/d\eta$. The boundary conditions for $Z(0)$, $Y(0)$, $X(0)$, $P(0)$, and $S(0)$ must be specified to start the integration.

The computer program FREEBL can be used to obtain solutions to the equivalent set of equations. (See Flow Chart 5-3 and Listing 5-3.) The variables in this program are identified in Table 5-4.

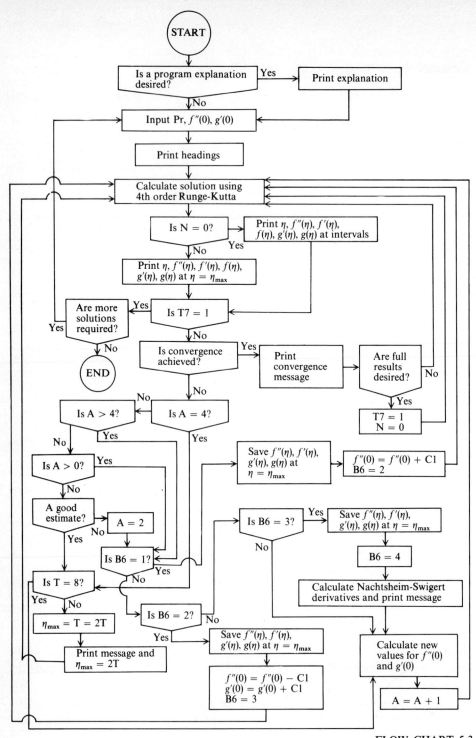

FLOW CHART 5-3

Table 5-4 VARIABLES IN THE FREEBL PROGRAM

Variable	Definition
A	Logical control variable
A9	Derivative $\Delta f'(\eta_{max})/\Delta f''(0)$
B6	Logical control variable
B9	Derivative $\Delta f'(\eta_{max})/\Delta \theta'(0)$
C1	Increment in $f''(0)$ and $\theta'(0)$
C9	Derivative $\Delta \theta(\eta_{max})/\Delta f''(0)$
D9	Derivative $\Delta \theta(\eta_{max})/\Delta \theta'(0)$
E	Eta—the independent variable
E1	Value of $f''(\eta_{max})$ for the second estimate of $f''(0)$ and $\theta'(0)$
E2	Value of $f'(\eta_{max})$ for the second estimate of $f''(0)$ and $\theta'(0)$
E3	Value of $\theta'(\eta_{max})$ for the second estimate of $f''(0)$ and $\theta'(0)$
E4	Value of $\theta(\eta_{max})$ for the second estimate of $f''(0)$ and $\theta'(0)$.
F	Nondimensional stream function and variable in Runge-Kutta integration scheme
F9	Derivative $\Delta f''(\eta_{max})/\Delta f''(0)$
G	Variable in Runge-Kutta integration scheme
G9	Derivative $\Delta \theta'(\eta_{max})/\Delta f''(0)$
H	Runge-Kutta step size
H1	Value of $f''(\eta_{max})$ for the third estimate of $f''(0)$ and $\theta'(0)$
H2	Value of $f'(\eta_{max})$ for the third estimate of $f''(0)$ and $\theta'(0)$
H3	Value of $\theta'(\eta_{max})$ for the third estimate of $f''(0)$ and $\theta'(0)$
H4	Value of $\theta(\eta_{max})$ for the third estimate of $f''(0)$ and $\theta'(0)$
H9	Derivative $\Delta f''(\eta_{max})/\Delta \theta'(0)$
I9	Derivative $\Delta \theta'(\eta_{max})/\Delta \theta'(0)$
J	Variable in Runge-Kutta integration scheme
J9	Function in Nachtsheim-Swigert iteration scheme
K1, 2, 3, 4	Variables in Runge-Kutta integration scheme
K9	Function in Nachtsheim-Swigert iteration scheme
L1, 2, 3, 4	Variables in Runge-Kutta integration scheme
M1, 2, 3, 4	Variables in Runge-Kutta integration scheme
N, N1	Logical control variables
P	Nondimensional temperature function and variable in Runge-Kutta integration scheme
P0	Initial value of P
P1	Prandtl number
R	Variable in Runge-Kutta integration scheme
S	Nondimensional temperature gradient and variable in Runge-Kutta integration scheme
S0	Initial value of S
T	Maximum value of the independent variable—eta
T1	Value of $f''(\eta_{max})$ for the first estimate of $f''(0)$ and $\theta'(0)$
T2	Value of $f'(\eta_{max})$ for the first estimate of $f''(0)$ and $\theta'(0)$
T3	Value of $\theta'(\eta_{max})$ for the first estimate of $f''(0)$ and $\theta'(0)$
T4	Value of $\theta(\eta_{max})$ for the first estimate of $f''(0)$ and $\theta'(0)$
T7, T8	Logical control variables
U1, 2, 3, 4	Variables in Runge-Kutta integration scheme
V1, 2, 3, 4	Variables in Runge-Kutta integration scheme
W	Variable in Runge-Kutta integration scheme
X	Corresponds to the second derivative of the nondimensional stream function
X0	Initial value of X
X8, X9	Functions in Nachtsheim-Swigert iteration scheme
Y	Corresponds to the first derivative of the nondimensional stream function
Y0	Initial value of Y
Y8	Function in Nachtsheim-Swigert iteration scheme
Y9	Negative of Y at η_{max}
Z	Corresponds to the nondimensional stream function
Z0	Initial value of Z
Z8	Function in Nachtsheim-Swigert iteration scheme
Z9	Negative of Z at η_{max}

Listing 5-3 FREEBL

```
1010 PRINT"DO YOU WISH A PROGRAM EXPLANATION?";
1030 INPUT I$
1050 IF I$="NO" THEN 1420
1060 REM
1070 REM PROGRAM EXPLANATION
1080 REM
1090 PRINT"THE FOLLOWING GIVES A SOLUTION OF THE NON-DIMENSIONAL VELOCITY"
1100 PRINT"AND TEMPERATURE PROFILES FOR LAMINAR FREE CONVECTION FROM A"
1110 PRINT"VERTICAL SURFACE.  THE ASSOCIATED BOUNDARY VALUE PROBLEM IS,"
1120 PRINT
1130 PRINT"      F***+3FF**-2F*F*+G=0"
1140 PRINT
1150 PRINT"      G**+3(PR)FG*=0"
1160 PRINT
1170 PRINT"WITH BOUNDARY CONDITIONS,"
1180 PRINT
1190 PRINT"      F(O)=F*(O)=0 ; G(O)=1"
1200 PRINT
1210 PRINT"      F*(INF)=0  ; G(INF)=0"
1220 PRINT
1230 PRINT"WHERE"
1240 PRINT"      G=(T-T(INF))/(TW-T(INF))"
1250 PRINT
1260 PRINT"      F*(ETA)=U/U(INF)"
1270 PRINT
1320 PRINT"HERE THE * REFERS TO DIFFERENTIATION WITH RESPECT TO THE"
1330 PRINT"INDEPENDENT VARIABLE ETA.  F IS A NONDIMENSIONAL STREAM"
1340 PRINT"FUNCTION AND G IS A NONDIMENSIONAL TEMPERATURE!"
1350 PRINT
1360 PRINT"A NACHTSHEIM-SWIGERT ITERATION SCHEME IS AUTOMATICALLY"
1370 PRINT"INCORPORATED INTO THE PROGRAM."
1380 PRINT
1390 REM
1400 REM SET ETA MAX,STEP SIZE AND INITIAL CONDITIONS
1410 REM
1420 LETT=2
1430 LETA=0
1440 LETH=.01
1450 LETN=1
1460 LETNI=0
1470 LETCI=.01
1480 PRINT"WHAT VALUE OF THE PRANDTL NUMBER DO YOU WISH?  PR=";
1490 INPUT P1
1500 PRINT
1550 PRINT"TYPE INITIAL GUESS OF SURFACE GRADIENT."
1560 PRINT"F**(O)=";
1570 INPUT XO
1580 PRINT
1590 PRINT"TYPE INITIAL GUESS OF SURFACE GRADIENT."
1600 PRINT"G*(O)=";
1610 INPUT SO
1620 PRINT
1630 PRINT"PRANDTL NUMBER=";P1
1650 PRINT"STEP SIZE=";H
1660 PRINT"ETA MAX=";T
1670 LETB6=1
1680 LETE=0
1690 LETNI=0
1700 LETPO=1
1710 LETYO=0
1720 LETZO=0
1730 LETS=SO
1740 LETP=PO
1750 LETX=XO
1760 LETY=YO
```

Listing 5-3   **FREEBL** (*continued*)

```
1770 LETZ=ZO
1775 IF A<4 THEN 1850
1780 PRINT
1790 PRINT"F**(0)=";XO
1800 PRINT"G*(0)=";SO
1810 PRINT
1820 PRINT"ETA","F**","F*","G*","G"
1830 PRINT
1840 PRINTE,X,Y,S,P
1850 REM
1860 REM FOURTH ORDER RUNGE-KUTTA NUMERICAL INTEGRATION SCHEME
1870 REM
1880 LETF=-3*Z*X+2*Y*Y-P
1890 IF ABS(F)>1E30 THEN 2760
1900 LETG=X
1910 LETJ=Y
1920 LETR=-3*PI*Z*S
1930 LETW=S
1940 IF NI=2 THEN 2090
1950 IF NI=3 THEN 2210
1960 IF NI=4 THEN 2330
1970 LETK1=H*F
1980 LETL1=H*G
1990 LETM1=H*J
2000 LETU1=H*R
2010 LETV1=H*W
2020 LETX=X+.5*K1
2030 LETY=Y+.5*L1
2040 LETZ=Z+.5*M1
2050 LETS=S+.5*U1
2060 LETP=P+.5*V1
2070 LETNI=2
2080 GOTO 1880
2090 LETK2=H*F
2100 LETL2=H*G
2110 LETM2=H*J
2120 LETU2=H*R
2130 LETV2=H*W
2140 LETX=X-.5*K1+.5*K2
2150 LETY=Y-.5*L1+.5*L2
2160 LETZ=Z-.5*M1+.5*M2
2170 LETS=S-.5*U1+.5*U2
2180 LETP=P-.5*V1+.5*V2
2190 LETNI=3
2200 GOTO 1880
2210 LETK3=H*F
2220 LETL3=H*G
2230 LETM3=H*J
2240 LETU3=H*R
2250 LETV3=H*W
2260 LETX=X-.5*K2+K3
2270 LETY=Y-.5*L2+L3
2280 LETZ=Z-.5*M2+M3
2290 LETS=S-.5*U2+U3
2300 LETP=P-.5*V2+V3
2310 LETNI=4
2320 GOTO 1880
2330 LETK4=H*F
2340 LETL4=H*G
2350 LETM4=H*J
2360 LETU4=H*R
2370 LETV4=H*W
2380 LETX=X-K3+(K1+2*K2+2*K3+K4)/6
2390 LETY=Y-L3+(L1+2*L2+2*L3+L4)/6
2400 LETZ=Z-M3+(M1+2*M2+2*M3+M4)/6
2410 LETS=S-U3+(U1+2*U2+2*U3+U4)/6
```

Listing 5-3 FREEBL (*continued*)

```
2420 LETP=P-V3+(V1+2*V2+2*V3+V4)/6
2430 LETE=E+H
2440 LETT8=T8+1
2450 LETN1=0
2460 REM
2470 REM PRINT SEQUENCE
2480 REM
2490 IF N>0 THEN 2540
2500 IF T9=T8 THEN 2520
2510 GOTO 1880
2520 PRINTE,X,Y,S,P
2530 LETT9=T8+20
2540 IF E<T THEN 1880
2545 IF A<4 THEN 2560
2550 PRINTE,X,Y,S,P
2560 IF T7=1 THEN 3630
2570 LETN=1
2580 REM
2590 REM CONVERGENCE CHECK
2600 REM
2610 IF ABS(X)>2E-4 THEN 2690
2620 IF ABS(Y)>2E-4 THEN 2690
2630 IF ABS(S)>2E-4 THEN 2690
2640 IF ABS(P)<2E-4 THEN 3490
2650 IF E>=T THEN 2690
2680 GOTO 1880
2690 IF A=4 THEN 2830
2700 IF A>4 THEN 2890
2710 IF A>0 THEN 2890
2720 IF ABS(Y)>5E-3 THEN 2740
2730 IF ABS(P)<5E-3 THEN 2760
2740 LETA=2
2750 GOTO 2890
2760 IF T=8 THEN 2880
2770 LETT=2*T
2780 PRINT
2790 PRINT"YOUR INITIAL ESTIMATE OF F**(O) AND G*(O) IS GOOD."
2800 PRINT"ETA MAX HAS BEEN INCREASED TO T=";T
2810 PRINT
2820 GOTO 1660
2830 LETA=0
2840 IF T=8 THEN 2950
2850 LETT=2*T
2860 PRINT
2870 GOTO 1660
2880 LETA=2
2890 IF B6=1 THEN 3120
2900 IF B6=2 THEN 3190
2910 IF B6=3 THEN 3270
2920 REM
2930 REM NACHTSHEIM-SWIGERT ITERATION SCHEME
2940 REM
2950 LETX8=A9*A9+C9*C9+F9*F9+G9*G9
2960 LETY8=B9*B9+D9*D9+H9*H9+I9*I9
2970 LETZ8=A9*B9+C9*D9+H9*F9+G9*I9
2980 LETY9=-Y
2990 LETZ9=-P
3000 LETR8=Y9*A9+Z9*C9-F9*X-G9*S
3010 LETR9=Y9*B9+D9*Z9-H9*X-I9*S
3020 LETX9=X8*Y8-Z8*Z8
3030 LETK9=(R8*Y8-R9*Z8)/X9
3040 LETJ9=(R9*X8-R8*Z8)/X9
3050 LETA=A+1
3060 LETX0=X0+K9
3070 LETS0=S0+J9
3080 GOTO 1680
```

Listing 5-3 **FREEBL** (*continued*)

```
3090 REM
3100 REM DETERMINATION OF NACHTSHEIM-SWIGERT ITERATION DERIVATIVES
3110 REM
3120 LETT1=X
3130 LETT2=Y
3140 LETT3=S
3150 LETT4=P
3160 LETB6=2
3170 LETX0=X0+C1
3180 GOTO 1680
3190 LETE1=X
3200 LETE2=Y
3210 LETE3=S
3220 LETE4=P
3230 LETB6=3
3240 LETX0=X0-C1
3250 LETS0=S0+C1
3260 GOTO 1680
3270 LETH1=X
3280 LETH2=Y
3290 LETH3=S
3300 LETH4=P
3310 LETB6=4.0
3320 LETS0=S0-C1
3330 LETA9=(E2-T2)/C1
3340 LETF9=(E1-T1)/C1
3350 LETG9=(E3-T3)/C1
3360 LETH9=(H1-T1)/C1
3370 LETI9=(H3-T3)/C1
3380 LETC9=(E4-T4)/C1
3390 LETB9=(H2-T2)/C1
3400 LETD9=(H4-T4)/C1
3410 PRINT
3420 REM
3425 GOTO 3480
3430 REM PROGRAM EXPLANATION CONTINUED
3440 REM
3450 PRINT"THE REQUIRED DERIVATIVES FOR THE NACHTSHEIM-SWIGERT ITERATION"
3460 PRINT"SCHEME HAVE NOW BEEN CALCULATED FOR ETA MAX=";T;".    AUTOMATIC"
3470 PRINT "ITERATION WILL NOW TAKE PLACE."
3480 GOTO2950
3490 PRINT
3500 PRINT"CONVERGENCE HAS BEEN ACHIEVED TO WITHIN A VALUE OF 2E-4."
3510 PRINT"DO YOU WISH THE FULL RESULTS?";
3530 INPUT S$
3540 PRINT
3550 IF S$="NO" THEN 3640
3560 LETN=0
3570 LETT8=0
3580 LETT9=20
3590 PRINT
3600 PRINT"PRANDTL NUMBER=";P1
3610 PRINT"STEP SIZE= H"
3620 GOTO 1680
3630 PRINT
3640 PRINT"DO YOU WISH TO OBTAIN ADDITIONAL SOLUTIONS?";
3660 INPUT R$
3670 LETT7=2
3680 PRINT
3690 IF R$="NO" THEN 3730
3700 LETT8=0
3710 LETT9=20
3720 GOTO 1420
3730 END
```

In the program FREEBL, a fixed step-size fourth-order Runge-Kutta integration scheme is used (Statements 1850 to 2430), along with a Nachtsheim-Swigert iteration scheme (Statements 2920 to 3480). The program is divided into four parts: an optional program explanation, Runge-Kutta integration scheme, Nachtsheim-Swigert iteration scheme, and various printing routines. Integration of the free-convection boundary-layer equations begins in Statement 1550 with a request for the initial estimate of the surface gradient $f''(0)$. The initial estimate of $\theta'(0)$ is requested in Statements 1580 to 1610. Values of f'', f', f, θ', and θ are initialized in Statements 1700 to 1770. The coupled system of five equivalent first-order ordinary differential equations appears in Statements 1880 to 1930.

During the initial passes through the program, $N = 0$ and Statement 2490 passes the program to Statement 2540 where if $E < T$ the program is returned to Statement 1880 and the next increment in eta is calculated. In this case no printing takes place. If at Statement 2540 $E > = T$, then the values of η, f'', f', θ', and θ are printed for the last value of $\eta = \eta_{max}$. Since for the initial passes through the program $T7 \neq 1$, the program passes through Statement 2560 to Statements 2580 to 2640 where the solution is checked for asymptotic convergence. If the convergence criteria are satisfied, the program passes to Statements 3490 to 3540, and results are printed. If convergence has not been achieved, the program passes to Statement 2690, where since $A = 0$ it passes through to Statements 2720 and 2730 where the appropriateness of the initial estimates of $f''(0)$ and $\theta'(0)$ are checked. If these estimates are sufficiently close to the correct values to yield acceptable values for $f'(\eta_{max})$ and $\theta(\eta_{max})$, the program is transferred to Statement 2760 where the value of η_{max} is increased to 4.0 in order to reduce the number of iterations required for convergence. The fundamental rationale for this procedure is discussed in Appendix B. If the initial estimates do not yield acceptable results, A is set equal to 2.0 in Statement 2740, and the program is transferred to Statement 2890 where, since $B6 = 1$, transfer to Statement 3120 occurs. In Statements 3120 to 3180 the values of f'', f', θ', and θ at η_{max} are saved as T1, T2, T3, and T4, respectively; B6 is set equal to 2.0, and $f''(0)$ incremented by $C1 = 0.01$. The program then returns to Statement 1680 from Statement 3180 for the second pass through the integration scheme. At Statements 3190 to 3260 the current values of f'', f', θ', and θ at η_{max} are stored as E1, E2, E3, and E4, respectively. Note that these results represent the changes which occur in f'', f', θ', and θ at η_{max} due to a change in $f''(0)$ while holding $\theta'(0)$ fixed. These values are subsequently used to calculate the necessary derivatives in the Nachtsheim-Swigert iteration scheme. In Statements 3230 to 3260 B6 is set equal to 3, X0 is returned to its original value, S0 is incremented by $C1 = 0.01$, and the program is returned to Statement 1680 for the third pass through the integration scheme. After this integration, current values of f'', f', θ', and θ at η_{max} are stored as H1, H2, H3, and H4, respectively, in Statements 3270 to 3300. Note that these results represent the changes in f'', f', θ', and θ due to a change in $\theta'(0)$ while holding $f''(0)$ fixed. In Statements 3310 to 3400, B6 is set equal to 4.0 and the convergence derivatives required are determined. In Statements 2950 to 3080 the necessary corrections to the initial estimates of $f''(0)$ and $\theta'(0)$ are determined and added to X0 and S0. The value of A is incremented to 3.0 in Statement 3050 and the program returned to Statement 1680 for the fourth pass through the integration scheme. On this pass, $A > 0$ and $B6 = 4$. Thus, the program transfers to Statement 2890 at Statement 2710 and passes through Statements 2890 to 2910 to the Nachtsheim-Swigert iteration scheme, Statements 2950 to 3080, where A is incremented to 4.0 and the corrections to X0 and S0 are again determined. The program is then returned to Statement 1680 for the fifth pass through the integration scheme. On this pass $A = 4$. Hence, Statement 2690 transfers the program to Statement 2830 where, since $T \neq 8$, the program passes through to Statement 2840 where A is set equal to zero, and the value of T is increased to 4. The program is then returned to Statement 1660 where the value of η_{max} is printed, and B1 is set equal to 1.0.

Since $B1 = 1.0$ and $A = 0$, the entire procedure described above is repeated with $T = 4.0$ and subsequently with $T = 8$. However, when $T = 8$, $A = 4$, $B1 = 4$, Statement 2690 transfers the program to Statement 2830 where the program is transferred directly to the Nachtsheim-Swigert iteration scheme, Statements 2950 to 3080. This procedure continues until convergence is achieved.

The boundary-value problem governing the free-convection boundary layer on an isothermal plate is now amenable to solutions by use of the program FREEBL. A typical run of the program for Prandtl number 0.72 is shown below. The solution is $f''(0) = 0.6760$ and $\theta'(0) = -0.5046$.

Using the results of $\theta'(0)$ for free convection, it is possible to calculate the local and average values of the convection heat transfer coefficients. Using Eq. (5-82) and Fourier's law gives

$$\frac{q}{A} = -k \left(\frac{\partial T}{\partial y} \right)_{y=0} = -k(T_w - T_\infty)Cx^{-1/4}\theta'(0) \qquad (5\text{-}88)$$

Also, by Newton's law of cooling

$$\frac{q}{A} = h_x(T_w - T_\infty)$$

Equating expressions for q/A and forming a local Nusselt number give

$$\frac{h_x x}{k} = -Cx^{3/4}\theta'(0) \qquad (5\text{-}89)$$

Computer Results 5-3   FREEBL

```
DO YOU WISH A PROGRAM EXPLANATION?? NO
WHAT VALUE OF THE PRANDTL NUMBER DO YOU WISH?   PR=? .72

TYPE INITIAL GUESS OF SURFACE GRADIENT.
F**(0)=? .6

TYPE INITIAL GUESS OF SURFACE GRADIENT.
G*(0)=? -.5

PRANDTL NUMBER= 0.72
STEP SIZE= 0.01
ETA MAX= 2


F**(0)= 0.632796
G*(0)=-0.572241
```

ETA	F**	F*	G*	G
0	0.632796	0	-0.572241	1
2.	-0.10419	0.139491	-0.267267	6.81414 E-2

```
ETA MAX= 4


F**(0)= 0.668511
G*(0)=-0.508928
```

ETA	F**	F*	G*	G
0	0.668511	0	-0.508928	1
4.01	-1.86359 E-2	7.01684 E-3	-2.53037 E-2	-4.35447 E-3

```
ETA MAX= 8


F**(0)= 0.678041
G*(0)=-0.500613
```

ETA	F**	F*	G*	G
0	0.678041	0	-0.500613	1
8.00999	-5.28176 E-3	-3.08335 E-2	-1.71296 E-4	9.59469 E-3

```
F**(0)= 0.675994
G*(0)=-0.504554
```

ETA	F**	F*	G*	G
0	0.675994	0	-0.504554	1
8.00999	-1.94463 E-4	-5.06714 E-4	-1.2796 E-4	1.04688 E-4

```
CONVERGENCE HAS BEEN ACHIEVED TO WITHIN A VALUE OF 2E-4.
DO YOU WISH THE FULL RESULTS?? NO
```

The average convection heat transfer coefficient is

$$\bar{h} = \frac{\int_0^L h_x \, dx}{L} = \frac{-Ck\theta'(0) \int_0^L x^{-1/4} \, dx}{L}$$

$$\bar{h} = -Ck\theta'(0) \frac{4}{3} \frac{L^{3/4}}{L} \tag{5-90}$$

Forming an average Nusselt number gives

$$\frac{\bar{h}L}{k} = -\frac{4}{3} \left[\frac{g\beta(T_w - T_\infty)L^3}{4v^2} \right]^{1/4} \theta'(0) \tag{5-91}$$

where $\overline{Nu} = \bar{h}L/k$ and $Gr = g\bar{\beta}(T_w - T_\infty)L^3/v^2$. It can be seen that the average Nusselt number for laminar free convection is equal to $\frac{4}{3}$, the local value at $x = L$. In terms of the Grashof number Gr, the equation is

$$\overline{Nu} = -0.9428Gr^{1/4}\theta'(0) \tag{5-92}$$

As shown in the sample solution following the program listing, when $Pr = 0.72$, the numerical results give a value of $\theta'(0) = -0.5046$. It follows that

$$\overline{Nu} = 0.4757Gr^{1/4} \tag{5-93}$$

Most empirical results for air in free convection, such as those given in Chap. 7, give larger values for the Nusselt number. These values result because it is very difficult to accurately measure heat transfer by free convection since stray air currents in the environment disturb the natural flow field.

An extension to this discussion involves free convection with a phase change. The species equation is used with the momentum and energy equations to form a set of three coupled equations. Solutions for this problem are presented in Ref. 5-8.

References

5-1   VALLENTINE, H. R.: "Applied Hydrodynamics," Butterworth, London, 1959.
5-2   PRANDTL, L.: Uber Flussigkeitbewegung bein sehr kleiner Reibung, *Proc. 3d Intern. Math. Congr.*, pp. 484–491, Heidelberg, 1904; also NACA TM 452, 1928.
5-3   SCHLICHTING, H.: "Boundary Layer Theory," 6th ed., McGraw-Hill, New York, 1968.
5-4   HANSEN, A. G.: "Similarity Analysis of Boundary Value Problems in Engineering," Prentice-Hall, Englewood Cliffs, N. J., 1964.
5-5   NACHTSHEIM, P. R., and P. SWIGERT: Satisfaction of Asymptotic Boundary Conditions in Numerical Solution of Systems of Non-linear Equations of Boundary Layer Type, NACA TN D-3004, 1965.
5-6   BLASIUS, H.: Grenzschichten in Flussigkeiten mit kleiner Reibung, *Z. Math. Phys.*, vol. 56, pp. 1–37, 1908; also NACA TM 1256.

5-7   REYNOLDS, W. C.: "Thermodynamics," 2d ed., McGraw-Hill, New York, 1968.

5-8   ADAMS, J. A., and R. L. LOWELL: Free Convection Organic Sublimation on a Vertical Semi-infinite Plate, *Int. J. Heat Mass Transfer*, vol. 11, pp. 1215–1224, 1968.

5-9   ECKERT, E. R. G.: Survey of Boundary Layer Heat Transfer at High Velocities and High Temperatures, *WADC Tech. Rept.* 59–624, 1960.

5-10 OSTRACH, S.: An analysis of Laminar Free-convection Flow and Heat Transfer about a Flat Plate Parallel to the Direction of the Generating Body Force, NACA TR-1111, 1953.

Problems[1]

5-1   Starting with Eqs. (5-15) and the similarity transformations, Eqs. (5-17), derive the Falkner-Skan equation (5-25). Derive the transformed boundary conditions as given by Eqs. (5-26).

5-2   Using the results in Table 5-2, prepare a graph of the local skin friction coefficient $C_{f,x}$ versus β.

5-3   Integrate Eq. (5-29), and obtain an expression for the total drag due to friction over a flat surface. What is the total drag on the surface if air flows over a flat surface of unit depth and length 0.6 m at a velocity of 1.5 m/s ($\mu = 1.93 \times 10^{-5}$ Pl, $\rho = 1.04$ kg/m³).

5-4   Using the results for I1 and I2 given in Table 5-2, calculate values of the momentum thickness Θ and the displacement thickness δ^* for $\beta = 0$, 0.5, and 1.0 for $\mathrm{Re}_x = 10^5$.

5-5   Using the full results given for the program FSKAN, calculate and plot $u = u(x, y)$ for various values of x. Specify your own fluid properties and laminar-flow conditions. What happens to the boundary-layer theory as the leading edge is approached, $x \to 0$?

5-6   Compare Eqs. (5-25) and (5-45) and their associated boundary conditions when $\beta = 0$, Pr = 1.0, and E = 0. Discuss the significance of the results.

5-7   Air flows over a heated flat surface of unit depth at a velocity of 5.0 ft/s ($\mu = 1.44 \times 10^{-5}$ lb/ft-s, $\rho = 0.06$ lb/ft³, $k = 0.0174$ Btu/h-ft-°F). The value of $T_w = 300$°F and $T_\infty = 100$°F. Calculate the average convection heat transfer coefficient for a plate length $L = 2.0$ ft, $L = 1.0$ ft, and $L = 0.1$ ft. What is the total heat transfer by convection for each surface?

5-8   Calculate the local wall shearing stress on a vertical isothermal surface in free convection for Pr = 0.72. Use the numerical results given in the sample solution in Sec. 5-7.

5-9   Starting with Eq. (5-63) and the similarity transformation given by Eqs. (5-65) and (5-66), go through the mathematical details that lead to Eq. (5-70). Show that the boundary conditions, Eq. (5-64), transform to Eq. (5-79).

---

[1] Problems whose numbers are followed by a superscript italic *c* should be analyzed by obtaining solutions on a digital computer.

5-10 Calculate the local heat transfer coefficient on a vertical plate of unit depth in free convection at $x = 0.3$ ft, given $T_w = 120°F$ and $T_\infty = 30°F$ $(g\beta/v^2 = 1.76 \times 10^6 \text{ ft}^{-3}\text{-°F}$, $k = 0.016$ Btu/h-ft-°F). If the plate is 1.0 ft long, calculate the average heat transfer coefficient and the total heat transfer from the surface. Treating the plate as a small gray surface ($\epsilon = 0.1$) in a large enclosure ($T_\infty = 30°F$), calculate the ratio of heat transfer by convection to heat transfer by radiation.

5-11^c Obtain solutions for Nu_x for laminar forced convection over a flat plate, using values $0.6 \le Pr \le 3.0$. Discuss the effect of the Prandtl number on the heat transfer. How do the results compare with values predicted by Eq. (5-53)?

5-12^c Obtain solutions for flow over a flat plate for various values of Eckert number $E > 0$. What value of the Eckert number produces a zero value of $\theta'(0)$? Is this value of E a function of the Prandtl number? What value of the PrE product produces a value of $\theta'(0) = -0.1$? What is the physical significance of this result?

5-13^c Obtain numerical solutions for free convection from an inclined plate for $0° < \alpha \le 20°$ when $T_w > T_\infty$. Discuss the effect of inclination on the heat transfer and shear at the wall compared to a vertical plate.

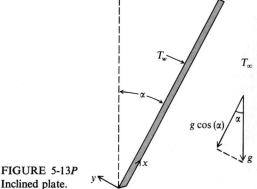

FIGURE 5-13P
Inclined plate.

5-14^c A thin vertical plate is initially in thermal equilibrium at $T = 15°C$. Suddenly radiation falls on each surface of the plate at a rate of $G = 3,000$ W/m². As the temperature rises, the plate loses energy by radiation and free convection to the surroundings at $T_\infty = 15°C$. Assume that the average value of the convection heat transfer coefficient at any instant is given by Eq. (5-93) with $T_w = f(t)$. Study the transient response of this plate for various plate properties and geometries. (Review Chap. 1.) Extra: Derive the boundary-layer energy equation for transient free convection on a vertical surface.

5-15^c In Ref. 5-9 it is suggested that local convective heat transfer rates on a flat plate with viscous dissipation can be estimated by using

$$Nu_x^* = 0.332 Re_x^{*1/2} Pr^{*1/3}$$

to calculate h_x. All properties are evaluated at

$$T^* = T_\infty + 0.5(T_w - T_\infty) + 0.22(T_{aw} - T_\infty)$$

Then $$q'' = h_x(T_w - T_{aw})$$

The procedure is to first calculate T_{aw} by use of Eqs. (5-59) and (5-58). Then calculate T^* and evaluate all fluid properties appearing in Nu_x, Re_x, and Pr at this temperature.

Obtain numerical solutions for $\theta'(0)$ on a flat plate with viscous dissipation, and compare the heat transfer rates found by using Fourier's law and the numerical results with those predicted by the above method.

5-16ᶜ   Many problems in the first four chapters where a value of $\bar{h}$ was given can now be solved for a variety of convective boundary conditions using the theoretical and numerical techniques presented in this chapter to determine the proper value of $\bar{h}$.

INTEGRAL METHODS FOR CONVECTION

6-1 Introduction

The boundary-layer analysis in Chap. 5 was based upon deriving the governing equations by use of a differential analysis. The differential analysis began by applying the basic conservation principles to a differential element within the boundary layer, and produced a set of governing partial differential equations. The analysis concentrated on a special type of laminar boundary-layer formation, called *similar flows*. This specialization allowed a mathematical transformation of the governing partial differential equations to ordinary nonlinear differential equations.

This chapter presents a method of analysis based upon an integral formulation of the governing equations. This alternate method of analysis is an approximate method, but it has the advantage that it can be applied to nonsimilar as well as similar boundary layers and to turbulent as well as laminar flow. In this chapter the integral method is applied to laminar and turbulent, external forced convection; laminar free convection; and laminar and turbulent, internal forced convection.

6-2   Momentum Integral Equation

An integral analysis as suggested by Pohlhausen (Ref. 6-1) and von Kármán (Ref. 6-2) is accomplished by applying the basic conservation principles of mass and momentum to a properly defined control volume, such as that shown in Fig. 6-1. The control volume has a finite height equal to the boundary-layer thickness δ, width dx, and unit depth. The boundary-layer thickness is determined by the condition that $u(x, y) = U(x)$, the free-stream velocity.

The mass flow into the control volume through the differential area dy (1) is $\rho u \, dy$ (1), and the total mass crossing plane 1 can be written

$$\dot{m}_x = \int_0^{\delta(x)} \rho u \, dy \, (1) \qquad (6\text{-}1)$$

where ρ is the fluid density in the control volume at plane 1, u is the velocity within the control volume at plane 1, and dy (1) is the differential area normal to the flow crossing plane 1. The momentum per unit time carried across the differential area is given by the product of the mass flow and velocity as $u\rho u \, dy$. The total rate of momentum carried across plane 1 due to the mass flow is then

$$\int_0^{\delta(x)} \rho u^2 \, dy$$

The corresponding quantities crossing plane 2 at $x + dx$ can be expressed in terms of the quantities at x by a first-order Taylor series expansion. The total mass flow is

$$\dot{m}_{x+dx} = \int_0^{\delta(x)} \left[\rho u + \frac{\partial}{\partial x} (\rho u) \, dx \right] dy \qquad (6\text{-}2)$$

and the total momentum per unit time is

$$\int_0^{\delta(x)} \left[\rho u^2 + \frac{\partial}{\partial x} (\rho u^2) \, dx \right] dy$$

There is more mass crossing plane 2 than plane 1 because the thickness of the boundary layer is growing. This added mass crosses the control volume through the top surface. We assume that no mass crosses the control volume boundary through the lower solid surface. Mass transfer can occur at the lower surface because of blowing or suction through a porous wall, or because of a phase change or chemical reaction at the surface.

In a steady-flow analysis, the conservation of mass principle requires that the mass entering the top surface be equal to

$$\dot{m}_{x+dx} - \dot{m}_x = \int_0^{\delta(x)} \left[\frac{\partial}{\partial x} (\rho u) \, dx \right] dy \qquad (6\text{-}3)$$

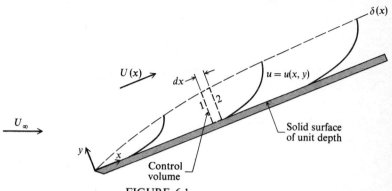

FIGURE 6-1
Boundary-layer control volume for an integral analysis.

The velocity of the mass crossing the top surface of the control volume is $U(x)$, where $U(x)$ is the free-stream velocity at the edge of the velocity boundary layer. The momentum per unit time carried across the top surface is then the product of a velocity and mass flow expressed as

$$U(x) \int_0^{\delta(x)} \left[\frac{\partial}{\partial x} (\rho u) \, dx \right] dy = U(x) \frac{\partial}{\partial x} \int_0^{\delta(x)} (\rho u \, dy) \, dx$$

Realizing that both δ and U are in general functions of x, we write $\delta = \delta(x)$ and $U = U(x)$. The net momentum rate crossing the surfaces of the control volume is

$$\int_0^{\delta} \left[\rho u^2 + \frac{\partial}{\partial x} (\rho u^2) \, dx \right] dy - \int_0^{\delta} \rho u^2 \, dy - U \frac{\partial}{\partial x} \int_0^{\delta} (\rho u \, dy) \, dx$$

This expression reduces to

$$\frac{\partial}{\partial x} \int_0^{\delta} (\rho u^2 \, dy) \, dx - U \frac{\partial}{\partial x} \int_0^{\delta} (\rho u \, dy) \, dx$$

The second term in the above expression can be expressed as the difference between two derivatives. The derivative of a product ab is written $d(ab) = a(db) + b(da)$, and therefore $a(db) = d(ab) - b(da)$. If we let $a = U$ and $b = \int_0^{\delta} \rho u \, dy$, it follows that

$$U \frac{\partial}{\partial x} \int_0^{\delta} (\rho u \, dy) \, dx = \frac{\partial}{\partial x} \left(U \int_0^{\delta} \rho u \, dy \right) dx - \frac{dU}{dx} \int_0^{\delta} (\rho u \, dy) \, dx$$

and the net momentum per unit time is

$$\frac{\partial}{\partial x} \int_0^{\delta} (\rho u^2 \, dy) \, dx - \frac{\partial}{\partial x} \int_0^{\delta} (\rho u U \, dy) \, dx + \frac{dU}{dx} \int_0^{\delta} (\rho u \, dy) \, dx$$

Newton's second law of motion states that the net momentum crossing the control volume per unit time is equal to the net forces acting on the control volume. The forces consist of surface forces and body forces. Body forces provide the driving force for fluid motion in free convection. In forced convection the body forces can be neglected if $Gr_x \ll Re^2$. Surface forces are caused by pressure and shearing stresses. The force due to pressure acting on the surface of plane 1 is $P \int_0^\delta dy(1)$ and on plane 2 is $[P + (dP/dx)dx] \int_0^\delta dy$. A shearing stress due to the velocity gradient at the wall acts on the lower surface of the control volume. For a Newtonian fluid, this resulting surface force is $-\tau_w\, dx = -\mu(\partial u/\partial y)_{y=0}\, dx$. There is no shearing stress on the upper surface of the control volume since $(\partial u/\partial y)_{y=\delta} = 0$.

Incompressible boundary-layer flow in forced convection is assumed. Applying Newton's second law of motion to the chosen control volume gives

$$-\frac{\tau_w}{\rho} - \frac{1}{\rho}\frac{dP}{dx}\int_0^\delta dy = -\frac{\partial}{\partial x}\int_0^\delta (U - u)u\, dy + \frac{dU}{dx}\int_0^\delta u\, dy \qquad (6\text{-}4)$$

The inviscid flow outside the boundary layer is considered a potential flow field. This was also done in Chap. 5. Writing Bernoulli's equation as

$$\frac{P}{\rho} + \frac{U^2}{2} = \text{constant} \qquad (6\text{-}5)$$

and differentiating yields

$$\frac{1}{\rho}\frac{dP}{dx} = -U\frac{dU}{dx} \qquad (6\text{-}6)$$

The boundary-layer approximation that the free-stream pressure gradient is impressed on the thin boundary layer also applies for an integral analysis. Thus the pressure gradient term in Eq. (6-6) is set equal to the pressure gradient term in Eq. (6-4). This gives

$$\frac{-\tau_w}{\rho} + U\frac{dU}{dx}\int_0^\delta dy = -\frac{\partial}{\partial x}\int_0^\delta U^2\left(1 - \frac{u}{U}\right)\frac{u}{U}\, dy + \frac{dU}{dx}\int_0^\delta u\, dy \qquad (6\text{-}7)$$

Introducing $\eta = y/\delta$ we can write this equation as follows:

$$\frac{-\tau_w}{\rho} + U\delta\int_0^1 \frac{dU}{dx}\, d\eta = -\frac{\partial}{\partial x}(\delta U^2)\int_0^1 \left[\frac{u}{U} - \left(\frac{u}{U}\right)^2\right] d\eta$$

$$+ \frac{dU}{dx}\delta U\int_0^1 \frac{u}{U}\, d\eta \qquad (6\text{-}8)$$

Finally, introducing the displacement thickness for incompressible flow, defined by

$$\delta^* = \delta\int_0^1 \left(1 - \frac{u}{U}\right) d\eta \qquad (6\text{-}9)$$

and the momentum thickness, given by

$$\Theta = \delta \int_0^1 \left[\frac{u}{U} - \left(\frac{u}{U}\right)^2\right] d\eta \qquad (6\text{-}10)$$

gives the following form of the momentum integral equation:

$$\frac{\tau_w}{\rho} = \frac{d}{dx}(U^2\Theta) + \delta^*U\frac{dU}{dx} \qquad (6\text{-}11)$$

Carrying out the indicated product derivative leads to

$$\frac{\tau_w}{\rho} = (2\Theta + \delta^*)U\frac{dU}{dx} + U^2\frac{d\Theta}{dx} \qquad (6\text{-}12)$$

This form of the momentum integral equation will be useful in the following sections.

6-3   Laminar Boundary Layers

In order to evaluate the displacement thickness δ^* and momentum displacement thickness Θ which appear in Eq. (6-12), it is necessary to know the velocity profile u/U as a function of η. In the integral method of analysis it is necessary to specify this function either by assuming a form of the profile or using results of experimental measurements. For a laminar boundary layer, a fourth-order polynomial provides a good approximation to the actual velocity profile. We assume

$$\frac{u}{U} = A\eta + B\eta^2 + C\eta^3 + D\eta^4 \qquad (6\text{-}13)$$

and evaluate the four constant coefficients by applying the following four boundary conditions

$$\eta = 1 \qquad u = U \qquad\qquad (6\text{-}14)$$

$$\eta = 1 \qquad \frac{\partial u}{\partial \eta} = 0 \qquad\qquad (6\text{-}15)$$

$$\eta = 1 \qquad \frac{\partial^2 u}{\partial \eta^2} = 0 \qquad\qquad (6\text{-}16)$$

$$\eta = 0 \qquad \nu\frac{\partial^2 u}{\partial \eta^2} = -\delta^2 U\frac{dU}{dx} \qquad (6\text{-}17)$$

Equation (6-14) expresses the physical condition for the edge of the velocity boundary layer. Equation (6-15) is based upon the fact that the shearing stress becomes zero at the edge of the boundary layer. Equation (6-16) expresses the asymptotic convergence of the velocity profile to the free-stream velocity. The final boundary condition is obtained from the governing differential momentum equation. If we use the boundary conditions $u = v = 0$ at $y = 0$, Eqs. (5-10b) and (6-6) give

$$\nu \frac{\partial^2 u}{\partial y^2} = -U \frac{dU}{dx} \qquad (6\text{-}18)$$

as a boundary condition at $y = 0$. Equation (6-13) also satisfies the no-slip condition at the surface, that is, $u = 0$ at $y = 0$.

   Application of the four boundary conditions gives $A = \lambda/6 + 2$, $B = -\lambda/2$, $C = \lambda/2 - 2$, and $D = -\lambda/6 + 1$, where we have defined

$$\lambda = \frac{\delta^2}{\nu} \frac{dU}{dx} \qquad (6\text{-}19)$$

By using Bernoulli's equation, λ can also be expressed by

$$\lambda = -\frac{dP}{dx} \frac{\delta}{\mu U/\delta}$$

The physical significance of λ can then be recognized as a ratio of pressure forces to viscous forces. It follows that the fourth-order polynomial which represents the velocity profile in laminar flow is given by

$$\frac{u}{U} = 2\eta - 2\eta^3 + \eta^4 + \frac{\lambda}{6}(\eta - 3\eta^2 + 3\eta^3 - \eta^4) \qquad (6\text{-}20)$$

   Combining Eqs. (6-20) and (6-9) leads to

$$\delta^* = \delta \left(\frac{3}{10} - \frac{\lambda}{120} \right) \qquad (6\text{-}21)$$

A combination of Eqs. (6-20) and (6-10) gives

$$\Theta = \delta \left(\frac{37}{315} - \frac{\lambda}{945} - \frac{\lambda^2}{9{,}072} \right) \qquad (6\text{-}22)$$

The velocity profile can also be used to express the wall shearing stress since

$$\tau_w = \mu \left(\frac{\partial u}{\partial y} \right)_{y=0} = \mu \left(\frac{\partial u}{\partial \eta} \frac{\partial \eta}{\partial y} \right)_{y=0}$$

$$= \frac{\mu U}{\delta} \left(\frac{\lambda}{6} + 2 \right) \qquad (6\text{-}23)$$

This equation leads to a nondimensional wall shearing stress given by

$$\frac{\tau_w \delta}{\mu U} = \frac{\lambda}{6} + 2 \qquad (6\text{-}24)$$

Boundary-layer separation occurs when $\tau_w = 0$. At this point, $\lambda = -12$.

Equation (6-12) can be put in nondimensional form by multiplying each side by Θ/vU. This gives

$$\frac{\tau_w \Theta}{\mu U} = \left(2 + \frac{\delta^*}{\Theta}\right)\frac{(dU/dx)\Theta^2}{v} + \frac{U\Theta}{v}\frac{d\Theta}{dx} \qquad (6\text{-}25)$$

We now define κ as follows

$$\kappa = \frac{(dU/dx)\Theta^2}{v} \qquad (6\text{-}26)$$

Equations (6-26) and (6-19) show that κ is connected with the momentum thickness the same way that λ is connected with the velocity boundary-layer thickness. Since $\kappa/\lambda = \Theta^2/\delta^2$, Eq. (6-22) gives

$$\frac{\kappa}{\lambda} = \left(\frac{37}{315} - \frac{\lambda}{945} - \frac{\lambda^2}{9{,}072}\right)^2 \qquad (6\text{-}27)$$

Following Ref. 6-3 we let $f_1(\kappa) = \delta^*/\Theta$ and use Eqs. (6-21) and (6-22) to write

$$f_1(\kappa) = \frac{\delta^*}{\Theta} = \frac{3/10 - \lambda/120}{37/315 - \lambda/945 - \lambda^2/9{,}072} \qquad (6\text{-}28)$$

A second function $f_2(\kappa)$ is obtained from the nondimensional wall shear stress parameter, Eq. (6-24), as

$$f_2(\kappa) = \frac{\tau_w \delta}{\mu U}\frac{\Theta}{\delta} = \frac{\tau_w \Theta}{\mu U}$$

$$f_2(\kappa) = \left(\frac{\lambda}{6} + 2\right)\left(\frac{37}{315} - \frac{\lambda}{945} - \frac{\lambda^2}{9{,}072}\right) \qquad (6\text{-}29)$$

Equation (6-25) can then be written

$$f_2(\kappa) = [2 + f_1(\kappa)]\kappa + \frac{U\Theta}{v}\frac{d\Theta}{dx} \qquad (6\text{-}30)$$

Using Eqs. (6-22) and (6-27) we calculate

$$\frac{d\Theta}{dx} = \delta\left[\frac{-d\lambda/dx}{945} - \frac{2\lambda(d\lambda/dx)}{9{,}072}\right] + \frac{d\delta}{dx}\sqrt{\frac{\kappa}{\lambda}}$$

Then

$$\Theta\frac{d\Theta}{dx} = \delta^2\sqrt{\frac{\kappa}{\lambda}}\left[\frac{-d\lambda/dx}{945} - \frac{2\lambda(d\lambda/dx)}{9{,}072}\right] + \delta\frac{d\delta}{dx}\frac{\kappa}{\lambda}$$

But from Eq. (6-19) $\delta^2 = \lambda v/(dU/dx)$, and thus by differentiation

$$\delta\frac{d\delta}{dx} = \frac{v}{2}\frac{(dU/dx)(d\lambda/dx) - \lambda(d^2U/dx^2)}{(dU/dx)^2}$$

Substituting the preceding two relationships into Eq. (6-30) gives

$$f_2(\kappa) = [2 + f_1(\kappa)]\kappa + \frac{U}{U'}\lambda\sqrt{\frac{\kappa}{\lambda}}\left[\frac{-d\lambda/dx}{945} - \frac{2\lambda(d\lambda/dx)}{9{,}072}\right]$$

$$+\frac{U}{2}\frac{U'(d\lambda/dx) - \lambda U''}{U'^2}\frac{\kappa}{\lambda} \qquad (6\text{-}31)$$

where the primes indicate differentiation with respect to x. Multiplying each side by U'/U and collecting like terms gives

$$\frac{U'}{U}[f_2(\kappa) - 2\kappa - \kappa f_1(\kappa)] + \frac{\kappa}{2}\frac{U''}{U'} = \frac{d\lambda}{dx}\left(\frac{-\lambda\sqrt{\kappa/\lambda}}{945} - \frac{2\lambda^2\sqrt{\kappa/\lambda}}{9{,}072} + \frac{1}{2}\frac{\kappa}{\lambda}\right) \qquad (6\text{-}32)$$

Multiplying each side of Eq. (6-32) by $7{,}560/\sqrt{\kappa/\lambda}$ and using the definitions of κ/λ, $f_1(\kappa)$, and $f_2(\kappa)$ lead to the following form of the integral equation:

$$\frac{U'}{U}(15{,}120 - 2{,}784\lambda + 79\lambda^2 + \tfrac{5}{3}\lambda^3) + \frac{U''}{U'}(444\lambda - 4\lambda^2 - \tfrac{5}{12}\lambda^3)$$

$$= \frac{d\lambda}{dx}(444 - 12\lambda - \tfrac{25}{12}\lambda^2) \qquad (6\text{-}33)$$

We now define

$$g_1(\lambda) = 15{,}120 - 2{,}784\lambda + 79\lambda^2 + \tfrac{5}{3}\lambda^3 \qquad (6\text{-}34)$$

$$g_2(\lambda) = 444\lambda - 4\lambda^2 - \tfrac{5}{12}\lambda^3 \qquad (6\text{-}35)$$

$$g_3(\lambda) = 444 - 12\lambda - \tfrac{25}{12}\lambda^2 \qquad (6\text{-}36)$$

and write the integral momentum equation as a nonlinear first-order differential equation, given by

$$\frac{d\lambda}{dx} = \frac{(U'/U)g_1(\lambda) + (U''/U')g_2(\lambda)}{g_3(\lambda)} \qquad (6\text{-}37)$$

To integrate this equation requires a value of λ at the stagnation point $x = 0$ as an initial condition. The velocity U is zero at the stagnation point, but in general U' is finite. If $d\lambda/dx$ is to be finite at $x = 0$, inspection of Eq. (6-37) shows that we must also require $g_1(\lambda) = 0$ at $x = 0$. There are three roots to Eq. (6-34) which will satisfy this requirement. These are $\lambda = 17.75$, $\lambda = 7.0523$, and $\lambda = -70.0$. Two of these roots can be rejected by physical reasoning. We have shown that separation of the boundary layer occurs at $\lambda = -12$. Larger negative values of λ occur in the separated flow region where the model of a steady-flow boundary layer is not valid, and we reject the root $\lambda = -70$. When $\lambda > +12$, the velocity profile u/U as given in Eq. (6-20) becomes greater than unity. This result is not physically possible for a fourth-order polynomial in steady, incompressible flow with the assumed boundary conditions, and we reject the root $\lambda = 17.75$. The root $\lambda = 7.0523$ is used to give an initial condition at $x = 0$.

Since the term $(U'/U)g_1(\lambda)$ is indeterminant at $x = 0$, L'Hopital's rule gives

$$\left[\frac{U'g_1(\lambda)}{U}\right]_{x=0} = \left[\frac{U'(dg_1/d\lambda)(d\lambda/dx)}{U'} + \frac{g_1 U''}{U'}\right]_{x=0} \qquad (6\text{-}38)$$

Using this expression in Eq. (6-37) and solving for $(d\lambda/dx)_{x=0}$ give

$$\left(\frac{d\lambda}{dx}\right)_{x=0} = \left[\frac{(U''/U')(g_1 + g_2)}{g_3 - dg_1/dx}\right]_{\lambda = 7.0523}$$

$$= 1.662 \frac{U''}{U'} \qquad (6\text{-}39)$$

This equation is used to calculate the initial gradient $d\lambda/dx$ at $x = 0$, and Eq. (6-37) is used to determine all subsequent gradients.

6-4   Solutions to the Integral Momentum Equation

The fourth-order Runge-Kutta scheme provides a numerical technique for integrating Eq. (6-37), subject to the initial conditions that $\lambda = 7.0523$ and $d\lambda/dx = 1.662U''/U'$ at $x = 0$. The free-stream velocity U and its first two derivatives must be specified as a function of x. When the velocity profile can be expressed as a polynomial, Eq. (6-37) can be solved to obtain λ as a function of x. The point along the surface at which $\lambda = -12$ indicates the point of boundary-layer separation. A change in body shape affects the free-stream velocity variation, and this result in turn affects the point of separation, as can be seen by obtaining solutions to Eq. (6-37) for various profiles $U = U(x)$.

In addition to the variation of λ with x, many other useful results are available from a solution to Eq. (6-37). The nondimensional free-stream velocity profile is given by $V = U/U_\infty = f(\xi)$, where $\xi = x/L$ and L is a characteristic body length. It follows that $dV/d\xi = (dU/dx)(dx/d\xi)(dV/dU) = (L/U_\infty)(dU/dx)$. Using the definition of λ, Eq. (6-19), a nondimensional boundary-layer thickness can be expressed as

$$\delta\sqrt{\frac{U_\infty}{\nu L}} = \sqrt{\frac{\lambda}{dV/d\xi}} \qquad (6\text{-}40)$$

Then from Eq. (6-21) we write

$$\delta^*\sqrt{\frac{U_\infty}{\nu L}} = \sqrt{\frac{\lambda}{dV/d\xi}}\left(\frac{3}{10} - \frac{\lambda}{120}\right) \qquad (6\text{-}41)$$

This nondimensional displacement thickness parameter gives a measure of the boundary-layer effect on the free-stream inviscid flow field. In a similar manner, from Eq. (6-22) we write

$$\Theta \sqrt{\frac{U_\infty}{\nu L}} = \sqrt{\frac{\lambda}{dV/d\xi}} \left(\frac{37}{315} - \frac{\lambda}{945} - \frac{\lambda^2}{9{,}072} \right) \qquad (6\text{-}42)$$

This nondimensional momentum thickness is a measure of the decrease in the momentum which the flow in the boundary layer undergoes because of the viscous effect of the fluid.

A final important result is the local wall shearing stress at the surface which causes the viscous drag. For a Newtonian fluid the wall shearing stress is given by $\tau_w = \mu (du/dy)_{y=0}$. Equations (6-24) and (6-40) lead to a nondimensional wall shear stress parameter given by

$$\left(\frac{du}{dy} \right)_{y=0} \sqrt{\frac{\nu L}{U_\infty^3}} = \frac{V}{\sqrt{\lambda/(dV/d\xi)}} \left(2 + \frac{\lambda}{6} \right) \qquad (6\text{-}43)$$

A computer program LBL and a typical solution are given below. (See Listing 6-1, Computer Results 6-1, and Flow Chart 6-1.) Table 6-1 identifies the computer variables. The results show lambda, displacement thickness parameter, and shear

Table 6-1   VARIABLES USED IN LBL

Variable	Definition
C1	Logic control variable
D	$[\lambda/(dU/d\xi)]^{1/2}$
D1	Nondimensional displacement thickness
F	Value of $d\lambda/d\xi$
F1	Initial value of $d\lambda/d\xi$
G1	$g_1(\lambda)$
G2	$g_2(\lambda)$
G3	$g_3(\lambda)$
H	Increment $\Delta \xi$
K1, K2, K3, K4	Runge-Kutta coefficients
M1	Nondimensional momentum thickness
S	Dependent variable λ
S1	Nondimensional wall shear stress parameter
T8, T9	Print control variables
V	Nondimensional free-stream velocity U/U_∞
V1	First derivative $dV/d\xi$
V2	Second derivative $d^2V/d\xi^2$
X	Independent variable ξ
X9	Maximum value of ξ

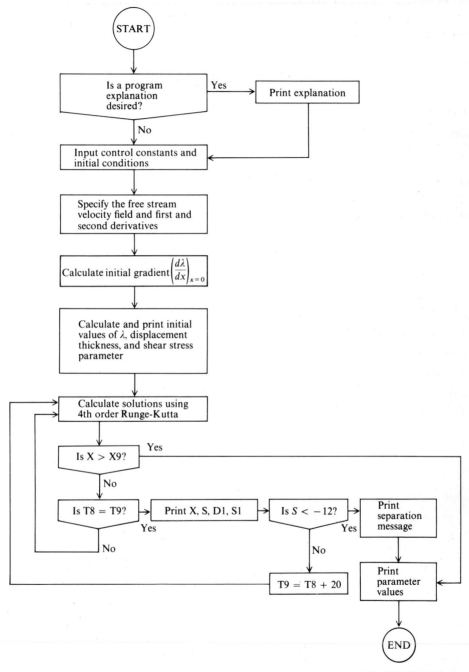

FLOW CHART 6-1

Listing 6-1   LBL

```
10    PRINT"TYPE I IF YOU WANT A PROGRAM EXPLANATION."
12    PRINT
13    PRINT"OTHERWISE, TYPE 2."
14    INPUTL
16    IF L=1 THEN 20
18    IF L=2 THEN 90
20    PRINT
22    PRINT"THIS PROGRAM USES A FOURTH ORDER RUNGE-KUTTA SCHEME"
24    PRINT"TO INTEGRATE THE FIRST ORDER, NON-LINEAR DIFFERENTIAL"
26    PRINT"EQUATION OBTAINED FROM THE MOMENTUM INTEGRAL EQUATION"
28    PRINT"FOR A LAMINAR BOUNDARY LAYER, GIVEN BY:"
30    PRINT
32    PRINT"   DS/DX = ((V'/V)G1 + (V''/V')G2)/G3"
34    PRINT
36    PRINT"WHERE G1, G2, AND G3 ARE FUNCTIONS OF S."
38    PRINT
90    PRINT"THE SOLUTION IS"
95    PRINT
100   REM SET VALUES OF NON-DIMENSIONAL PARAMETERS
150   PRINT
200   REM INITIALIZE CONTROL CONSTANTS
210   LETC1=0
240   LETT8=0
250   LETT9=20
260   LETX9=1
270   LETH=0.002
1000  REM SET PROGRAM VALUES AND INITIAL CONDITIONS
1005  LETS=7.0523
1010  LETX=0
1018  LETV=(X-X^3)
1020  LETV1=(1-3*X^2)
1025  LETV2=-6*X
1050  REM SPECIFY INITIAL CONDITION
1075  REM CALCULATE INITIAL GRADIENT DS/DX.
1080  LETF1=1.662*(V2/V1)
1090  LETD=SQR(S/V1)
1094  LETD1=SQR(S/V1)*(3/10-S/120)
1096  LETM1=SQR(S/V1)*(37/315-S/945-S^2/9072)
1098  LETS1=(V/SQR(S/V1))*(S/6+2)
1100  PRINT"X","S","D1","S1"
1110  PRINT
1120  PRINTX,S,D1,S1
1130  REM
1132  LETV=(X-X^3)
1134  LETV1=(1-3*X^2)
1136  LETV2=-6*X
1140  LETD=SQR(S/V1)
1145  LETD1=SQR(S/V1)*(3/10-S/120)
1150  LETM1=SQR(S/V1)*(37/315-S/945-S^2/9072)
1155  LETS1=(V/SQR(S/V1))*(S/6+2)
1156  REM FOURTH ORDER RUNGE-KUTTA INTEGRATION SCHEME
1158  REM
1160  REM CALCULATE G1, G2, AND G3
1175  LETG1=15120-2784*S+79*S^2+5/3*S^3
1180  LETG2=444*S-4*S^2-5/12*S^3
1185  LETG3=444-12*S-25/12*S^2
1186  IF X>0 THEN 1190
1187  LETF=F1
1188  GOTO 1195
```

Listing 6-1 **LBL** (*continued*)

```
1190 LETF=(G1*(V1/V)+G2*(V2/V1))/G3
1195 IF C1=1 THEN 1300
1200 IF C1=2 THEN 1380
1210 IF C1=3 THEN 1460
1220 LETK1=H*F
1225 LETX=X+0.5*H
1230 LETS=S+0.5*K1
1240 LETC1=1
1290 GOTO 1130
1300 LETK2=H*F
1350 LETS=S-0.5*K1+0.5*K2
1360 LETC1=2
1370 GOTO 1130
1380 LETK3=H*F
1385 LETX=X+0.5*H
1430 LETS=S-0.5*K2+K3
1440 LETC1=3
1450 GOTO 1130
1460 LETK4=H*F
1510 LETS=S-K3+(K1+2*K2+2*K3+K4)/6
1520 LETC1=0
1540 LETT8=T8+1
1618 REM
1620 REM PRINT SEQUENCE
1622 REM
1625 IF X>=X9 THEN 2000
1630 IF T8=T9 THEN 1670
1635 GOTO 1130
1670 PRINTX,S,D1,S1
1675 IF S<-12 THEN 1695
1680 LETT9=T8+20
1690 GOTO 1130
1695 PRINT
1696 PRINT"SEPARATION OF THE BOUNDARY LAYER HAS OCCURRED."
2000 PRINT
2005 PRINT"THE PARAMETERS USED IN THIS SOLUTION ARE"
2006 PRINT
2010 PRINT"DELTA X INCREMENT="H
2020 PRINT
2030 PRINT"D1 IS A NON-DIMENSIONAL DISPLACEMENT THICKNESS."
2040 PRINT
2050 PRINT"S1 IS A NON-DIMENSIONAL WALL SHEAR STRESS."
3000 END
```

Computer Results 6-1   LBL

```
TYPE 1 IF YOU WANT A PROGRAM EXPLANATION.

OTHERWISE, TYPE 2.
? 1

THIS PROGRAM USES A FOURTH ORDER RUNGE-KUTTA SCHEME
TO INTEGRATE THE FIRST ORDER, NON-LINEAR DIFFERENTIAL
EQUATION OBTAINED FROM THE MOMENTUM INTEGRAL EQUATION
FOR A LAMINAR BOUNDARY LAYER, GIVEN BY:

    DS/DX = ((V'/V)G1 + (V''/V')G2)/G3

WHERE G1, G2, AND G3 ARE FUNCTIONS OF S.

THE SOLUTION IS
```

X	S	DI	SI
0	7.0523	0.640617	0
0.04	7.03844	0.641836	4.76499 E-2
0.08	6.99623	0.645528	9.42271 E-2
0.12	6.92402	0.651811	0.138668
0.16	6.81884	0.660889	0.179925
0.2	6.6762	0.673073	0.216978
0.24	6.48972	0.688811	0.248839
0.28	6.25047	0.708731	0.274562
0.32	5.94607	0.733721	0.293251
0.36	5.55914	0.765046	0.304061
0.4	5.06491	0.804552	0.306202
0.44	4.42697	0.855022	0.298928
0.48	3.58975	0.920861	0.281515
0.52	2.46342	1.0095	0.253187
0.56	0.890419	1.13471	0.212932
0.6	-1.27223	1.23865	0.172164
0.64	-4.80808	1.55892	9.88006 E-2
0.68	-14.2353	2.53792	-2.24361 E-2

```
SEPARATION OF THE BOUNDARY LAYER HAS OCCURRED.

THE PARAMETERS USED IN THIS SOLUTION ARE

DELTA X INCREMENT= 0.002

DI IS A NON-DIMENSIONAL DISPLACEMENT THICKNESS.

SI IS A NON-DIMENSIONAL WALL SHEAR STRESS.
```

stress parameter as a function of $\xi = x/L$. For the example shown, the free-stream velocity is expressed by

$$\frac{U}{U_\infty} = \xi - \xi^3 \qquad (6\text{-}44)$$

Notice that the value of lambda S continually decreases from its initial value. Separation occurs at $\xi \approx 0.676$. The shear stress parameter S1 increases from zero at the stagnation point to a maximum in the vicinity of $\xi = 0.4$ and then decreases to zero at the point of boundary-layer separation. The displacement thickness D1 continually increases from the stagnation point to the separation point.

6-5   A Modified Integral Analysis[1]

One disadvantage in using Eq. (6-37) is that the second derivative U'' appears explicitly in the differential equation. The value of this second derivative is difficult to specify if the velocity field around the body under investigation is obtained from experimental measurements.

The following modifications suggested by Holstein and Bohlen (Ref. 6-4) avoids this difficulty. We define $Z = \Theta^2/v$ and write $\kappa = (dU/dx)Z$. If we use the equality $(\Theta/v)(d\Theta/dx) = \frac{1}{2}(dZ/dx)$ and the definitions of $f_1(\kappa)$ and $f_2(\kappa)$, Eq. (6-25) may be written

$$f_2(\kappa) = [2 + f_1(\kappa)]\kappa + U\frac{1}{2}\frac{dZ}{dx} \qquad (6\text{-}45)$$

Solving for the derivative gives

$$\frac{dZ}{dx} = \frac{1}{U}[2f_2(\kappa) - 4\kappa - 2\kappa f_1(\kappa)] = \frac{F(\kappa)}{U} \qquad (6\text{-}46)$$

The full expression for $F(\kappa)$ in terms of λ is

$$F(\kappa) = 2\left(\frac{37}{315} - \frac{\lambda}{945} - \frac{\lambda^2}{9{,}072}\right)\left[2 - \frac{116\lambda}{315} + \left(\frac{2}{945} + \frac{1}{120}\right)\lambda^2 + \frac{2\lambda^3}{9{,}072}\right] \qquad (6\text{-}47)$$

Since $U = 0$ at $x = 0$ for a blunt-nose body, Eq. (6-46) requires that $F(\kappa) = 0$ at $x = 0$ for $(dZ/dx)_{x=0}$ to be finite. The three roots of Eq. (6-47) are the same as for Eq. (6-34), and we choose $\lambda = 7.0523$ at $x = 0$.

Again there is a singular point in the governing differential equation at $x = 0$. Using L'Hopital's rule we evaluate

$$\lim_{x\to 0}\frac{dZ}{dx} = \lim_{x\to 0}\frac{F(\kappa)}{U} = \lim_{x\to 0}\frac{F'(\kappa)}{dU/dx} \qquad (6\text{-}48)$$

where

$$F'(\kappa) = dF(\kappa)/dx = \frac{dF(\kappa)}{d\kappa}\frac{d\kappa}{dx}$$

Since $\kappa = Z(dU/dx)$, we can write

$$\frac{d\kappa}{dx} = \frac{dU}{dx}\frac{dZ}{dx} + Z\frac{d^2U}{dx^2} \qquad (6\text{-}49)$$

---

[1] This section may be omitted without loss of continuity of material presentation.

Substituting these relations into Eq. (6-48) gives

$$\left(\frac{dZ}{dx}\right)_{x=0} = \left\{\frac{[(dU/dx)(dZ/dx) + Z(d^2U/dx^2)][dF(\kappa)/d\kappa]}{dU/dx}\right\}_{x=0} \tag{6-50}$$

Initially at $x = 0$, $(Z)_{x=0} = \kappa_0/(dU/dx)_{x=0}$. Evaluating κ_0 at $\lambda_0 = 7.0523$ by use of Eq. (6-27) gives $(Z)_{x=0} = 0.0770/(dU/dx)_{x=0}$. The value of the derivative $[dF(\kappa)/d\kappa]_{x=0}$ appearing in Eq. (6-50) can be obtained numerically and is equal to -5.53. (See Ref. 6-23.) Solving for $(dZ/dx)_{x=0}$ from Eq. (6-50) then gives

$$\left(\frac{dZ}{dx}\right)_{x=0} = \frac{0.0770}{6.53}\left[\frac{d^2U/dx^2}{(dU/dx)^2}\right]_{x=0}(-5.53)$$

$$\left(\frac{dZ}{dx}\right)_{x=0} = -0.0652\left[\frac{d^2U/dx^2}{(dU/dx)^2}\right]_{x=0} \tag{6-51}$$

This equation gives the initial value of dZ/dx to be used in the numerical integration scheme at $x = 0$. It should be pointed out that differentiating Eq. (6-47) to obtain $dF(\kappa)/d\lambda$ and evaluating the expression at $\lambda_0 = 7.0523$ also give -0.0652.

Two special cases of interest lead to simple analytical solutions. For a flat plate no stagnation point exists and $U = $ constant. It follows that $dU/dx = 0$ and thus $\lambda = 0$ and $\kappa = 0$. Equation (6-46) then becomes a linear equation given by

$$\frac{dZ}{dx} = \frac{F(0)}{U} = \frac{0.4698}{U} \tag{6-52}$$

Integration gives $Z = 0.4698x/U$. By definition, $Z = \Theta^2/\nu$ from which $\Theta/x = 0.686/\text{Re}_x^{1/2}$, where the local Reynolds number $\text{Re}_x = Ux/\nu$. From Eq. (6-22)

$$\frac{\delta}{x} = \frac{5.84}{\text{Re}_x^{1/2}} \tag{6-53}$$

and Eq. (6-21) gives

$$\frac{\delta^*}{x} = \frac{1.75}{\text{Re}_x^{1/2}} \tag{6-54}$$

We can compare the above approximate integral results with the numerical solutions obtained by the method discussed in Sec. 5-3. These results are $\delta^*/x = 1.720784/\text{Re}_x^{1/2}$ and $\Theta/x = 0.664112/\text{Re}_x^{1/2}$. See Eqs. (5-31) and (5-32), and Table 5-2.

The second case of special interest is two-dimensional stagnation flow where $U = U_\infty x$. Thus, $dU/dx = U_\infty = $ constant and $d^2U/dx^2 = 0$. This inviscid flow field will produce similar boundary-layer profiles with $\beta = 1.0$. (See Chap. 5.) From boundary-layer theory (Ref. 6-3) it is shown that Θ and δ^* are independent of x for stagnation flow. It follows that $\lambda = $ constant since dU/dx is a constant, and $\delta = $ constant, i.e., independent of x. Thus, κ and $F(\kappa)$ are constants with $\kappa = \kappa_0 = 0.0770$, and $\Theta = (Zv)^{1/2} = (\kappa_0 v / U_\infty)^{1/2}$, or $\Theta(U_\infty/v)^{1/2} = 0.278$. The displacement thickness is

$$\delta^* \left(\frac{U}{v}\right)^{1/2} = \kappa_0^{1/2} f_1(\kappa_0) = 0.641$$

From the definition of $f_2(\kappa)$ one obtains

$$\frac{\tau_w}{\mu U} \left(\frac{v}{U_\infty}\right)^{1/2} = \frac{f_2(\kappa_0)}{\kappa_0^{1/2}} = 1.19$$

for the shearing stress at the wall. Note that τ_w is also independent of x for two-dimensional stagnation flow. The corresponding solution obtained from a differential analysis and the program FSKAN given in Chap. 5 gives $\Theta(U_\infty/v)^{1/2} = 0.2923$, $\delta^*(U/v)^{1/2} = 0.6479$, and $(\tau_w/\mu U)(v/U_\infty)^{1/2} = 1.233$.

The numerical method of solution is somewhat different for Eq. (6-46). The initial value of λ gives the initial value of $F(\kappa)$. Then the integration of Eq. (6-46) gives Z. A new κ for $x + \Delta x$ is then calculated as $\kappa = Z(dU/dx)$. However, before Eq. (6-46) can be integrated again, a new value of $F(\kappa)$ at $x + \Delta x$ is needed. Thus, after each κ is found, the corresponding value of λ must be determined from Eq. (6-27). The value of λ is then used to calculate a new $F(\kappa)$ using Eq. (6-47), and the process is repeated.

The advantage of this modified approach is that only the value of U'' at $x = 0$ as required in Eqs. (6-51) is needed for the calculations. The second derivative does not appear explicitly in the governing equation. Notice the governing equation (6-46) can also be written in terms of $V = U/U_\infty$ and $\xi = x/L$, given by

$$\frac{dZ}{d\xi} = \frac{F(\kappa)}{V} \qquad (6\text{-}55)$$

where

$$\kappa = Z \frac{dV}{d\xi} \qquad (6\text{-}56)$$

The inviscid velocity profile is again given as $V = U/U_\infty = f(\xi)$. A computer program using the modified approach discussed above, called LBLMOD, is listed in Listing 6-2 below. A table of X(I) versus Z(I) is initially read into the computer which gives κ as a function of λ. (See Statements 40 to 85.) Every time a new value of κ is calculated in Statement 1138, the corresponding value of λ is obtained from the stored table by linear interpolation in Statements 1142 to 1150. Equation (6-46) is contained in Statements 1175 to 1190. The remainder of the program is similar to the LBL program.

Listing 6-2   LBLMOD

```
5     DIM X(50),Z(50)
10    PRINT"TYPE 1 IF YOU WANT A PROGRAM EXPLANATION."
12    PRINT
13    PRINT"OTHERWISE, TYPE 2."
14    INPUTL
16    IF L=1 THEN 20
18    IF L=2 THEN 39
20    PRINT
22    PRINT"THIS PROGRAM USES A FOURTH ORDER RUNGE-KUTTA SCHEME"
24    PRINT"TO INTEGRATE THE SET OF GOVERNING EQUATIONS   FOR AN"
26    PRINT"INTEGRAL BOUNDARY LAYER ANALYSIS   FOR LAMINAR,"
27    PRINT"INCOMPRESSIBLE FLOW.  THE SET OF EQUATIONS IS GIVEN BY:"
28    PRINT
30    PRINT"   DZ/DX = F(K)/V   ;   K = Z(DV/DX)"
32    PRINT
34    PRINT"WHERE Z, K, AND F(K) ARE RELATED TO THE MOMENTUM AND"
36    PRINT"DISPLACEMENT THICKNESS.  V IS THE FREE STREAM VELOCITY RATIO"
37    PRINT"EXPRESSED AS A FUNCTION OF X, THE DISTANCE ALONG THE"
38    PRINT"SURFACE MEASURED FROM THE STAGNATION POINT."
39    REM CALCULATE AND STORE VALUES OF KAPPA AND LAMBDA.
40    FOR I=1 TO 50
50    LET S=-12.5+(I-1)/2
60    LET K=((37/315-S/945-(1/9072)*S^2)^2)*S
70    LET X(I)=K
80    LET Z(I)=S
85    NEXT I
86    PRINT
90    PRINT"THE SOLUTION IS"
95    PRINT
100   REM SET VALUES OF NON-DIMENSIONAL PARAMETERS
120   LET K9=0.18E-3
130   LET S=7.0523
140   LET K=0.0770
150   PRINT
160   LET F1=(3/10-S/120)/((37/315)-(S/945)-(1/9072)*S^2)
170   LET F2=(2+S/6)*((37/315)-(S/945)-(1/9072)*S^2)
200   REM INITIALIZE CONTROL CONSTANTS
210   LET C1=0
240   LET T8=0
250   LET T9=20
260   LET X9=3
270   LET H=0.002
1000  REM SET PROGRAM VALUES AND INITIAL CONDITIONS
1010  LET X=0
1018  LET V=(X-X^3)
1020  LET V1=(1-3*X^2)
1025  LET V2=-6*X
1050  REM SPECIFY INITIAL CONDITION
1056  IF V1=0 THEN 2050
1060  LET Z0=K/V1
1070  LET Z=Z0
1075  REM CALCULATE INITIAL GRADIENT DZ/DX.
1080  LET F=-0.10836*V2/(V1^2)
1090  LET M1=SQR(K/V1)
1094  LET D1=F1*SQR(K/V1)
1096  LET S1=(F2*(3/10-S/120)*V*SQR(S*V1))/(K*F1)
1100  PRINT"X","S","D1",S1"
1110  PRINT
1120  PRINTX,S,D1,S1
1130  IF X=0 THEN 1195
1132  LET V=(X-X^3)
1134  LET V1=(1-3*X^2)
1136  LET V2=-6*X
1138  LET K=Z*V1
```

Listing 6-2 **LBLMOD** (*continued*)

```
1140 LET I=1
1142 IF X(I)>K THEN 1150
1144 IF I>=50 THEN 1150
1146 LET I=I+1
1148 GOTO 1142
1150 LET S=Z(I-1)+(K-X(I-1))*(Z(I)-Z(I-1))/(X(I)-X(I-1))
1152 REM A CORRECTED VALUE OF LAMBDA NOW EXISTS.
1154 REM
1156 REM FOURTH ORDER RUNGE-KUTTA INTEGRATION SCHEME
1158 REM
1160 REM CALCULATE F1(K), F2(K), AND F(K).
1175 LET F1=(3/10-S/120)/((37/315)-(S/945)-(1/9072)*S^2)
1180 LET F2=(2+S/6)*((37/315)-(S/945)-(1/9072)*S^2)
1185 LET K5=2*F2-4*K-2*K*F1
1190 LET F=K5/V
1191 LET M1=SQR(K/V1)
1192 LET D1=F1*SQR(K/V1)
1193 LET S1=(F2*(3/10-S/120)*V*SQR(S*V1))/(K*F1)
1195 IF C1=1 THEN 1300
1200 IF C1=2 THEN 1380
1210 IF C1=3 THEN 1460
1220 LET K1=H*F
1225 LET X=X+0.5*H
1230 LET Z=Z+0.5*K1
1240 LET C1=1
1290 GOTO 1130
1300 LET K2=H*F
1350 LET Z=Z-0.5*K1+0.5*K2
1360 LET C1=2
1370 GOTO 1130
1380 LET K3=H*F
1385 LET X=X+0.5*H
1430 LET Z=Z-0.5*K2+K3
1440 LET C1=3
1450 GOTO 1130
1460 LET K4=H*F
1510 LET Z=Z-K3+(K1+2*K2+2*K3*K4)/6
1520 LET C1=0
1540 LET T8=T8+1
1560 IF K<=X(1) THEN 1580
1565 IF K<=X(50) THEN 1620
1570 PRINT"K IS TOO LARGE, K="K,"LAMBDA="S
1575 GOTO 3000
1580 PRINT"K IS TOO SMALL, LAMBDA="S
1585 GOTO 1695
1618 REM
1620 REM PRINT SEQUENCE
1622 REM
1625 IF X>=X9 THEN 2000
1630 IF T8=T9 THEN 1670
1635 GOTO 1130
1670 PRINTX,S,D1,S1
1675 IF S<-12 THEN 1695
1680 LET T9=T8+20
1690 GOTO 1130
1695 PRINT
1696 PRINT"SEPARATION OF THE BOUNDARY LAYER HAS OCCURRED."
2000 PRINT
2005 PRINT"THE PARAMETERS USED IN THIS SOLUTION ARE"
2006 PRINT
2010 PRINT"DELTA X INCREMENT="H
2020 GOTO 3000
2050 PRINT
2060 PRINT"SINCE DV/DX=0, SEE TEXT FOR DISCUSSION."
3000 END
```

One can obtain the values of momentum thickness, displacement thickness, and wall shearing stress from a numerical solution. From Eq. (6-26)

$$\Theta = \left(\frac{\kappa \nu}{U'}\right)^{1/2}$$

or

$$\Theta\left(\frac{U_\infty}{\nu L}\right)^{1/2} = \left(\frac{\kappa}{dV/d\xi}\right)^{1/2} \qquad (6\text{-}57)$$

Then from Eq. (6-28) $\delta^* = \Theta f_1(\kappa)$, or

$$\delta^*\left(\frac{U_\infty}{\nu L}\right)^{1/2} = f_1(\kappa)\left(\frac{\kappa}{dV/d\xi}\right)^{1/2} \qquad (6\text{-}58)$$

From the definitions of κ and λ it follows that $(\kappa/\lambda)^{1/2} = \Theta/\delta$, and the nondimensional shearing stress is

$$\frac{\tau_w \delta}{\mu U} = \frac{f_2(\kappa)}{(\kappa/\lambda)^{1/2}}$$

By using Eq. (6-21) and the above equations we can write

$$\left(\frac{du}{dy}\right)_{y=0}\left(\frac{\nu L}{U_\infty{}^3}\right)^{1/2} = \frac{f_2(\kappa)\left(\dfrac{3}{10} - \dfrac{\lambda}{120}\right) V[\lambda(dV/d\xi)]^{1/2}}{\kappa f_1(\kappa)} \qquad (6\text{-}59)$$

A comparison of results obtained from the programs LBL and LBLMOD for the inviscid velocity field given by $U/U_\infty = \xi - \xi^3$ is shown in Fig. 6-2. The comparison is good except for small differences near the separation point. As mentioned by Schlichting (Ref. 6-3), experience has shown that the approximate integral analysis gives good results in regions of accelerated potential flow which provide a favorable pressure gradient. For adverse pressure gradients the approximate solutions are less accurate as the separation point is approached.

6-6   Integral Energy Equation

To formulate the integral energy equation for two-dimensional forced convection, a control volume is defined with height δ_t, width dx, and unit depth as shown in Fig. 6-3. Steady, incompressible boundary-layer flow is assumed. In addition, it is assumed that changes in kinetic and potential energy terms are negligible, and no thermal energy is generated due to viscous dissipation caused by fluid friction. A review of Eqs. (5-38) and (5-39) shows that the kinetic energy terms canceled out in the differential analysis. The conservation of energy principle for this control volume is then

$$q + \dot{m}_{in} h_{in} = \dot{m}_{out} h_{out} \qquad (6\text{-}60)$$

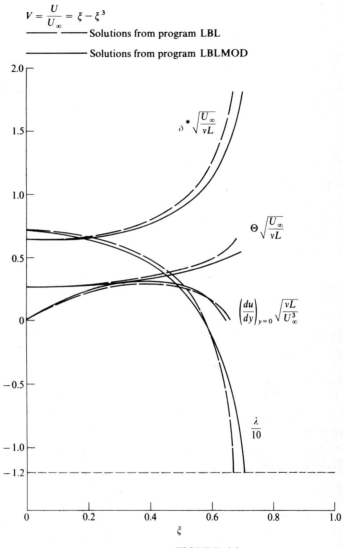

FIGURE 6-2
Comparison of numerical solutions.

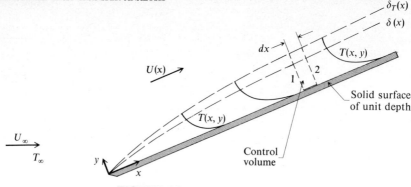

FIGURE 6-3
Thermal boundary layer-control volume for an integral analysis.

where q is the rate of heat transfer crossing the control volume surface, $\dot{m}$ is the rate of mass flow crossing the boundary, and h is the static enthalpy of the fluid at the indicated control volume boundary.

At the solid surface the heat transfer crossing the control volume boundary can be expressed by Fourier's law of conduction,

$$q = -k \, dx(1) \left(\frac{\partial T}{\partial y} \right)_{y=0} \qquad (6-61)$$

The specific heat at constant pressure is defined for the fluid in the boundary layer as (see Ref. 1-1)

$$c_p = \left(\frac{\partial h}{\partial T} \right)_p$$

For incompressible flow we can neglect the effect of pressure and write $dh = c_p \, dT$. The enthalpy crossing the differential area $dy(1)$ at plane 1 is $h_{\text{in}} = c_p T$. The enthalpy along the top surface of the control volume is constant and equal to $c_p T_\infty$. The enthalpy crossing a differential area dy at plane 2 in terms of plane 1 is expressed by means of a first-order Taylor series expansion about h_{in}:

$$h_{\text{out}} = c_p T + \frac{\partial}{\partial x} (c_p T) \, dx$$

If we use the expressions for mass flow in Sec. 6-2 and integrate over the control volume areas, the energy equation (6-60) becomes

$$-k \frac{\partial T}{\partial y} \, dx + \rho c_p \int_0^{\delta_t} uT \, dy + \rho c_p T_\infty \frac{\partial}{\partial x} \left(\int_0^{\delta_t} u \, dy \right) dx$$

$$= \rho c_p \int_0^{\delta_t} uT \, dy + \rho c_p \frac{\partial}{\partial x} \left(\int_0^{\delta_t} uT \, dy \right) dx \qquad (6-62)$$

for a constant-density (incompressible) fluid with constant specific heat. This expression reduces to

$$\frac{\partial}{\partial x} \int_0^{\delta_t} (T_\infty - T) u \ dy = \alpha \left(\frac{\partial T}{\partial y} \right)_{y=0} \qquad (6\text{-}63)$$

where $\alpha = k/\rho c_p$ is the thermal diffusivity of the fluid in the boundary layer. Equation (6-63) is the integral energy equation for boundary-layer flow. Before the indicated integration can be carried out, both $u = u(y)$ and $T = T(y)$ must be specified within the thermal boundary layer.

A fourth-order polynomial can be used to represent the temperature profile in a laminar, thermal boundary layer. Defining $\gamma = y/\delta_t$, and $\theta = (T - T_w)/(T_\infty - T_w)$ we write

$$\theta = C_1\gamma + C_2\gamma^2 + C_3\gamma^3 + C_4\gamma^4 \qquad (6\text{-}64)$$

The four boundary conditions needed to evaluate the four constants are

$$\gamma = 1 \qquad \theta = 1$$

$$\gamma = 1 \qquad \frac{\partial \theta}{\partial \gamma} = 0$$

$$\gamma = 1 \qquad \frac{\partial^2 \theta}{\partial \gamma^2} = 0 \qquad (6\text{-}65)$$

$$\gamma = 0 \qquad \frac{\partial^2 \theta}{\partial \gamma^2} = 0$$

The first condition specifies that $T = T_\infty$ at $y = \delta_t$. The second specifies no heat transfer at the edge of the thermal boundary layer. This is analogous to specifying zero shearing stress at the edge of a velocity boundary layer. The third condition results from the asymptotic convergence of the temperature profile at $y = \delta_t$. The final condition comes from the governing differential energy equation (5-41). For no viscous dissipation this becomes

$$u \frac{\partial T}{\partial x} + v \frac{\partial T}{\partial y} = \alpha \frac{\partial^2 T}{\partial y^2} \qquad (6\text{-}66)$$

At $y = 0$, $u = v = 0$; and it follows that

$$\frac{\partial^2 T}{dy^2} = 0 = \frac{\partial^2 \theta}{\partial \gamma^2}$$

Use of the thermal boundary conditions in Eq. (6-62) in the temperature profile results in the following expression:

$$\theta = 2\gamma - 2\gamma^3 + \gamma^4 \qquad (6\text{-}67)$$

By comparing Eq. (6-67) with Eq. (6-20) it can be seen that the velocity and tempera-ture profiles are identical for a flat plate ($\lambda = 0$) if $\delta = \delta_t$. This special condition is discussed later.

The integral equation (6-63) can be written in terms of θ and γ as follows:

$$\frac{\partial}{\partial x}\left[U \int_0^\infty (1 - \theta)\frac{u}{U}\,\delta_t\,dy\right] = 2\frac{\alpha}{\delta_t} \qquad (6\text{-}68)$$

This leads to the definition of an energy thickness

$$\Gamma = \delta_t \int_0^\infty \frac{u}{U}(1 - \theta)\,d\gamma \qquad (6\text{-}69)$$

and Eq. (6-68) is written

$$\frac{d}{dx}(U\Gamma) = U\frac{d\Gamma}{dx} + \Gamma\frac{dU}{dx} = 2\frac{\alpha}{\delta_t} \qquad (6\text{-}70)$$

First, consider the ratio $\zeta = \delta_t/\delta < 1.0$. Equation (6-69) is then written

$$\Gamma = \delta_t \int_0^{\delta_t/\delta} \frac{u}{U}(1 - \theta)\,d\gamma + \delta_t \int_{\delta_t/\delta}^\infty \frac{u}{U}(1 - \theta)\,d\gamma$$

but when $y > \delta_t$, $(\gamma > 1.0)$, $\theta = 1.0$, and the second integral vanishes. We only need to evaluate

$$\Gamma = \delta_t \int_0^1 \frac{u}{U}(1 - \theta)\,d\gamma \qquad (6\text{-}71)$$

Substituting Eqs. (6-67) and (6-20) into Eq. (6-69) and carrying out the integration give

$$\Gamma = \delta_t\left[\frac{2}{15}\zeta - \frac{3}{140}\zeta^3 + \frac{1}{180}\zeta^4 \right.$$
$$\left. + \frac{\lambda}{6}\left(\frac{1}{15}\zeta - \frac{1}{14}\zeta^2 + \frac{9}{280}\zeta^3 - \frac{1}{180}\zeta^4\right)\right] \qquad \zeta < 1.0 \qquad (6\text{-}72)$$

or
$$\Gamma = \delta_t\left[H_1(\zeta) + \frac{\lambda}{6}I_1(\zeta)\right] \qquad \zeta < 1.0 \qquad (6\text{-}73)$$

For $\zeta > 1.0$ we express Eq. (6-69) as

$$\Gamma = \delta_t \int_0^{\delta/\delta_t} \frac{u}{U}(1 - \theta)\,d\gamma + \delta_t \int_{\delta/\delta_t}^1 \frac{u}{U}(1 - \theta)\,d\gamma \qquad (6\text{-}74)$$

When $y > \delta$, $u/U = 1.0$, and Eq. (6-74) is written

$$\Gamma = \delta_t \int_0^{1/\zeta} \frac{u}{U}(1 - \theta)\,d\gamma + \delta_t \int_{1/\zeta}^1 (1 - \theta)\,d\gamma \qquad (6\text{-}75)$$

Integration gives

$$\Gamma = \delta_t \left[\frac{3}{10} - \frac{3}{10}\frac{1}{\zeta} + \frac{2}{15}\frac{1}{\zeta^2} - \frac{3}{140}\frac{1}{\zeta^4} + \frac{1}{180}\frac{1}{\zeta^5} \right.$$

$$\left. + \frac{\lambda}{6}\left(\frac{3}{10} - \frac{19}{20}\frac{1}{\zeta} + \frac{29}{30}\frac{1}{\zeta^2} - \frac{69}{140}\frac{1}{\zeta^4} + \frac{499}{2,520}\frac{1}{\zeta^5} \right) \right] \qquad \zeta > 1.0 \qquad (6\text{-}76)$$

or $\qquad \Gamma = \delta_t \left[H_2(\zeta) + \frac{\lambda}{6} I_2(\zeta) \right] \qquad \zeta > 1.0 \qquad (6\text{-}77)$

We now nondimensionalize Eq. (6-70) by multiplying each side by Γ/ν to obtain

$$U\frac{\Gamma}{\nu}\frac{d\Gamma}{dx} + \frac{(dU/dx)\Gamma^2}{\nu} = 2\frac{\alpha}{\nu}\frac{\Gamma}{\delta\zeta} \qquad (6\text{-}78)$$

The ratio of kinematic viscosity to thermal diffusivity ν/α is the Prandtl number of the fluid in the boundary layer.

The parameter N (nu) is defined as

$$N = \frac{(dU/dx)\Gamma^2}{\nu} \qquad (6\text{-}79)$$

Again, N is connected with the energy thickness the same way that λ is connected with the velocity boundary-layer thickness. Since $N/\lambda = \Gamma^2/\delta^2$, Eq. (6-73) gives

$$\frac{\Gamma^2}{\delta^2} = \frac{N}{\lambda} = \left\{ \zeta \left[H_1(\zeta) + \frac{\lambda}{6} I_1(\zeta) \right] \right\}^2 \qquad \zeta < 1.0 \qquad (6\text{-}80)$$

and Eq. (6-77) gives

$$\frac{\Gamma^2}{\delta^2} = \frac{N}{\lambda} = \left\{ \zeta \left[H_2(\zeta) + \frac{\lambda}{6} I_2(\zeta) \right] \right\}^2 \qquad \zeta > 1.0 \qquad (6\text{-}81)$$

6-7 Solutions to the Integral Energy Equation

For a flat plate, $\lambda = 0$ since $dU/dx = 0$. Thus, Eq. (6-78) becomes

$$\frac{d\Gamma}{dx} = \frac{d}{dx}[\delta_t H(\zeta)] = \frac{2}{\delta_t}\frac{\alpha}{U} \qquad (6\text{-}82)$$

where $H(\zeta) = H_1(\zeta) + (\lambda/6) I_1(\zeta)$ for $\zeta < 1.0$, and $H(\zeta) = H_2(\zeta) + (\lambda/6) I_2(\zeta)$ for $\zeta > 1.0$. For $\zeta < 1.0$ this relationship becomes

$$U\frac{d}{dx}\left[\delta\zeta\left(\frac{2}{15}\zeta - \frac{3}{140}\zeta^3 + \frac{1}{180}\zeta^4 \right) \right] = \frac{2\alpha}{\delta\zeta} \qquad (6\text{-}83)$$

Since $\zeta < 1.0$ we can usually neglect the terms involving ζ^4 and ζ^5, and write

$$\frac{2}{15} U \frac{d}{dx} (\delta\zeta^2) = \frac{2\alpha}{\delta\zeta} \qquad (6\text{-}84)$$

Differentiating gives

$$\frac{U}{15} \left(2\delta\zeta \frac{d\zeta}{dx} + \zeta^2 \frac{d\delta}{dx} \right) = \frac{\alpha}{\delta\zeta}$$

and rearranging leads to

$$\frac{U}{15} \left(2\delta^2\zeta^2 \frac{d\zeta}{dx} + \zeta^3\delta \frac{d\delta}{dx} \right) = \alpha \qquad (6\text{-}85)$$

From Eq. (6-53), $\delta(d\delta/dx) = [(5.84)^2/2]v/U$, and $\delta^2 = (5.84)^2 vx/U$.
    Thus, Eq. (6-85) becomes

$$\zeta^3 + 4x\zeta^2 \frac{d\zeta}{dx} = \frac{15}{17} \frac{\alpha}{v} \qquad (6\text{-}86)$$

This nonlinear equation can be written as a linear equation by letting $y = \zeta^3$:

$$y + \frac{4}{3} x \frac{dy}{dx} = \frac{15}{17} \frac{\alpha}{v} \qquad (6\text{-}87)$$

The solution is

$$y = \zeta^3 = Cx^{-3/4} + \frac{15}{17} \frac{1}{\mathrm{Pr}} \qquad (6\text{-}88)$$

The initial condition is $\delta_t = \zeta = 0$ at $x = x_0$, the start of the point of heating or cooling of the surface relative to the free stream. This condition gives

$$\zeta = \frac{\delta_t}{\delta} = \frac{\mathrm{Pr}^{-1/3}}{1.0426} \left[1 - \left(\frac{x_0}{x} \right)^{3/4} \right]^{1/3} \qquad (6\text{-}89)$$

When the entire surface is heated or cooled, $x_0 = 0$, and this approximate analytical solution predicts that ζ is independent of x. A feeling for the accuracy of this solution is obtained when it is realized that the similarity solution for $x_0 = 0$ in Chap. 5 shows that $\delta = \delta_t$ when $\mathrm{Pr} = 1.0$, $E = 0$, and $\theta = 0$. Equation (6-89) gives $\delta = 1.0426 \, \delta_t$ for $\mathrm{Pr} = 1.0$.

The Prandtl number is inversely proportional to $\zeta = \delta_t/\delta$. When $\zeta < 1.0$, $\mathrm{Pr} > 1.0$, and the velocity boundary-layer thickness is greater than the thermal boundary-layer thickness, and vice versa. Recall that the Prandtl number is a ratio of the diffusion of a momentum disturbance into the free stream to the diffusion of a thermal disturbance. In fact, the kinematic viscosity v is sometimes called the momentum diffusivity.

Expressing the heat transfer at the surface by Fourier's law gives

$$q = -k\left(\frac{\partial T}{\partial y}\right)_{y=0} = -k\frac{\partial \theta}{\partial \gamma}\frac{\partial \gamma}{\partial y}\left(\frac{\partial T}{\partial \theta}\right)_{y=0}$$

$$q = -\frac{2k}{\delta_t}(T_\infty - T_w)$$

We can also express q in terms of a local convection heat transfer coefficient h_x by using Newton's law of cooling:

$$q = h_x(T_w - T_\infty) = \frac{2k}{\delta\zeta}(T_w - T_\infty)$$

Using Eqs. (6-89) and (6-53) we obtain

$$h_x = 0.357\frac{k}{x}\,\mathrm{Re}_x^{1/2}\mathrm{Pr}^{1/3}\left[1 - \left(\frac{x_0}{x}\right)^{3/4}\right]^{-1/3} \tag{6-90}$$

A local Nusselt number $\mathrm{Nu}_x = h_x x/k$ is then given by

$$\mathrm{Nu}_x = 0.357\mathrm{Re}_x^{1/2}\mathrm{Pr}^{1/3}\left[1 - \left(\frac{x_0}{x}\right)^{3/4}\right]^{-1/3} \tag{6-91}$$

Three important nondimensional parameters for incompressible forced convection appear in this equation. Viscous dissipation has been neglected, and thus the Eckert number does not appear. The pressure gradient parameter does not appear since Eq. (6-91) is limited to a flat plate. The fluid properties appearing in Re, Pr, and Nu are evaluated at the mean film temperature, given by

$$T_f = \frac{T_w + T_\infty}{2} \tag{6-92}$$

The constant in Eq. (6-91) is larger than the value 0.332 obtained by the numerical solution in Chap. 5. [See Eq. (5-53).] An improved value of Nu_x can be obtained from the integral method by obtaining solutions to the full nonlinear equation given by Eq. (6-83) for $\zeta < 1.0$. Carrying out the differentiation gives

$$U\left[\frac{d\zeta}{dx}\delta^2\left(\frac{2}{15}\zeta^2 - \frac{3}{70}\zeta^4 + \frac{1}{72}\zeta^5\right)\right.$$
$$\left. + \delta\frac{d\delta}{dx}\left(\frac{1}{15}\zeta^3 - \frac{3}{280}\zeta^5 + \frac{1}{360}\zeta^6\right)\right] = \alpha \tag{6-93}$$

Using the known expressions for δ^2 and $\delta(d\delta/dx)$ gives

$$x\frac{d\zeta}{dx}\left(\frac{4}{15}\zeta^2 - \frac{6}{70}\zeta^4 + \frac{1}{36}\zeta^5\right) + \frac{1}{15}\zeta^3 - \frac{3}{280}\zeta^5 + \frac{1}{360}\zeta^6 = \frac{1}{17.05}\frac{1}{\mathrm{Pr}}$$

Then

$$\frac{d\zeta}{dx} = \frac{1}{17.05 \, \text{Pr} \, \text{G1}x} - \frac{\text{G2}}{\text{G1}x} \qquad (6\text{-}94)$$

where

$$\text{G1} = \frac{4}{15} \zeta^2 - \frac{6}{70} \zeta^4 + \frac{1}{36} \zeta^5 \qquad (6\text{-}95)$$

and

$$\text{G2} = \frac{1}{15} \zeta^3 - \frac{3}{280} \zeta^5 + \frac{1}{360} \zeta^6 \qquad (6\text{-}96)$$

A similar equation with different functions G1 and G2 applies when $\zeta > 1.0$.

Equation (6-94), along with initial values of ζ and $d\zeta/dx$ at $x = 0$, constitutes a first-order, nonlinear initial-value problem. A numerical solution can be obtained by making a few changes to the program RKINT listed in Chap. 1. The proper initial conditions can be obtained from the analytical solution, Eq. (6-89). For $x_0 = 0$, these are $\zeta = 1/1.0426\text{Pr}^{1/3}$ and $d\zeta/dx = 0$ at $x = 0$. Program ZETA (Table 6-2 and Listing 6-3) can be used to obtain solutions to Eq. (6-94). Results for various Prandtl numbers are given in Table 6-3. It can be seen that the ratio $\text{Nu}/\text{Re}^{1/2}\text{Pr}^{1/3}$ is relatively constant over a large Prandtl number range. The results also agree more closely with the solution for a flat plate obtained in Chap. 5 than the results given by Eq. (6-91).

For fluids with high Prandtl numbers such as oils, the variation of the dynamic viscosity μ with temperature must be included in the mathematical model to obtain results which agree with experimental measurements. Many empirical results exist in the literature which give relationships for the Nusselt number and attempt to account for this property variation. Some of these are presented in the next chapter.

Liquid metals such as sodium, bismuth, and mercury have Prandtl numbers of the order 0.01. Since $\delta_t >> \delta$ for very low Prandtl numbers, it follows that the free-stream velocity exists throughout most of the thermal boundary layer. An

Table 6-2 VARIABLES USED IN PROGRAM ZETA

Variable	Definition
C1	Logical control variable
F	Derivative $d\zeta/dx$
G1	Right side of Eq. (6-95)
G2	Right side of Eq. (6-96)
H	Increment in x
K1, K2, K3, K4	Coefficients required for Runge-Kutta scheme
P5	Prandtl number
T8, T9	Print control variables
X	Independent variable x
X9	Maximum value of x
Y	Dependent variable ζ
Y0	Initial value of ζ, Eq. (6-89)

Listing 6-3   ZETA

```
10    PRINT"TYPE 1 IF YOU WANT A PROGRAM EXPLANATION."
12    PRINT
13    PRINT"OTHERWISE, TYPE 2."
14    INPUTL
16    IF L=1 THEN 20
18    IF L=2 THEN 39
20    PRINT
22    PRINT"THIS PROGRAM USES A FOURTH ORDER RUNGE-KUTTA SCHEME"
24    PRINT"TO INTEGRATE THE GOVERNING DIFFERENTIAL EQUATION"
26    PRINT"DESCRIBING THE THERMAL BOUNDARY LAYER ON A FLAT PLATE"
27    PRINT"FOR STEADY, LAMINAR, INCOMPRESSIBLE FLOW GIVEN BY:"
28    PRINT
30    PRINT"   DY/DX = 1/(17.05*PR*G1*X) - G2/(G1*X)"
32    PRINT
34    PRINT"WHERE PR IS THE PRANDTL NUMBER, X IS THE DISTANCE ALONG"
36    PRINT"THE PLATE, AND G1 AND G2 ARE FUNCTIONS OF Y."
38    PRINT
39    PRINT"THE SOLUTION IS"
40    PRINT

110   LETP5=100
120   IF P5<1 THEN 1700
130   LETH=0.01
200   REM INITIALIZE CONTROL CONSTANTS
210   LETC1=0
240   LETT8=0
250   LETT9=10
260   LETX9=1.01
1000  REM SET PROGRAM VALUES AND INITIAL CONDITIONS
1030  REM INITIALIZE INDEPENDENT VARIABLE
1040  LETX=0
1050  REM SPECIFY INITIAL CONDITION
1060  LETY0=1/(1.0426*P5^.3333)
1070  LETY=Y0
1075  REM CALCULATE INITIAL GRADIENT USING GOVERNING EQN.
1080  LETF=0
1100  PRINT "DISTANCE","ZETA","NU/(RE^.5)(PR^.33333)"
1110  PRINT
1120  PRINTX,Y,2/(5.84*Y*P5^0.333)
1130  REM
1140  REM FOURTH ORDER RUNGE-KUTTA INTEGRATION SCHEME
1150  REM
1160  IF X=0 THEN 1190
1175  LETG1=(4/15)*Y^2-(6/70)*Y^4+(1/36)*Y^5
1180  LETG2=(1/15)*Y^3-(3/280)*Y^5+(1/360)*Y^6
1185  LETF=1/(17.05*P5*G1*X)-G2/(G1*X)
1190  IF C1=1 THEN 1300
1200  IF C1=2 THEN 1380
1210  IF C1=3 THEN 1460
1220  LETK1=H*F
1240  LETX=X+0.5*H
1270  LETY=Y+0.5*K1
1280  LETC1=1
1290  GOTO 1160
1300  LETK2=H*F
1350  LETY=Y-0.5*K1+0.5*K2
1360  LETC1=2
1370  GOTO 1160
1380  LETK3=H*F
1400  LETX=X+0.5*H
1430  LETY=Y-0.5*K2+K3
1440  LETC1=3
1450  GOTO 1160
1460  LETK4=H*F
```

Listing 6-3 ZETA (*continued*)

```
1510 LETY=Y-K3+(K1+2*K2+2*K3+K4)/6
1520 LETC1=0
1540 LETT8=T8+1
1550 REM
1560 REM PRINT SEQUENCE
1570 REM
1580 IF X>X9 THEN 2000
1590 IF T8=T9 THEN 1610
1600 GOTO 1160
1610 PRINTX,Y,2/(5.84*Y*P5^0.333)
1620 LETT9=T8+10
1660 GOTO 1160
1700 PRINT
1710 PRINT"FOR PR<1, EQUATIONS MUST BE CHANGED. SEE TEXT."
1720 GOTO 3000
2000 PRINT
2005 PRINT"THE PARAMETERS USED IN THIS SOLUTION ARE"
2006 PRINT
2010 PRINT"PRANDTL NUMBER="P5
3000 END
```

approximate solution can be obtained by assuming a uniform velocity equal to the free-stream velocity throughout the entire thermal boundary layer. This is called *slug flow*. Since $u/U = 1.0$, Eq. (6-68) can be integrated directly to give

$$\delta_t \, d\delta_t = \frac{20}{3} \frac{\alpha}{U} \, dx \qquad (6\text{-}97)$$

The solution to this differential equation, if one assumes that $\delta_t = 0$ at $x = 0$, is

$$\delta_t = \left(\frac{40\alpha x}{3U}\right)^{1/2} \qquad (6\text{-}98)$$

The convection heat transfer coefficient is $h_x = 2k/\delta_t$, and this leads to

$$\mathrm{Nu}_x = 0.548 \mathrm{Re}_x^{1/2} \mathrm{Pr}^{1/2} = 0.548 \mathrm{Pe}^{1/2} \qquad (6\text{-}99)$$

The product RePr is called the *Peclet number* Pe.

The foundation has been laid for an integral analysis of both similar and non-similar boundary-layer flow over a curved, heated or cooled surface in the presence

Table 6-3 FORCED CONVECTION WITH $\mathrm{Pr} \geq 1.0$

	Pr = 1	Pr = 2	Pr = 5	Pr = 10	Pr = 100
$\mathrm{Nu}/\mathrm{Re}^{1/2}\mathrm{Pr}^{1/3}$	0.3427	0.3480	0.3523	0.3543	0.3572
$\zeta = \delta_t/\delta$	0.999	0.781	0.569	0.449	0.207

of a free-stream pressure gradient. The integral energy equation is given by Eq. (6-78) with Γ given by Eq. (6-72) when $\zeta < 1.0$ and Eq. (6-76) when $\zeta > 1.0$. Many integral solutions exist in the literature, for example Refs. 6-6 through 6-9, which present further analysis and results.

6-8 Turbulent External Convection

In turbulent flow it is customary to define momentum and energy diffusivities. This is done to account for the additional transport of momentum and energy caused by the turbulent mixing of the flow. The transport properties are then represented by the sum of the microscopic properties such as observed in laminar flow and the macroscopic eddy diffusivities.

Fourier's heat conduction law for laminar flow can be rewritten as

$$\frac{q''}{\rho c_p} = -\alpha \frac{dT}{dy}$$

In turbulent flow we add the *eddy diffusivity of heat* ε_H and write

$$\frac{q''}{\rho c_p} = -(\alpha + \varepsilon_H) \frac{dT}{dy} \qquad (6\text{-}100)$$

In a similar manner, Newton's law of viscosity for laminar flow is given as $\tau = \mu(du/dy)$ and can be rewritten as

$$\frac{\tau}{\rho} = v \frac{du}{dy}$$

In turbulent flow we add the *momentum eddy diffusivity* ε_M and write

$$\frac{\tau}{\rho} = (v + \varepsilon_M) \frac{du}{dy} \qquad (6\text{-}101)$$

The true values of ε_H and ε_M are still subject to much doubt, especially for liquids. For gases, experiments seem to indicate that $\varepsilon_H = \varepsilon_M$ when the Prandtl number is close to unity. Also, by definition, $v = \alpha$ when the molecular Prandtl number is unity. Restricting ourselves to a small Prandtl number range near unity, the combination of Eqs. (6-100) and (6-101) gives

$$\frac{q''}{c_p \tau} du = -dT \qquad (6\text{-}102)$$

Since the transport properties have been assumed equal and since the integral momentum and energy equations are similar when $dP/dx = 0 = dU/dx$, we assume that the ratio of energy transport to momentum transport is constant. (Compare Eqs. (6-4) and (6-63) to see when they have a similar form.) Thus we write

$$\frac{q''}{\tau} = \text{constant} = \left(\frac{q''}{\tau}\right)_w \qquad (6\text{-}103)$$

for turbulent boundary-layer flow of a gas over a flat plate.

Since the specific heat is not a strong function of temperature, it can be considered constant within a turbulent boundary layer, and Eq. (6-102) can be integrated directly to obtain

$$\frac{q_w'' U_\infty}{c_p \tau_w} = T_w - T_\infty \qquad (6\text{-}104)$$

where U_∞ is the free-stream velocity, and T_∞ is the free-stream temperature. Using the definition of the local heat transfer coefficient as given by Newton's law of cooling, we write

$$T_w - T_\infty = \frac{q''}{h_x} \qquad (6\text{-}105)$$

Combining Eqs. (6-104) and (6-105) results in $h_x U_\infty = c_p \tau_w$.

In turbulent flow the velocity profile throughout most of the boundary layer is more uniform than in a laminar boundary layer because of the momentum exchange between the fluid layers. Near the wall, however, the velocity gradient du/dy is larger than in a laminar boundary layer. This characteristic causes a greater wall shearing stress in turbulent flow. For fluids with Prandtl numbers close to unity and with equal momentum and energy boundary conditions, the turbulent nondimensional temperature profile has the same profile characteristics as the nondimensional velocity profile.

In fluid mechanics, a local friction factor is defined in terms of the wall shearing stress as

$$C_{f_x} = \frac{\tau_w}{\frac{1}{2}\rho U_\infty^2} \qquad (6\text{-}106)$$

The heat transfer at the surface can be related to the shearing stress at the surface by combining Eqs. (6-104) and (6-106). This gives

$$\frac{h_x}{\rho U_\infty c_p} = \frac{C_{f_x}}{2} \qquad (6\text{-}107)$$

This nondimensional equation is known as *Reynolds' analogy* for turbulent flow over a flat plate with $dP/dx = 0$. It is strictly valid only for Pr = 1.0. The nondimensional parameter $h_x/\rho U_\infty c_p$ is called the *Stanton number*. It is actually the ratio Nu/RePr.

Reynolds' analogy allows the convection heat transfer coefficient to be calculated by using empirical correlations which give the turbulent flow friction factor C_{f_x} as a function of the Reynolds number.

An empirical result often used for the local friction coefficient for turbulent flow over a flat plate is given by Ref. 6-3 as

$$C_{f_x} = 0.0576 \text{Re}_x^{-1/5} \qquad (6\text{-}108)$$

for $5 \times 10^5 < \text{Re}_x < 1 \times 10^7$. It follows that

$$\text{St}_x = 0.0288 \text{Re}_x^{-1/5}$$

and
$$\text{Nu}_x = 0.0288 \text{Re}_x^{4/5} \qquad \text{for } \text{Pr} = 1.0 \qquad (6\text{-}109)$$

Experimental results have shown that the Prandtl number range can be extended by modifying the above equation to give

$$\text{Nu}_x = 0.0288 \text{Pr}^{1/3} \text{Re}_x^{4/5} \qquad 0.6 < \text{Pr} < 50 \qquad (6\text{-}110)$$

Assuming that turbulent flow exists over the entire surface and integrating over the surface to calculate $\bar{h}$ gives the average Nusselt number

$$\overline{\text{Nu}_L} = 0.036 \text{Pr}^{1/3} \text{Re}_L^{4/5} \qquad 0.6 < \text{Pr} < 50 \qquad (6\text{-}111)$$

Equation (6-12) can be used to obtain an expression for $\delta(x)$ in turbulent flow if the correct velocity profile is known. However, because of the complicated nature of turbulent flow, it is difficult to specify a single continuous function to represent the velocity profile across the entire boundary layer. Based upon experimental observation, a turbulent velocity boundary layer can be described as consisting of an intermediate unstable layer (buffer layer) which separates a laminar sublayer adjacent to the wall from the outer fully turbulent region which makes up the major portion of the boundary layer. Martinelli (Ref. 6-21) described a universal turbulent velocity profile in terms of these three distinct regions.

For a flat plate with no pressure gradient, the velocity profile in the fully developed turbulent portion of the boundary layer can be approximated by the profile

$$\frac{u}{U_\infty} = \left(\frac{y}{\delta}\right)^{1/7} \qquad (6\text{-}112)$$

This velocity profile is not valid within the laminar sublayer and cannot be used to calculate the wall shearing stress. Thus, it becomes necessary to use an empirical relation for τ_w as a function of $\delta(x)$ in Eq. (6-12) as well as in Eq. (6-112). Kays (Ref. 6-13) develops the following expression from experimental data:

$$\tau_w = 0.0228 \, \rho U_\infty^2 \left(\frac{\nu}{U_\infty} \delta\right)^{1/4} \qquad (6\text{-}113)$$

Use of Eqs. (6-113) and (6-112) in Eq. (6-12) for $dU/dx = 0$ leads to the following equation

$$0.0228\left(\frac{v}{U_\infty \delta}\right)^{1/4} = \frac{7}{72}\frac{d\delta}{dx} \qquad (6\text{-}114)$$

which reduces to $\delta^{1/4}\,d\delta = 0.235(v/U_\infty)^{1/4}\,dx$. Integrating and letting $\delta = 0$ at $x = 0$ give

$$\frac{\delta}{x} = 0.376\mathrm{Re}_x^{-1/5} \qquad (6\text{-}115)$$

Comparing Eqs. (6-115) and (6-53) shows that a turbulent velocity boundary layer on a flat plate grows at a faster rate than a laminar boundary layer. However, the heat transfer by convection and the heat transfer by surface friction are both greater in turbulent flow. This result is due to the turbulent mixing of momentum and energy which more than compensates for the insulating effect of the increased boundary-layer thickness. Thus, turbulence is an advantage in promoting heat transfer, but it also increases the drag between the fluid and the solid surface.

6-9   Integral Analysis of Free Convection

In Sec. 6-2 it was shown that the net momentum per unit time crossing the surfaces of a boundary-layer control volume is given by

$$\frac{d}{dx}\int_0^\delta (\rho u^2\,dy)\,dx - U\,\frac{d}{dx}\int_0^\delta (\rho u\,dy)\,dx$$

For free-convection control volumes as shown in Fig. 6-4, the second term vanishes since $U = 0$. The forces acting on the free-convection control volume of unit depth, differential height, and finite width are shear forces and body forces. Using Newton's second law, equating the time rate of change of momentum to the forces, gives [see Eq. (5-64)]

$$\frac{d}{dx}\int_0^\delta \rho u^2\,dy = -\tau_w + \int_0^\delta g_x \rho_\infty \bar{\beta}(T - T_\infty)\,dy \qquad (6\text{-}116)$$

The integral energy equation for a boundary-layer control volume is exactly the same in free convection as in forced convection and is given by Eq. (6-63):

$$\frac{\partial}{\partial x}\int_0^{\delta_t} (T_\infty - T)u\,dy = \alpha\left(\frac{\partial T}{\partial y}\right)_{y=0} \qquad (6\text{-}63)$$

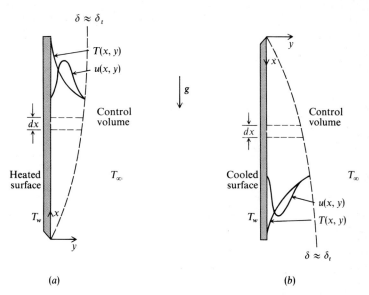

FIGURE 6-4
Boundary layers in free convection.

The integral equations of momentum and energy for a free-convection boundary layer are coupled equations since the temperature appears as a dependent variable in the momentum equation. An approximate laminar velocity profile can be specified as a fourth-order polynomial given by

$$\frac{u}{U_{\text{ref}}} = C_1\eta + C_2\eta^2 + C_3\eta^3 + C_4\eta^4 \qquad (6\text{-}117)$$

The free-stream velocity cannot be used as the reference velocity U_{ref} since $U = 0$, but U_{ref} can be an arbitrary function with dimensions of velocity. Four boundary conditions which fix the four constant coefficients are

$$
\begin{aligned}
\eta &= 1 & u &= 0 \\[2mm]
\eta &= 1 & \frac{\partial u}{\partial \eta} &= 0 \\[2mm]
\eta &= 1 & \frac{\partial^2 u}{\partial \eta^2} &= 0 \\[2mm]
\eta &= 0 & \frac{\partial^2 u}{\partial \eta^2} &= -\delta^2 g_x \frac{\bar{\beta}(T_w - T_\infty)}{\nu}
\end{aligned}
\qquad (6\text{-}118)
$$

The first three boundary conditions are the same as those previously used for Eq. (6-13). The last condition comes from the differential energy equation at $y = 0$. From Chap. 5, this is given by

$$u \frac{\partial u}{\partial x} + v \frac{\partial u}{\partial y} = g_x \bar{\beta}(T - T_\infty) + v \frac{\partial^2 u}{\partial y^2} \qquad (6\text{-}119)$$

Since $u = v = 0$ at $y = 0$, this leads to

$$\left(\frac{\partial^2 u}{\partial y^2} \right)_{y=0} = \frac{-g_x \bar{\beta}(T_w - T_\infty)}{v} \qquad (6\text{-}120)$$

Thus

$$\left(\frac{\partial^2 u}{\partial \eta^2} \right)_{\eta=0} = \frac{-\delta^2 g_x \bar{\beta}(T_w - T_\infty)}{v} \qquad (6\text{-}121)$$

gives the required boundary condition at $\eta = 0$.

Applying the boundary conditions to evaluate the four constants in Eq. (6-117) leads to

$$\frac{u}{U_{ref}} = \frac{\delta^2 g_x \bar{\beta}(T_w - T_\infty)}{6v U_{ref}} \eta(1 - 3\eta + 3\eta^2 - \eta^3) \qquad (6\text{-}122)$$

We now define $V = \dfrac{u}{U_{ref}} = u/[\partial^2 g_x \bar{\beta}(T_w - T_\infty)/6v]$ as a nondimensional velocity and write

$$V = \eta(1 - 3\eta + 3\eta^2 - \eta^3) \qquad (6\text{-}123)$$

We must now determine a temperature profile for free convection. Because of the appearance of the temperature difference $T_w - T_\infty$ in the buoyant force, we define $\theta = (T - T_\infty)/(T_w - T_\infty)$. A nondimensional temperature profile for free convection suggested by Eckert (Ref. 6-10) is

$$\theta = (1 - \eta)^2 \qquad (6\text{-}124)$$

The energy equation (6-63) then becomes

$$\frac{\partial}{\partial x} \left(U_1 \int_0^1 \theta V \delta \, d\eta \right) = \frac{2\alpha}{\delta} \qquad (6\text{-}125)$$

where $U_1 = U_{ref}$ and the momentum equation can be written

$$\frac{\partial}{\partial x} \left(U_1^2 \int_0^1 V^2 \delta \, d\eta \right) = \frac{-\tau_w}{\rho} + (T_w - T_\infty) \int_0^1 g_x \bar{\beta} \theta \delta \, d\eta \qquad (6\text{-}126)$$

Using the expressions for θ and V in these equations and integrating give

$$\frac{1}{42}\frac{\partial}{\partial x}(U_1\delta) = \frac{2\alpha}{\delta} \tag{6-127}$$

and

$$\frac{1}{252}\frac{\partial}{\partial x}(U_1{}^2\delta) = -\frac{\nu U_1}{\delta} + \frac{1}{3}g_x\bar{\beta}(T_w - T_\infty)\delta \tag{6-128}$$

One way to obtain an approximate solution to the above two equations was suggested by Eckert (Ref. 6-10). A condition that agrees with experimental observation is to let $\delta = C_1 x^m$ and $U_1 = C_2 x^n$, where m and n are exponents to be determined. The two equations then become

$$\frac{1}{42}(m + n)C_1 C_2 x^{m+n-1} = \frac{2\alpha}{C_1}x^{-m}$$

and

$$\frac{1}{252}(m + 2n)C_2{}^2 C_1 x^{m+2n-1} = \frac{-\nu C_2 x^{n-m}}{C_1} + \frac{1}{3}g_x\bar{\beta}(T_w - T_\infty)C_1 x^m$$

Equating exponents on each side of the two equations gives $m = \frac{1}{4}$ and $n = \frac{1}{2}$. Thus

$$\frac{3}{168}C_1 C_2 x^{-1/4} = \frac{2\alpha}{C_1}x^{-1/4}$$

$$\frac{5}{1{,}008}C_1 C_2{}^2 x^{1/4} = \frac{-\nu C_2}{C_1}x^{1/4} + \frac{1}{3}g_x\bar{\beta}(T_w - T_\infty)C_1 x^{1/4}$$

It follows that

$$C_1 = (336)^{1/4}\left(\frac{35}{63} + \frac{\nu}{\alpha}\right)^{1/4}\left[\frac{g_x\bar{\beta}(T_w - T_\infty)}{\nu^2}\right]^{1/4}\left(\frac{\nu}{\alpha}\right)^{-1/2}$$

$$C_2 = \frac{112}{(336)^{1/2}}\nu\left(\frac{35}{63} + \frac{\nu}{\alpha}\right)^{-1/2}\left[\frac{g_x\bar{\beta}(T_w - T_\infty)}{\nu^2}\right]^{1/2}$$

Then $\delta = C_1 x^{1/4} = (336)^{1/4}\left(\frac{35}{63} + \frac{\nu}{\alpha}\right)^{1/4}\left[\frac{g_x\bar{\beta}(T_w - T_\infty)}{\nu^2}\right]^{-1/4}\left(\frac{\nu}{\alpha}\right)^{-1/2}x^{1/4}$

$$\frac{\delta}{x} = 4.27\text{Pr}^{-1/2}\left(\frac{35}{63} + \text{Pr}\right)^{1/4}\text{Gr}_x{}^{-1/4} \tag{6-129}$$

The Grashof number, $\text{Gr}_x = g_x\bar{\beta}(\bar{T}_w - T_\infty)x^3/\nu^2$, has been introduced in Eq. (6-129). It is a nondimensional ratio of body force to viscous force. The Grashof and Prandtl numbers, in addition to the Nusselt number, are the important parameters that arise in the study of free convection. For free convection of gases on a vertical plate, when the value of the Grashof number lies between 10^4 and 10^9 a laminar, free-convective boundary layer exists. At smaller Grashof numbers the buoyant force is not large enough to cause a convective, boundary-layer flow to form. At larger Grashof numbers the free-convective boundary layer is turbulent.

To calculate the Nusselt number, we write

$$-k\left(\frac{\partial T}{\partial y}\right)_{y=0} = h_x(T_w - T_\infty)$$

This gives $h_x = 2k/\delta$ and $\mathrm{Nu}_x = 2x/\delta$. The approximate Nusselt number for laminar free convection is then

$$\mathrm{Nu}_x = 0.468\mathrm{Pr}^{1/2}\left(\frac{35}{63} + \mathrm{Pr}\right)^{-1/4} \mathrm{Gr}_x^{1/4} \qquad (6\text{-}130)$$

In calculating the values for Nu, Pr, and Gr, the properties are evaluated at a mean film temperature $T_f = (T_w + T_\infty)/2$.

For a flat surface that is slightly inclined at an angle α with the vertical, the above results can be used if the Grashof number is defined as $\mathrm{Gr}_x = g_x\bar{\beta}(T_w - T_\infty)x^3 \cos \alpha/v^2$. On the other hand, if the surface is nearly horizontal, the empirical results given in Table 7-4 should be used since boundary-layer flow will not exist.

An integral method may also be used to obtain approximate solutions for turbulent free convection of air at $\mathrm{Gr}_x > 10^9$. Eckert (Ref. 6-12) obtained the following expression for the average Nusselt number:

$$\overline{\mathrm{Nu}_L} = 0.024\left(\frac{\mathrm{Pr}^{1.17}}{1 + 0.494\mathrm{Pr}^{2/3}} \mathrm{Gr}_L\right)^{2/5} \qquad (6\text{-}131)$$

An equation that is easier to use is

$$\overline{\mathrm{Nu}_L} = 0.0210(\mathrm{GrPr})^{2/5}$$

which agrees closely with Eckert's results. Since this expression is based upon the assumption that the flow is turbulent over the entire surface, it is more accurate for $\mathrm{Gr}_x > 10^{10}$ since the surface covered by a laminar boundary layer will then be a small fraction of the total surface. Because of the nature of a turbulent boundary layer, the local heat transfer coefficient is nearly constant in turbulent free convection.

6-10   Free Convection Due to Condensation

A special free-convective process of much practical importance is condensation. Condensation of vapor on a cooler surface may be classified as either *filmwise* condensation or *dropwise* condensation. When the liquid condensate is in contact with a wetting surface, a thin film is formed which can be analyzed as a boundary layer. If the surface is nonwetting, liquid beads form from selected nucleation sites. The heat transfer rate is greater for dropwise condensation as compared to filmwise condensation since the diameter of the liquid beads is smaller than the film thickness which offers resistance to conduction through the film.

However, filmwise condensation can be predicted analytically, and it serves as a model which gives a conservative estimate of the heat transfer during condensation. Also, dropwise condensation cannot yet be consistently obtained because of the lack of a permanent promoter to produce a nonwetting surface and the presence of excessive noncondensable gases in the condensing vapor.

Figure 6-4b can be used as a model to represent laminar filmwise condensation. Applying the first law of thermodynamics to the control volume shown gives

$$q + (\dot{m}h)_{in} = (\dot{m}h)_{out} \qquad (6\text{-}132)$$

Using the boundary-layer assumption that $dT/dy >> dT/dx$ allows us to assume that the temperature in the liquid film is not a function of x. The net energy added to the control volume of width dx is the product $\dot{m}_c h_{fg}$, where $\dot{m}_c$ is the flow rate of additional condensate added to the boundary layer and $h_{fg} = h_g - h_f$ is the latent heat of condensation. Equation (6-132) can then be written

$$k\, dx \left(\frac{dT}{dy}\right)_{y=0} = \dot{m}_c h_{fg} \qquad (6\text{-}133)$$

The mass flow added to the boundary layer $\dot{m}_c$ must be expressed in terms of the velocity profile. Using the integral method with $\eta = y/\delta$, we assume

$$V = \frac{u}{U_{ref}} = C_1 + C_2 \eta + C_3 \eta^2 \qquad (6\text{-}134)$$

and apply the following boundary conditions:

$$\eta = 0 \qquad V = 0$$

$$\eta = 1 \qquad \frac{dV}{d\eta} = 0 \qquad\qquad (6\text{-}135)$$

$$\eta = 0 \qquad \frac{dV}{d\eta} = \frac{(\rho - \rho_v) g_x \delta^2}{\mu\, U_{ref}}$$

The third boundary condition comes from a force balance applied to the control volume. The forces acting on the control volume are a surface force due to shear at the wall and a body force due to buoyancy. Equating these two forces gives

$$\mu \left(\frac{du}{dy}\right)_{y=0} dx = (\rho - \rho_v) g_x \delta\, dx \qquad (6\text{-}136)$$

where $\rho_v = \rho_\infty$, the density of the vapor.

Evaluating the three constants in Eq. (6-134) gives the following velocity profile within the thin condensate film:

$$V = \frac{u}{U_{ref}} = \frac{g_x \delta^2 (\rho - \rho_v)}{\mu U_{ref}} \left(\eta - \frac{\eta^2}{2} \right) \tag{6-137}$$

Choosing $U_{ref} = g_x \delta^2 (\rho - \rho_v)/\mu$ gives

$$V = \eta - \frac{\eta^2}{2} \tag{6-138}$$

The mass flow crossing the top surface of the control volume is

$$\dot{m}_x = \int_0^\delta \rho u \, dy = \int_0^\delta \rho \frac{g_x(\rho - \rho_v)}{\mu} \left(\delta y - \frac{y^2}{2} \right) dy$$

$$= \frac{\rho g_x (\rho - \rho_v)\delta^3}{3\mu} \tag{6-139}$$

The mass flow leaving the control volume across the lower surface is expressed by using a first-order Taylor series expansion.

$$\dot{m}_{x+dx} = \dot{m}_x + \frac{d}{dx}(\dot{m}_x) \, dx$$

The mass flow of condensate entering the control volume between x and $x + dx$ due to additional condensation of the free-stream vapor is then

$$\dot{m}_c = \frac{d}{dx} \left[\frac{\rho g_x (\rho - \rho_v)\delta^3}{3\mu} \right] dx = \frac{\rho g_x (\rho - \rho_v)\delta^2 \, d\delta}{\mu} \tag{6-140}$$

Since the velocity in the falling condensate film is usually small, we can neglect the effect of convection on the temperature profile in the film and assume a linear temperature variation between T_w and T_v. Equation (6-133) can now be written

$$k \, dx \frac{T_v - T_w}{\delta} = \frac{\rho g_x (\rho - \rho_v)\delta^2 \, d\delta}{\mu} h_{fg} \tag{6-141}$$

Integrating and specifying $\delta = 0$ at $x = 0$ gives

$$\delta = \left[\frac{4\mu k(T_v - T_w)x}{\rho g_x h_{fg}(\rho - \rho_v)} \right]^{1/4} \tag{6-142}$$

To form a Nusselt number, we write

$$-k \, dx \frac{T_v - T_w}{\delta} = h_x \, dx(T_w - T_v)$$

It follows that

$$h_x = \frac{k}{\delta} = \left[\frac{\rho g_x h_{fg}(\rho - \rho_v)k^3}{4\mu(T_v - T_w)x}\right]^{1/4} \tag{6-143}$$

By integration, the average convection heat transfer coefficient becomes $\bar{h} = \frac{4}{3}h_{x=L}$. The local Nusselt number is

$$\mathrm{Nu}_x = \frac{h_x x}{k} = \left[\frac{\rho g_x h_{fg}(\rho - \rho_v)x^3}{4\mu k(T_v - T_w)}\right]^{1/4}$$

This equation can be written

$$\mathrm{Nu}_x = 0.707\left[\mathrm{Gr}_x \mathrm{Pr}\frac{h_{fg}}{c_p(T_v - T_w)}\right]^{1/4} \tag{6-144}$$

where $\mathrm{Gr}_x = g_x(1 - \rho_v/\rho)x^3/v^2$, $\mathrm{Pr} = \mu c_p/k$, and $h_{fg}/c_p(T_v - T_w)$ is an additional nondimensional parameter that arises because of the two-phase nature of the flow. It is the ratio of the change in enthalpy that occurs with a phase change to the enthalpy change that would occur between T_v and T_w without a phase change.

In condensers one is often interested in condensation on horizontal tubes rather than vertical surfaces since the tubes contain the coolant that removes the thermal energy given up during the condensation process. Nusselt recommended the following equation for laminar filmwise condensation on horizontal tubes of diameter D,

$$\overline{\mathrm{Nu}} = \frac{\bar{h}D}{k} = 0.725\left[\mathrm{Gr}_D \mathrm{Pr}\frac{h_{fg}}{c_p(T_v - T_w)}\right]^{1/4} \tag{6-145}$$

where $\mathrm{Gr}_D = g_x(1 - \rho_v/\rho)D^3/v^2$. The total heat transfer from a surface can be determined, once $\bar{h}$ is known, by using Newton's law of cooling, $q = \bar{h}A_s(T_w - T_v)$. The saturation temperature T_v is a function of pressure and can be found in standard thermodynamic tables.

6-11   Internal Convection

In this section we consider steady, incompressible, constant-property flow of fluids through circular tubes of constant cross-sectional area. The coordinate system is indicated in Fig. 6-5. Both laminar and turbulent flow are considered. Laminar flow exists when the Reynolds number $\mathrm{Re} = VD_H/v$ is less than 2,300. The hydraulic diameter D_H is defined as four times the flow cross-sectional area divided by the wetted perimeter, and it is the characteristic length for internal convection. For a circular tube, $D_H = D$, the tube diameter.

When fluid enters a tube, a boundary layer forms in the entrance region. Equations of motion for the entrance region are partial differential equations similar to

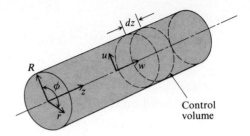

FIGURE 6-5
Cylindrical coordinate system.

those discussed in Chap 5. In internal convection, the entrance-boundary-layer-type flow soon disappears as the boundary layers grow to form a viscous flow region throughout the entire pipe cross section. This region is called *fully developed flow*. In fully developed flow of a constant-property fluid in a constant-area tube, the velocity profile is independent of axial location. The entrance length L_e required to establish a fully developed flow can be estimated by using the relationship $L_e/D = 0.05 Re_D Pr$. (See Ref. 6-13.) When a temperature difference exists between the tube wall and the fluid, a thermal boundary layer also grows into a *fully developed thermal region*. The growth is quicker than the velocity profile for $Pr < 1$ and slower for $Pr > 1$.

Fully developed laminar flow in circular tubes is often called *Hagen-Poiseuille flow*. The solution to the differential equations of motion gives a parabolic velocity profile across the tube diameter. To show this result using the integral method, we can specify a velocity profile given by

$$V = \frac{w}{U_0} = C_1 + C_2\eta + C_3\eta^2 \qquad (6\text{-}146)$$

where U_0 is the centerline velocity and $\eta = r/R$. The three boundary conditions are

$$\eta = 0 \qquad V = 1.0$$

$$\eta = 0 \qquad \frac{dV}{d\eta} = 0 \qquad (6\text{-}147)$$

$$\eta = 1 \qquad V = 0$$

$$V = 1 - \eta^2 \qquad (6\text{-}148)$$

which give the expected parabolic profile.

A force balance on the control volume shown in Fig. 6-5 requires equating the wall shearing force to the net pressure force acting on the control volume surface.

This gives

$$\pi R^2 \frac{dp}{dz} \, dz = 2\pi R\mu \, dz \left(\frac{dw}{dr}\right)_{r=R}$$

which leads to

$$\left(\frac{dV}{d\eta}\right)_{\eta=1} = \frac{R^2}{2\mu U_0} \frac{dp}{dz} \qquad (6\text{-}149)$$

Combining Eqs. (6-148) and (6-149) gives

$$U_0 = -R^2 \frac{dP/dz}{4\mu} \qquad (6\text{-}150)$$

for the centerline velocity, where a pressure drop $dP/dz < 0$ gives a positive velocity.

The average velocity in laminar, constant-property flow is given by

$$U_m = \frac{\displaystyle\int_0^R w(2\pi r) \, dr}{\pi R^2} \qquad (6\text{-}151)$$

Integration gives $U_m = U_0/2$.

In turbulent convection one can separate the time average values of local temperature and local velocity from the corresponding fluctuating components. At any instant of time, the instantaneous properties are written as a sum of the time average value plus the instantaneous deviation from this value. For example, the local axial velocity is written as $w_i = w + w'$, and the local temperature is written as $T_i = T + T'$.

The Reynolds averaging technique is to substitute the instantaneous values of velocity and temperature into the momentum and energy equations and average all fluctuating quantities over time. By definition, the time average values $\hat{u}'$, $\hat{v}'$, $\hat{w}'$, and $\hat{T}'$ will equal zero. The hat has been used to indicate the time average of the fluctuating components. However, terms like $\widehat{u'v'}$, $\widehat{u'T'}$, $\widehat{w'T'}$, etc., are not necessarily zero.

Because of the complexity of fully developed turbulent flow, it is necessary to use experimentally determined velocity profiles to analyze fluid flow and heat transfer in internal forced convection. One such profile, called *Kármán universal turbulent velocity profile*, is discussed in Ref. 6-13. This profile requires two or three separate algebraic equations since the flow is divided into separate regions of a laminar sub-layer, buffer layer, and a turbulent core.

A different profile for steady, axisymmetric, incompressible turbulent flow inside a smooth tube was presented by Pai in Ref. 6-14. The axial velocity w is given by

$$\frac{w}{U_0} = 1 - \frac{n-s}{n-1}\eta^2 - \frac{s-1}{n-1}\eta^{2n} \qquad (6\text{-}152)$$

and the momentum eddy diffusivity is given by

$$\frac{\varepsilon_M}{\nu} = \frac{s(n-1)}{n-s+n(s-1)\eta^{2n-2}} - 1.0 \qquad (6\text{-}153)$$

The constants n and s are discussed below. The advantages offered by this profile are that it is continuously differentiable across the tube, it fits the physical boundary conditions, it is given in simple polynomial form, and it reduces to the well-known results for laminar flow when the turbulent effects are eliminated.

For turbulent flow in tubes, the momentum eddy diffusivity is defined by

$$\varepsilon_M = \frac{\widehat{u'v'}}{-\partial w/\partial r} \qquad (6\text{-}154)$$

and the energy eddy diffusivity is defined as

$$\varepsilon_H = \frac{\widehat{u'T'}}{-\partial T/\partial r} \qquad (6\text{-}155)$$

In turbulent flow, nondimensional velocities can be obtained by use of a reference shear velocity given by $(g_c\tau_w/\rho)^{1/2}$. We define

$$\overline{w} = \frac{w}{(g_c\tau_w/\rho)^{1/2}} \qquad (6\text{-}156)$$

and introduce

$$\overline{y} = \frac{\widehat{u'w'}}{g_c\tau_w/\rho} \qquad (6\text{-}157)$$

where $\overline{y}$ is a turbulent shear stress parameter and $\widehat{u'w'}$ is the time average of the product of the radial and axial fluctuating velocity components. The appropriate boundary conditions are

$$\begin{aligned} \eta = 1 \qquad \overline{w} = 0 \\ \eta = 1 \qquad \overline{y} = 0 \end{aligned} \qquad (6\text{-}158)$$

Using the Pai profile to calculate the average velocity gives

$$U_m = \frac{\int_0^R w\,dA}{A} = \frac{U_0}{\pi R^2}\int_0^R \left(1 - \frac{n-s}{n-1}\eta^2 - \frac{s-1}{n-1}\eta^{2n}\right)2\pi r\,dr$$

Since $\eta = r/R$, the integral reduces to

$$U_m = 2U_0\int_0^1 \left(\eta - \frac{n-s}{n-1}\eta^3 - \frac{s-1}{n-1}\eta^{2n+1}\right)d\eta \qquad (6\text{-}159)$$

The results of the integration can be put in the following form:

$$\frac{U_m}{U_0} = \frac{1}{2}\frac{n+s}{n+1} \qquad (6\text{-}160)$$

It can be seen that when $n = 0$ and $s = 1.0$, Eq. (6-160) reduces to the results for laminar flow, $U_m = U_0/2$.

The constant s is a ratio of the actual wall shearing stress to the wall shearing stress in laminar flow with centerline velocity U_0:

$$s = \frac{\tau_w}{(\tau_L)_w} \qquad (6\text{-}161)$$

In laminar flow the value of s is unity by definition. The wall shearing stress can also be expressed in terms of the *Fanning friction factor f* which is defined by

$$f = \frac{\tau_w}{\frac{1}{2}\rho U_m{}^2} \qquad (6\text{-}162)$$

In laminar flow of a Newtonian fluid,

$$(\tau_L)_w = -\mu \left(\frac{\partial w}{\partial r}\right)_{r=R} \qquad (6\text{-}163)$$

where the velocity gradient can be obtained from Eq. (6-148). This relationship gives

$$(\tau_L)_w = \frac{4\mu U_m}{R} = \frac{4\mu U_0}{D} \qquad (6\text{-}164)$$

Combining Eqs. (6-161), (6-162), and (6-164) gives

$$s = \frac{f}{8}\,\mathrm{Re}_0\left(\frac{U_m}{U_0}\right)^2 \qquad (6\text{-}165)$$

where $\mathrm{Re}_0 = \rho U_0 D/\mu$. Finally, use of Eq. (6-160) yields

$$s = \frac{f}{8}\,\mathrm{Re}_0\left[\frac{\frac{1}{2}(n+s)}{n+1}\right]^2 \qquad (6\text{-}166)$$

Solving Eq. (6-160) for n yields

$$n = \frac{s - 2U_m/U_0}{2U_m/U_0 - 1} \qquad (6\text{-}167)$$

Combining Eqs. (6-166) and (6-167) gives

$$n = \frac{(f/16)\mathrm{Re} - 1}{1 - \frac{1}{2}U_0/U_m} \qquad (6\text{-}168)$$

where $\mathrm{Re} = \rho U_m D/\mu$.

Equations (6-165) and (6-168) show that s and n may be determined by measuring f, Re, and U_0/U_m. Values for n and s as a function of the Reynolds number have been evaluated by Brodkey in Ref. 6-15. Further tabulation of n and s values is given in Ref. 6-16. They are reproduced in Table 6-4. These values give utility to the use of the Pai velocity profile for heat- and momentum-transfer calculations.

Haberstroh and Baldwin (Ref. 6-16) numerically integrated the governing energy equation for fully developed flow with a constant wall heat flux by using the Pai profile to specify the velocity distribution. Their results have been reproduced in Fig. 6-6. For Reynolds numbers less than 2,300 the result is Nu = 4.36, which agrees with analytical solutions for laminar flow with a constant wall heat flux. (See Ref. 6-13.)

The results for turbulent flow give satisfactory agreement with other available correlations for $0.01 \le \text{Pr} \le 10$ and $10^4 \le \text{Re} \le 10^7$. This range of Prandtl numbers

Table 6-4 VALUES OF THE FLOW PARAMETERS n AND s FOR USE WITH THE PAI VELOCITY PROFILE

Re	n	s
Laminar	—	1.00
2,000	9.25	1.041
2,100	5.74	1.108
2,200	4.35	1.251
2,300	3.22	1.447
2,400	2.57	1.741
2,500	2.45	2.07
2,600	2.64	2.31
2,700	2.83	2.49
2,800	3.02	2.64
2,900	3.18	2.76
3,000	3.31	2.86
3,200	3.58	3.06
3,600	4.10	3.41
4,000	4.57	3.72
4,500	5.15	4.10
5,000	5.72	4.46
8,000	9.00	6.43
10,000	10.93	7.56
12,000	12.70	8.60
15,000	15.18	10.13
30,000	26.51	17.60
50,000	39.91	26.62
70,000	52.18	35.04
100,000	69.27	46.97
300,000	165.12	116.40
500,000	247.00	177.83
700,000	322.00	235.20
1,000,000	426.40	316.30
10,000,000	2,608.00	2,151.00

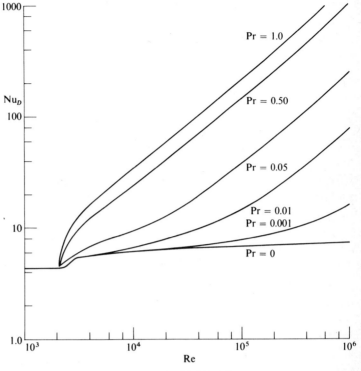

FIGURE 6-6
Nusselt numbers for internal convection.

includes liquid metals, gases, and water. Organic liquids with high Prandtl numbers usually have temperature-dependent properties which require modified mathematical models or empirical correlations to obtain the functional relationship between the controlling nondimensional parameters.

Reynolds' analogy, which was discussed in Sec. 6-8 for external, turbulent convection, can also be developed for turbulent, internal forced convection. If we use the same reasoning and assumptions, Eq. (6-102) applies to internal convection. To integrate Eq. (6-102), we specify limits of integration between the values at the tube wall and the mean velocity and temperature values in the turbulent flow. The mean temperature for internal convection is called the *bulk temperature* and is defined as

$$T_B = \frac{\displaystyle\int_0^R \rho c_p u T (2\pi r)\, dr}{\displaystyle\int_0^R \rho c_p u (2\pi r)\, dr} \qquad (6\text{-}169)$$

To evaluate the bulk temperature requires a knowledge of u and T as a function of r. The bulk temperature is also used as the reference temperature for defining the local convection heat transfer coefficient for internal convection by writing $q = h_z A_s(T_w - T_B)$. Equation (6-102) is now written as

$$\int_0^{U_m} \frac{q''}{c_p \tau} \, du = - \int_{T_w}^{T_B} dT \qquad (6\text{-}170)$$

In external flow, Reynolds' analogy was developed for a flat plate where $dP/dx = 0$. The relationship $q''/\tau = q_w''/\tau_w = $ constant, Eq. (6-103), was assumed to apply for a flat plate. This argument does not apply to internal convection since there is no flow when $dP/dz = 0$. However, for turbulent flow in a circular tube there is a linear relationship between shearing stress and radius. To show this relationship, we again use a force balance on a control volume as shown in Fig. 6-5 to obtain

$$\tau(2\pi r) + \frac{dP}{dz} \pi r^2 = 0 \qquad (6\text{-}171)$$

where τ is an apparent turbulent shearing stress acting on the surface of the control volume parallel to the turbulent axial flow. From this equation it follows that

$$\frac{\tau}{\tau_w} = \frac{r}{R} \qquad (6\text{-}172)$$

To obtain Reynolds' analogy for internal, turbulent convection, we assume that the local heat flux follows a similar linear variation:

$$\frac{q''}{q_w''} = \frac{r}{R} \qquad (6\text{-}173)$$

From the above two equations we conclude that

$$\frac{q''}{\tau} = \frac{q_w''}{\tau_w} = \text{constant} \qquad (6\text{-}174)$$

The validity of this approximation has been tested by comparison with experimental measurements. It is a good approximation for turbulent flow of fluids with Prandtl numbers near unity. It is not a good approximation for laminar, internal convection, even when the Prandtl is near unity. Physically, this fact can be explained because turbulence produces similar velocity and temperature profiles which do not normally occur in laminar, internal convection.

Integration of Eq. (6-170) gives

$$\frac{q_w'' U_m}{\tau_w c_p} = T_w - T_B \qquad (6\text{-}175)$$

From Eq. (6-171)

$$\tau_w = \frac{-(dP/dz)R}{2} = \frac{\Delta P D}{4L} \qquad (6\text{-}176)$$

where L is the tube length and ΔP is the pressure difference across the tube length.

The Fanning friction factor, Eq. (6-162), is used to express the pressure drop through a pipe (Ref. 6-17):

$$\Delta P = f \frac{L}{D} \frac{\rho U_m^2}{2} \qquad (6\text{-}177)$$

It follows that $\tau_w = (f/8)\rho U_m^2$. The average wall heat flux is expressed by $q_w'' = \bar{h}(T_w - T_B)$. These two expressions along with Eq. (6-175) give the Stanton number

$$\text{St} = \frac{\bar{h}}{\rho U_m c_p} = \frac{f}{8} \qquad (6\text{-}178)$$

The friction factor f is a function of tube roughness and the Reynolds number. Moody diagrams, which show the variation of the friction factor, are available in most elementary books on fluid mechanics. For smooth tubes, several empirical correlations are available. Reference 6-13 gives

$$f = 0.316\text{Re}^{-1/4} \qquad 5 \times 10^3 < \text{Re} < 3 \times 10^4$$
$$f = 0.184\text{Re}^{-1/5} \qquad 3 \times 10^4 < \text{Re} < 1 \times 10^6$$

Equation (6-178) gives good results only for turbulent flow of gases which have Prandtl numbers near unity. To increase the valid Prandtl number range of Eq. (6-178), the same modification used in Sec. 6-8 for external, turbulent convection has been suggested. That is, we write

$$\text{St}\,\text{Pr}^{2/3} = \frac{f}{8} \qquad (6\text{-}179)$$

For $5 \times 10^3 < \text{Re} < 3 \times 10^4$ this equation gives

$$\overline{\text{Nu}} = 0.0395\text{Re}^{3/4}\text{Pr}^{1/3} \qquad (6\text{-}180)$$

and for $3 \times 10^4 < \text{Re} < 1 \times 10^6$

$$\overline{\text{Nu}} = 0.023\text{Re}^{4/5}\text{Pr}^{1/3} \qquad (6\text{-}181)$$

These results give reasonable accuracy for $0.5 < \text{Pr} < 100$. Various modifications and analogies have been suggested to expand the results of this simple analysis to larger Prandtl number ranges. (See Refs. 6-18 to 6-21.)

In fully developed turbulent flow the velocity and temperature profiles are uniform across much of the tube cross section because of the turbulent mixing. Thus the ratios U_m/U_0 and T_B/T_0 are much closer to unity in turbulent flow than in laminar flow. It is possible to assume uniform, one-dimensional velocity and temperature profiles to estimate heat transfer rates in turbulent flow. This assumption eliminates the need to know the temperature profile as required in Eq. (6-169) in order to determine the bulk temperature. The first law of thermodynamics applied to turbulent flow through a tube gives

$$q = \dot{m}(h_{\text{out}} - h_{\text{in}}) \qquad (6\text{-}182)$$

For a constant wall temperature this equation becomes

$$\bar{h} A_s (T_w - T_B) = \rho A_c U_m c_p (T_{\text{out}} - T_{\text{in}}) \qquad (6\text{-}183)$$

where A_c is the tube cross-sectional flow area, and A_s is the tube surface area.

Suppose that T_w and T_{in} are known and T_{out} is to be determined. The mean bulk temperature can be estimated to be $T_B = (T_{\text{in}} + T_{\text{out}})/2$. The procedure is to assume a value for T_{out}, calculate T_B, and then use Eq. (6-183) to see if an equality exists. The value of $\bar{h}$ comes from an equation such as Eq. (6-181), where the properties appearing in Nu, Re, and Pr are evaluated at the mean bulk temperature. The process is repeated until the correct outlet temperature is found. Notice that if T_{in}, T_{out}, and T_w can all be measured, the value of $\bar{h}$ can be determined directly from Eq. (6-183) without any additional knowledge of heat transfer. This procedure provides an experimental technique for determining average values of $\bar{h}$ in a complex heat exchanger configuration. Heat exchangers are discussed in the next chapter.

References

6-1   POHLHAUSEN, K.: Zur näherungsweisen Integration der Differentialgleichung der laminaren Reibungsschicht, *ZAMM*, vol. 1, pp. 252–268, 1921.

6-2   VON KÁRMÁN, T., and C. B. MILLIKAN: On the Theory of Laminar Boundary Layer Involving Separation, NACA Rept. 504, 1934.

6-3   SCHLICHTING, H.: "Boundary-layer Theory," 6th ed., McGraw-Hill, New York, 1968.

6-4   HOLSTEIN, H., and T. BOHLEN: Ein einfaches Verfahren zur Berechnung laminaren Reibungsschichten, die dem Näherungsansatz von K. Pohlhausen genugen, Lilienthal-Bericht, S10, pp. 5–16, 1940.

6-5   MILNE, J. S.: A Complete Computer Solution to the Pohlhausen Laminar Boundary Layer Equation for Steady Incompressible Flow, "Bulletin of Mechanical Engineering Education," vol. 9, p. 97, Pergamon, New York, 1970.

6-6   ECKERT, E. R. G., and J. N. B. LIVINGOOD: Method for Calculation of Laminar Heat Transfer in Air Flow around Cylinders of Arbitrary Cross-section, NACA Rept. 1118, 1953.

6-7   SQUIRE, H. B.: Heat Transfer Calculations for Aerofoils, ARC RM 1986, 1942.

6-8   DIENEMANN, W.: "Berechnung des Warme ubergranges an laminar umstromten Korpern mit konstanter und ortsveranderlicher Wandtemperatur," dissertation, Braunschweig, 1951, *ZAMM*, vol. 33, pp. 89–109, 1953.

6-9 SPALDING, D. B., and W. M. PUN: A Review of Methods for Predicting Heat Transfer Coefficients for Laminar Uniform-property Boundary Layer Flows, *Intern. J. Heat Mass Transfer*, vol. 5, pp. 239–250, 1962.

6-10 ECKERT, E. R. G., and R. M. DRAKE, JR.: "Heat and Mass Transfer," McGraw-Hill, New York, 1959.

6-11 MC ADAMS, W. H.: "Heat Transmission," 3d ed., McGraw-Hill, New York, 1954.

6-12 ECKERT, E. R. G., and T. W. JACKSON: Analysis of Turbulent Free Convection Boundary Layer on Flat Plate, NACA Rept. 1015, July 1950.

6-13 KAYS, W. M.: "Convective Heat and Mass Transfer," McGraw-Hill, New York, 1966.

6-14 PAI, S. I.: On Turbulent Flow in Circular Pipe, *J. Franklin Inst.*, vol. 256, no. 4, pp. 337–352, 1953.

6-15 BRODKEY, R. S.: Limitations on a Generalized Velocity Distribution, *AIChE J.*, vol. 9, no. 4, pp. 448–451, 1963.

6-16 HABERSTROH, R. D., and L. V. BALDWIN: Application of a Simplified Velocity Profile to the Prediction of Pipe Flow Heat Transfer, *J. Heat Transfer*, vol. 90, ser. C, August 1968.

6-17 SHAMES, I. H.: "Mechanics of Fluids," McGraw-Hill, New York, 1962.

6-18 LYON, R. N.: Liquid Metal Heat Transfer Coefficients, *Chem. Eng. Progr.*, vol. 47, p. 75, 1951.

6-19 VON KÁRMÁN, T.: Analogy between Fluid Friction and Heat Transfer, *Trans. ASME*, vol. 61, pp. 705–710, November 1936.

6-20 DEISSLER, R. G.: Analytical and Experimental Investigation of Fully Developed Turbulent Flow of Air in a Smooth Tube with Heat Transfer and Variable Fluid Properties, NACA TN 2629, 1952.

6-21 MARTINELLI, R. C.: Heat Transfer to Molten Metals, *Trans. ASME*, vol. 69, pp. 947–959, November 1947.

6-22 COLBURN, A. P.: A Method of Correlating Forced Convection Heat Transfer Data and a Comparison with Fluid Friction, *AIChE J.*, vol. 29, p. 174, 1933.

6-23 HOUGHTON, E. L., and R. P. BOSWELL: "Further Aerodynamics for Engineering Students," Spottiswoode, Ballantyne, London, 1969.

Problems[1]

6-1 Assume a velocity profile in the form of a third-order polynomial given by

$$\frac{u}{U} = C_1 + C_2\eta + C_3\eta^2 + C_4\eta^3$$

where $\eta = y/\delta$.

(a) Evaluate the four constants by applying appropriate boundary conditions for a laminar boundary layer on a flat plate.

(b) Calculate the ratio δ^*/δ for the above profile.

(c) Calculate the ratio Θ/δ for the above profile.

(d) Calculate a value for the nondimensional wall shear stress parameter $\tau_w \delta/\mu U$.

---

[1] Problems whose numbers are followed by a superscript italic c should be analyzed by obtaining solutions on a digital computer.

6-2   Derive Eq. (6-20) by evaluating the four constants in Eq. (6-13).

6-3   Derive Eq. (6-21).

6-4   Derive Eq. (6-22).

6-5   Assume a linear boundary-layer profile given by $V = u/U_\infty = y/\delta$. Use this profile in the integral momentum equation for laminar, incompressible flow over a flat plate, Eq. (6-8), and calculate the value δ/x as a function of the Reynolds number $\text{Re} = Vx/\gamma$.

6-6^c The velocity profile for the inviscid laminar flow around a cylinder in cross flow can be approximated by

$$V = \frac{U}{U_\infty} = 2\left(\xi - \frac{1}{3!}\xi^3 + \frac{1}{5!}\xi^5 + \cdots\right)$$

where $\xi = x/D$.

(a) Determine the separation point.

(b) Calculate the variation of δ^* and Θ with respect to ξ.

(c) Calculate the maximum local shearing stress τ_w and the point where it occurs.

6-7^c Milne (Ref. 6-5) gives the following dimensional velocity profile around a 9.74-cm cylinder in water with a kinematic viscosity $\nu = 0.0101 \text{ cm}^2/\text{s}$. The free-stream velocity $V_\infty = 19.2 \text{ cm/s}$, and x is the distance on the cylinder from the stagnation point.

$$U = 7.151x - 0.04497x^3 - 0.00033x^5$$

(a) Calculate the separation point.

(b) Calculate the boundary-layer thickness at the separation point.

(c) Calculate the variation of δ^* and Θ with respect to x.

6-8^c A free-stream velocity field is given by

$$V = \frac{U}{U_\infty} = 1 - \xi$$

where $\xi = x/L$.

(a) Find the separation point for the surface which produces this free-stream velocity field.

(b) Arbitrarily add higher-order terms, and see if you can increase the value of ξ at which separation occurs.

(c) Compare the variation of the local shearing stress of any of the solutions obtained in part (b) with the solution to part (a).

6-9   The momentum equation for steady, two-dimensional, axially symmetric flow is (Ref. 6-2)

$$U^2 \frac{d\Theta}{dx} + 2(\Theta + \delta^*)U \frac{dU}{dx} + \frac{U^2\Theta}{r}\frac{dr}{dx} = \frac{\tau_w}{\rho}$$

(a) Show that by defining $z = \Theta^2/\nu$ the equation can be written

$$\frac{dz}{dx} = \frac{1}{U}\left[F(\kappa) - 2\kappa \frac{1}{r}\left(\frac{dr}{dx}\right)\frac{U}{U'}\right]$$

$$\kappa = zU'$$

(*b*) For a blunt nose, at the stagnation point (see Fig. 6-9P)

$$\lim_{x \to 0} \left(\frac{1}{r} \frac{dr}{dx} \frac{U}{U'} \right) = 1$$

Show that it must follow that $\lambda_0 = 4.716$ at $x = 0$, where λ is defined in Eq. (6-19).

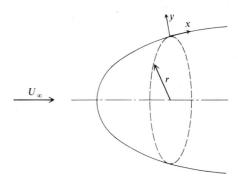

FIGURE 6-9P
Body of revolution.

6-10   Using the velocity profile specified by a cubic polynomial, calculate the value of λ where separation will occur for flow with a pressure gradient. How does this value compare with the value of λ for a fourth-order velocity profile? Physically, how much difference will there be in the predicted location of the separation point on the surface obtained by using the two different velocity profiles?

6-11   Using the results given in Fig. 6-2 for a gas ($\nu = 0.2 \times 10^{-3}$ ft^2/s) with a free-stream velocity of 10 ft/s over a surface of length 0.5 ft, calculate the following:
(*a*) the Reynolds number
(*b*) the maximum local shearing stress if $\mu = 1.3 \times 10^{-5}$ lb/ft-s
(*c*) the total wall shearing force up to the point of boundary-layer separation
(*d*) the boundary-layer thickness at the point of separation
(*e*) the boundary-layer thickness at the point where the favorable free-stream pressure gradient ends and the adverse pressure gradient begins

6-12^c Consider the following two inviscid flow fields for laminar, incompressible forced convection over a surface:

$$V = \frac{U}{U_\infty} = 4\xi - \xi^2$$

$$V = \frac{U}{U_\infty} = 1 + 4\xi - \xi^2$$

Obtain computer solutions for λ, δ^*, and Θ as a function of ξ, as well as the wall shearing stress. Compare the results.

6-13^c For $\zeta > 1.0$, that is Pr < 1.0, make the necessary modifications to the ZETA computer program to obtain the values of Nu/Re$^{1/2}$Pr$^{1/3}$ for Prandtl numbers of 0.01, 0.1, and

1.0. Compare with the results obtained from the ZETA program for $Pr > 1.0$. For a Prandtl number 0.01, compare the computer results with the approximate results assuming slug flow given by Eq. (6-99).

6-14   For laminar free convection on a vertical surface, assume a third-order velocity profile given by

$$\frac{u}{U_{ref}} = C_1 + C_2\eta + C_3\eta^2 + C_4\eta^3$$

rather than the fourth-order profile given by Eq. (6-117). Derive an expression for $Nu = f(Gr_x, Pr)$.

6-15   Consider laminar, fully developed flow in a circular tube. Assume that a second-degree polynomial is chosen to represent the temperature profile. Let

$$\theta = \frac{T_w - T}{T_w - T_B} = C_1 + C_2\eta + C_3\eta^2$$

(*a*) List three appropriate boundary conditions, and evaluate C_1, C_2, and C_3.

(*b*) Use the first law of thermodynamics, and show that

$$\left(\frac{\partial\theta}{\partial\eta}\right)_{\eta=1} = \frac{\dot{m}c_p(dT_B/dz)}{2\pi k(T_w - T_B)}$$

(*c*) Show that

$$\theta = \frac{U_0 R^2(dT/dz)}{8\alpha(T_w - T_B)}(\eta^2 - 1)$$

(*d*) Calculate the Nusselt number $\overline{Nu} = \bar{h}D/k$.

6-16   Using the turbulent boundary-layer profile given by Eq. (6-112) along with the wall shearing stress given by Eq. (6-113), derive Eq. (6-115). Show that Eq. (6-112) gives unrealistic results if it is incorrectly used to calculate the wall shearing stress.

6-17   For laminar, internal, fully developed, constant-property flow without internal generation, the Nusselt number is given by $Nu = 3.66$ for constant wall temperature and $Nu = 4.36$ for uniform wall heat flux. If Reynolds' analogy is incorrectly used to calculate the Nusselt number for the flow, what answer is obtained? Discuss why Reynolds' analogy fails, even when $Pr = 1.0$.

6-18   An empirical equation for laminar flow in a tube with constant wall temperature is given by

$$Nu_d = \frac{3.66 + 0.0668(d/L)Re_d\,Pr}{1 + 0.04[(d/L)Re_d\,Pr]^{2/3}}$$

The corresponding analytical solution for laminar flow with constant properties is $Nu_d = 3.66$. Compare the two solutions and explain the differences. When are they approximately equal?

7

EMPIRICAL METHODS

7-1 Empirical Results

The analysis of convective heat transfer presented in this book has concentrated on boundary-layer-type flow. There are many practical applications where a boundary-layer analysis adequately describes the convective behavior. However, there are also many examples of flow patterns which cannot be described by the boundary-layer model. Such regions are found in the wake of separated flows around nonslender bodies, in free convection of gases at Grashof numbers less than 10^4, in free convection from horizontal surfaces, and in most internal convective situations. Many empirical correlations exist in the literature which allow estimations of local and average convection heat transfer coefficients. Some of these are tabulated in this section.

External convection around cylinders and spheres occurs frequently in practice. The boundary-layer analysis in the previous chapter can be applied up to the point of boundary-layer separation, that is, the point where $(\partial u/\partial y)_{y=0} = 0$. In laminar flow the separation point on a cylinder in cross flow occurs at $\theta \approx 80°$. The separation point in turbulent flow occurs near $\theta \approx 130°$ and is indicated in Fig. 7-1. When transition to turbulence occurs before separation, the point of boundary-layer separation is located at larger values of θ because of the increased exchange of momentum within the turbulent boundary layer which helps to overcome the adverse pressure

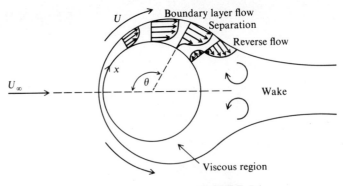

FIGURE 7-1
Cross flow around a cylinder.

gradient. This phenomenon reduces the size of the wake and has a significant effect on the heat transfer rate between the cylinder surface and the fluid. Thus, the variation of the local convection heat transfer coefficient is a strong function of θ. The variations of Nu_θ versus θ as measured by W. H. Giedt, Ref. 7-1, are shown in Fig. 7-2 for flow over a circular cylinder in cross flow.

The influence of transition, turbulence, and separation on the local Nusselt number can be seen in Fig. 7-2. For curve 1, the Reynolds number is below the critical Reynolds number where transition from laminar to turbulent flow occurs. The maximum value of the local convection heat transfer coefficient occurs at the stagnation point and then decreases with increasing θ because of the thickening of the boundary layer. Boundary-layer separation occurs at $\theta \approx 80°$. The increase in h_θ for $\theta > 80°$ is due to the vortex motion in the wake of the separated flow. Curve 2 for $Re = 10^5$ is similar to curve 1 except that the wake begins to have a greater influence on h_θ at the rear of the cylinder. Curve 3 shows the effect of boundary-layer transition from laminar to turbulent flow before separation. Transition occurs at $\theta \approx 85°$, and a sudden increase in h_θ occurs because of the effect of turbulence. Separation does not occur until $\theta \approx 130°$ where the second minimum in h_θ is noted. The delay in boundary-layer separation which reduces the wake region reduces the form drag of the cylinder. Curve 4 clearly shows the effects of transition and separation. Notice that the maximum value of h_θ now occurs within the turbulent boundary layer rather than at the stagnation point.

For bodies with strong curvature, it is often not practical to obtain a mathematical expression for the variation of the local convection heat transfer coefficient over the entire surface. When this is true, $\bar{h}$ and hence $\overline{Nu}$ must be obtained experimentally. Some of the many empirical correlations for the average Nusselt number as a function

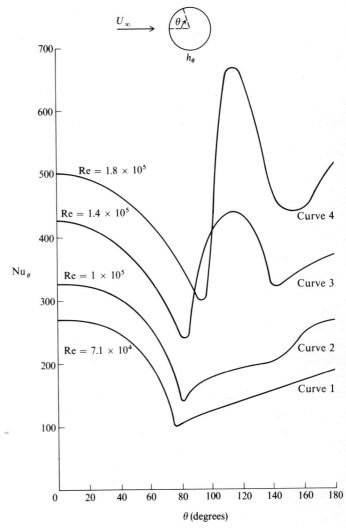

FIGURE 7-2
Local Nusselt number around a cylinder in cross flow.

of Re and Pr for external forced convection over bodies of various shapes are given in Table 7-1.

Correlations for external and internal free convection are given in Tables 7-2 and 7-3, respectively. The type of convective pattern is a strong function of the surface boundary conditions as well as the Grashof and Prandtl numbers. When making estimates of free-convective heat transfer rates, one must be aware that combined effects of forced and free convection exist when the Grashof number is the same order of magnitude as the square of the Reynolds number.

Correlations for both laminar and turbulent, internal forced convection are also numerous. The complications added by the effects of surface roughness, property variation with temperature, entrance-length effects, and variable cross-sectional flow area limit the accuracy of the empirical results. Table 7-4 gives some of the more common correlations.

Table 7-1   EXTERNAL FORCED CONVECTION
[Properties evaluated at the film temperature $T_f = (T_w + T_\infty)/2$]

Configuration	Limitations	Correlation	Comments
Cylinder in cross flow	$10^{-1} < \text{Re} < 10^5$	$\overline{\text{Nu}}_D = (0.35 + 0.56\text{Re}^{0.52})\text{Pr}^{0.3}$	Liquids only
	$0.4 < \text{Re} < 4$	$\overline{\text{Nu}}_D = 0.891\text{Re}^{0.33}$	Gases only
	$4.0 < \text{Re} < 40$	$\overline{\text{Nu}}_D = 0.821\text{Re}^{0.385}$	Gases only
	$40 < \text{Re} < 4 \times 10^3$	$\overline{\text{Nu}}_D = 0.615\text{Re}^{0.466}$	Gases only
	$4 \times 10^3 < \text{Re} < 4 \times 10^4$	$\overline{\text{Nu}}_D = 0.174\text{Re}^{0.618}$	Gases only
	$4 \times 10^4 < \text{Re} < 4 \times 10^5$	$\overline{\text{Nu}}_D = 0.0239\text{Re}^{0.805}$	Gases only (Refs. 7-25 and 7-26)
Sphere in cross flow	$17 < \text{Re} < 7 \times 10^4$	$\overline{\text{Nu}}_D = 0.37\text{Re}^{0.6}$	Gases only
	$1.0 < \text{Re} < 2 \times 10^3$	$\overline{\text{Nu}}_D = (0.97 + 0.68\text{Re}^{0.5})\text{Pr}^{0.3}$	Liquids only (Refs. 6-11 and 7-27)
Square bar in cross flow	$5 \times 10^3 < \text{Re} < 10^5$	$\overline{\text{Nu}}_L = 0.092\text{Re}^{0.675}$	Gases only (Ref. 7-24)

Table 7-2 EXTERNAL FREE CONVECTION
[Properties evaluated at the film temperature $T_f = (T_w + T_\infty)/2$]

Configuration	Limitations	Correlation	Comments
Horizontal cylinder	$10^3 < GrPr < 10^9$ $10^9 < GrPr < 10^{12}$ $Pr << 1$	$\overline{Nu}_D = 0.53(GrPr)^{1/4}$ $\overline{Nu}_D = 0.126(GrPr)^{1/3}$ $\overline{Nu}_D = 0.53(GrPr^2)^{1/4}$	Laminar flow Turbulent flow Liquid metals (Refs. 6-11 and 7-29)
Sphere	$10^3 < GrPr < 10^7$ $1 < Gr < 10^5$	$\overline{Nu}_D = 0.51(GrPr)^{1/4}$ $\overline{Nu}_D = 2 + 0.43(GrPr)^{1/4}$	Laminar flow Non-boundary- layer flow (Refs. 6-11 and 7-28)
Horizontal surface, heated surface up or cooled surface down	$10^5 < GrPr < 2 \times 10^7$ $2 \times 10^7 < GrPr < 3 \times 10^{10}$	$\overline{Nu}_L = 0.54(GrPr)^{1/4}$ $\overline{Nu}_L = 0.14(GrPr)^{1/3}$	Laminar flow Turbulent flow (Ref. 6-11)
Horizontal surface, heated surface down or cooled surface up	$3 \times 10^5 < GrPr < 3 \times 10^{10}$	$\overline{Nu}_L = 0.27(GrPr)^{1/4}$	Laminar flow (Ref. 6-11)
Vertical surface	$10^3 < GrPr < 10^9$ $10^9 < GrPr < 10^{12}$	$\overline{Nu}_L = 0.59(GrPr)^{1/4}$ $\overline{Nu}_L = 0.13(GrPr)^{1/3}$	Laminar flow Turbulent flow (Ref. 6-11)

Table 7-3 INTERNAL FREE CONVECTION
(Properties evaluated at the mean bulk temperature)

Configuration	Limitations	Correlation	Comments
T_2 ↓ / L / T_1 ↑ Horizontal plates	$0.02 < \text{Pr} < 8{,}750$ $3 \times 10^5 < \text{GrPr} < 10^9$	$\overline{\text{Nu}}_L = 0.069 \text{Gr}^{1/3}\text{Pr}^{0.074}$	$T_1 > T_2$ (Ref. 7-31)
L / H Vertical plates	$2 \times 10^4 < \text{Gr} < 2 \times 10^5$ $2 \times 10^5 < \text{Gr} < 1 \times 10^7$ $\text{GrPr} < 10^3$ $10^3 < \text{GrPr} < 10^7$	$\overline{\text{Nu}}_L = 0.18 \text{Gr}^{1/4}(H/L)^{-1/9}$ $\overline{\text{Nu}}_L = 0.065 \text{Gr}^{1/3}(H/L)^{-1/9}$ $\overline{\text{Nu}}_L = 1.0$ $\overline{\text{Nu}}_L = 0.280(\text{GrPr})^{1/4}(H/L)^{-1/4}$	Air Air Liquids Liquids (Refs. 7-24 and 7-30)

Table 7-4 INTERNAL FORCED CONVECTION
(Properties evaluated at mean bulk temperature of fluid, except where indicated elsewhere by use of a subscript $w \equiv$ wall temperature)

Configuration	Limitations	Correlation	Comments
D / T_w Circular tubes of constant cross section	$\text{Re} > 10^4$ $0.6 < \text{Pr} < 160$ $L/D > 60$ Constant fluid properties	$\overline{\text{Nu}}_D = 0.023\text{Re}^{0.8}\text{Pr}^{0.4}$ $\overline{\text{Nu}}_D = 0.023\text{Re}^{0.8}\text{Pr}^{0.3}$	Heating fluid Cooling fluid (Ref. 7-32)
	$\text{Re} > 10^4$ $0.7 < \text{Pr} < 16{,}700$ Constant T_w	$\overline{\text{Nu}}_D = 0.023\text{Re}^{0.8}\text{Pr}^{0.33}(\mu/\mu_w)^{0.14}$	Fully developed turbulent flow (Ref. 7-33)
	$10 < L/D < 400$	$\overline{\text{Nu}}_D = 0.036\text{Re}^{0.8}\text{Pr}^{0.33}C$ where $C = (D/L)^{0.055}$	Flow in entrance region (Ref. 7-34)
	Constant wall heat flux $10^2 < \text{RePr} < 10^4$ $L/D > 60$	$\overline{\text{Nu}}_D = 0.625(\text{RePr})^{0.4}$	Liquid metals (Ref. 7-35)
	Constant wall temperature $\text{RePr} > 10^2$ $L/D > 60$	$\overline{\text{Nu}}_D = 5 + 0.025(\text{RePr})^{0.8}$	Liquid metals (Ref. 7-36)
	$\text{RePr } D/L > 10$ $\text{Re} < 2{,}300$	$\overline{\text{Nu}}_D = 1.86(\text{RePr})^{1/3}C$ where $C = (D/L)^{0.33}(\mu/\mu_w)^{0.14}$	Laminar flow (Ref. 7-33)
	Constant wall temperature $\text{Re} < 2{,}300$ $\text{RePr } D/L > 10$	$\overline{\text{Nu}}_D = [3.66 + 0.0668\text{Gz}/(1 + 0.04\text{Gz}^{2/3})]$ $\times (\mu_B/\mu_w)^{0.14}$ where $\text{Gz} = \text{RePr}D/L$	Laminar flow (Ref. 7-37)

The numerical results for internal, turbulent forced convection obtained by Haberstroh in Ref. 6-16, and reproduced in Fig. 6-6, were also presented in the form of an empirical correlation. This relationship was given by

$$\overline{\mathrm{Nu}} = 6(1 - \mathrm{Pr}^{0.8}) + (0.0189\mathrm{Pr}^{0.973} - 8 \times 10^{-5})\mathrm{Re}^{0.814\mathrm{Pr}^{-0.035}}$$

This equation predicts the calculated results with a maximum deviation of 8 percent for $0.01 < \mathrm{Pr} < 10$ and $10^4 < \mathrm{Re} < 10^7$.

The correlations presented in this section are useful for estimating heat transfer rates and determining required heat transfer surface areas. However, the correlations are limited to specific ranges of Prandtl, Reynolds, and/or Grashof numbers and must be used with caution.

7-2   One-dimensional Flow

When appropriate analytical, numerical, or empirical relations are available for the Nusselt number, it is possible to calculate the convection heat transfer coefficient without detailed knowledge of the temperature and velocity distributions. For internal forced convection, this procedure leads to a simplified analysis based upon the assumption of a one-dimensional temperature profile $T = T(z)$ and velocity profile $u = u(z)$. If the tube has a constant cross-sectional area, the velocity for a steady, constant-property flow can be assumed constant. Analysis based upon one-dimensional flow is commonly used to estimate the performance of heat exchangers.

The flow of a fluid through a single pipe is an example of a simple heat exchanger. Most heat exchangers are designed for turbulent-flow operation to reduce the size and cost. The increased pressure drop due to turbulence is offset by greater heat transfer rates. Fully developed turbulent profiles of velocity and temperature in a fluid with constant properties are more one-dimensional in nature than the parabolic profiles which occur in laminar flow. This fact is due to the exchange of momentum and energy between the fluid layers caused by turbulent mixing. Thus, a simple thermodynamic analysis based upon the assumption of one-dimensional, steady flow is useful.

Consider the heat exchanger shown in Fig. 7-3. The pipe-wall temperature is maintained constant at T_w. This often occurs physically when condensation of a fluid occurs on the outside surface of the pipe. Let the constant-property fluid inside the constant-diameter pipe be increased from T_1 to T_2 under conditions of steady, single-phase, fully developed turbulent flow. The problem is to calculate the required length of the pipe for known values of T_1 and T_2. Assuming one-dimensional velocity and temperature profiles, the first law of thermodynamics applied to the indicated control volume within the pipe, assuming negligible changes in kinetic and potential energy, gives (Ref. 7-2)

$$q = \dot{m}\, dh = \dot{m}c_p\, dT \qquad (7\text{-}1)$$

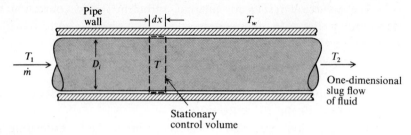

FIGURE 7-3
A simple heat exchanger.

where dh is the change in enthalpy across the differential element, and $\dot{m}$ is the mass flow rate of the fluid. The convective heat transfer rate between the pipe wall and the fluid is q. The rate of heat transfer can also be expressed by Newton's law of cooling given by

$$q = h_z \, dA_s \, (T_w - T) \qquad (7\text{-}2)$$

where dA_s is the surface area of the differential control volume in contact with the pipe, that is, $dA_s = \pi D \, dx$. The local convection heat transfer coefficient for internal convection between the pipe and the fluid is h_z, and the fluid temperature throughout the differential control volume is $T = T(z)$.

Combining Eqs. (7-1) and (7-2) gives

$$h_z \pi D(T_w - T) \, dx = \dot{m} c_p \, dT \qquad (7\text{-}3)$$

Integration gives

$$L = \int_0^L dx = \frac{\dot{m}}{\pi D} \int_{T_1}^{T_2} \frac{c_p \, dT}{h_z (T_w - T)} \qquad (7\text{-}4)$$

The value of h_z is obtained from the proper Nusselt number correlation given in the previous section. Since the fluid properties are constant and the flow is fully developed, Re and Pr are constants. It follows that under these conditions the heat transfer coefficient for internal forced convection is constant, $h_z = \bar{h}$. Also, since $T_w =$ constant, Eq. (7-4) becomes

$$L = \frac{-\dot{m} c_p}{\pi D \bar{h}} \ln \frac{T_w - T_2}{T_w - T_1} \qquad (7\text{-}5)$$

For heating of a gas in turbulent flow, the correlation for $\bar{h}$ given in Table 7-4 is

$$\bar{h} = \frac{0.023k}{D} \left(\frac{\rho V D}{\mu}\right)^{0.8} \left(\frac{\mu c_p}{k}\right)^{0.4} \qquad (7\text{-}6)$$

or

$$\bar{h} = \frac{0.023k}{D} \left(\frac{4\dot{m}}{\pi D \mu}\right)^{0.8} \left(\frac{\mu c_p}{k}\right)^{0.4} \qquad (7\text{-}7)$$

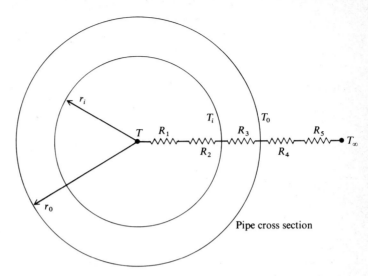

FIGURE 7-4
Resistance network for heat exchanger.

Equations (7-5) and (7-7) can be used to calculate the required heat exchanger length for a given turbulent-flow situation. Other correlations for different fluids in either laminar or turbulent flow may be used in a similar manner to obtain $\bar{h}$. However, the one-dimensional flow assumption is much less accurate for laminar flow because of the parabolic nature of the velocity profile.

7-3   Overall Conductance

Often the wall temperature of the pipe may not be known, but the temperature of the fluid surrounding the external surface of the pipe may be available. The steady-state, one-dimensional heat transfer from the internal fluid to the environment may be estimated by considering the series resistance network shown in Fig. 7-4. The first resistance $R_1 = 1/\bar{h}A$ is the resistance to convection between the internal fluid and the pipe. R_2 and R_4 are added resistances which may be significant if fouling caused by foreign film deposits is present on the tube surfaces. This resistance due to fouling can be written $R = 1/\bar{h}_f A$, where $1/\bar{h}_f$ is called a *fouling factor*. It is a function of velocity, temperature, fluid properties, and length of heat exchanger service. Values of the fouling factor (h-ft^2-°F/Btu) usually fall between 5×10^{-4} and 5×10^{-3}. (See Refs. 7-3 and 7-4.) The resistance R_3 is the conductive resistance within the pipe wall. The resistance to steady-state, radial, one-dimensional conduction was shown to be $R_3 = \ln(r_o/r_i)/2\pi kL$ in Sec. 4-10. The final resistance R_5 is the equivalent resistance to energy transfer by convection and radiation between the exterior pipe surface and the environment. This equivalent resistance was discussed in Sec. 1-8.

An overall conductance U is defined based upon the total resistance between the fluid at temperature T and the surroundings at temperature T_∞. For negligible scale resistance due to fouling and negligible thermal radiation, the overall conductance for the pipe shown in Fig. 7-4 is defined by

$$UA_s = \frac{1}{\sum R} = \frac{1}{\dfrac{1}{\bar{h}_i A} + \dfrac{\ln (r_o/r_i)}{2\pi kL} + \dfrac{1}{\bar{h}_o A}} \qquad (7\text{-}8)$$

The surface area A_s is a reference area, such as the inside surface area of the pipe. This choice of A_s leads to

$$U = \frac{1}{\bar{h}_i + \dfrac{r_i}{k} \ln \dfrac{r_o}{r_i} + \dfrac{r_i}{r_o} \dfrac{1}{\bar{h}_o}} \qquad (7\text{-}9)$$

If thermal radiation is significant, $\bar{h}_o$ can be replaced by $\bar{h}_r$. (See Sec. 1-8.) The added resistance due to scale fouling gives a modified conductance U_f expressed as

$$\frac{1}{U_f} = \frac{1}{U} + \frac{1}{\bar{h}_{f,i}} + \frac{r_i}{r_o \bar{h}_{f,o}} \qquad (7\text{-}10)$$

The heat transfer between T and T_∞ across the differential control volume is then given by

$$q = U_f \, dA_s \, (T_\infty - T) \qquad (7\text{-}11)$$

The expression for calculating the unknown length of the pipe required to obtain a given temperature change in the fluid inside the pipe is given by

$$L = \frac{\dot{m}}{\pi D} \int_{T_1}^{T_2} \frac{c_p \, dT}{U_f (T_\infty - T)} \qquad (7\text{-}12)$$

For constant properties and constant T_∞, integration gives

$$L = \frac{-\dot{m} c_p}{\pi D U_f} \ln \frac{T_\infty - T_2}{T_\infty - T_1} \qquad (7\text{-}13)$$

To analytically consider the effects of variable properties, two-dimensional velocity and temperature profiles must be used in the mathematical model. For liquids, the viscosity decreases with increasing temperature, but the reverse is true for gases. When a liquid coolant is used to cool a heated pipe, the viscosity near the heated surface will be substantially lower than the viscosity at the pipe centerline. This reduction in viscosity near the pipe surface can result in an increase in the heat transfer coefficient by several percent, as compared to the value for constant properties. A similar increase occurs when a hot gas flows through a cool pipe. Conversely, the heat transfer between a cold surface being heated by a hot liquid or a hot surface being cooled by a gas is reduced when the fluid viscosity is temperature dependent. Analytical solutions for variable-property fluids are given in Ref. 7-5.

In practice, empirical expressions, such as Eq. (7-7), based upon mean property values can be used to express $\bar{h} = f(\text{Re}, \text{Pr}, k/D)$. The effect of variable properties on the heat transfer is then accounted for in an approximate manner by multiplying by a correction factor such as $(\mu/\mu_w)^n$, where n is an empirically determined constant exponent, μ is the average fluid viscosity, and μ_w is the fluid viscosity evaluated at the wall temperature. Variable-property correlations are indicated in Table 7-4.

EXAMPLE 7-1   For the single-pipe heat exchanger shown in Fig. 7-3, let the inlet temperature of the gas coolant be $T_1 = 20°C$ and the outlet temperature be $T_2 = 120°C$. The coolant properties at the mean temperature of $T_m = 70°C$ are $c_p = 0.9$ kJ/kg-°C, $\mu = 1.7 \times 10^{-5}$ Pl $= 1.7 \times 10^{-5}$ N-s/m², and $k = 0.02$ W/m-°C. The pipe-wall temperature is constant at $T_w = 150°C$. For a coolant flow rate of 9.0 kg/h through a pipe of inside diameter 1.25 cm, calculate the length of pipe required.

$$\bar{h} = \frac{0.023k}{D}\left(\frac{4\dot{m}}{\pi D \mu}\right)^{0.8}\left(\frac{\mu c_p}{k}\right)^{0.4}$$

$$\bar{h} = 36.8(1.5 \times 10^4)^{0.8}(0.765)^{0.4} = 73.3 \text{ W/m}^2\text{-°C}$$

$$L = \frac{-\dot{m}c_p}{\pi D \bar{h}} \ln \frac{T_w - T_2}{T_w - T_1} = 11.5 \text{ m}$$

If a mean temperature difference between the wall and gas $T_w - T_m$ is incorrectly used to calculate the length, one obtains

$$q = \bar{h}A_s(T_w - T_m) = \dot{m}c_p(T_2 - T_1)$$

$$L = \frac{\dot{m}c_p(T_2 - T_1)}{\bar{h}\pi D(T_w - T_m)} = 9.8 \text{ m}$$

////

EXAMPLE 7-2   Consider the heat exchanger analyzed in Example 7-1. If the length of the pipe is increased to 15 m and all other parameters remain the same except the outlet temperature, calculate the outlet temperature.

$$L = \frac{-\dot{m}c_p}{\pi D \bar{h}} \ln \frac{T_w - T_2}{T_w - T_1} = 15$$

$$\frac{T_w - T_2}{T_w - T_1} = \exp(-1.92) = 0.147$$

$$T_2 = 130.9°C$$

////

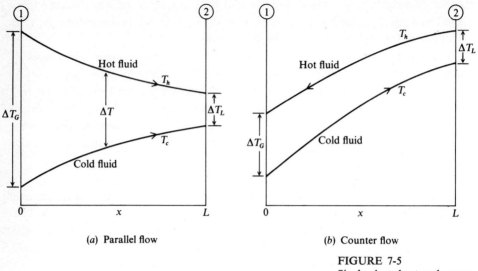

(a) Parallel flow (b) Counter flow

FIGURE 7-5
Single-phase heat exchangers.

7-4   Single-phase Heat Exchangers

Heat exchangers can be classified as either *recuperators* or *regenerators*. A recuperator consists of a heat transfer surface which separates two fluids flowing simultaneously through the heat exchanger. Heat transfer occurs directly from the hot fluid to the cold fluid across the surface. In a regenerator, thermal energy obtained from the hot fluid is stored in a structural matrix, and later this energy is transferred to the cold fluid. The heat exchanger matrix is alternately traversed by the hot and cold streams.

The temperature change which occurs in a fluid during its path through a recuperator is directly proportional to the heat transfer and inversely proportional to the fluid heat capacity $C = \dot{m}c_p$. Often the flow rates are such that significant temperature changes occur in both fluids when there is no phase change involved. When this is true, the values of T_w and T_∞ cannot be assumed constant as was done in the previous section.

The flow configurations that occur in a recuperator are classified as parallel flow, counterflow, single-pass cross flow, or multipass cross flow. The temperature changes that occur in a single-phase, parallel-flow or counterflow heat exchanger can be conveniently displayed on a two-dimensional plot as shown in Fig. 7-5. In this figure, ΔT_G indicates the greatest temperature difference, and ΔT_L indicates the lowest temperature difference between the two fluids.

The two concentric pipes shown in Fig. 7-6 can operate as either a parallel-flow or a counterflow heat exchanger, with one fluid inside the inner tube and one in the

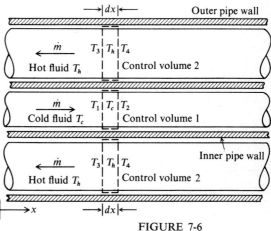

FIGURE 7-6
Counterflow heat exchanger.

annulus. Consider counterflow operation as indicated in the figure. Applying the conservation of energy principle to the control volume for the cold fluid flowing through the inner tube and assuming steady, incompressible flow with one-dimensional velocity and temperature profiles gives

$$q_c = \dot{m}(dh)_c = (\dot{m}c_p)_c \, dT_c = C_c \, dT_c \qquad (7\text{-}14)$$

Applying the conservation of energy principle to the fluid in the annulus gives

$$q_h = \dot{m}(dh)_h = (\dot{m}c_p)_h \, dT_h = C_h \, dT_h \qquad (7\text{-}15)$$

Now consider a differential section of the tube wall as a closed thermodynamic system. Neglecting axial conduction along the tube wall and neglecting radiation heat transfer allows one to write $-q_h = q_c$. Using the above equations gives

$$-C_h \, dT_h = C_c \, dT_c \qquad (7\text{-}16)$$

The negative sign on q_h is required since dT_h is negative, $(T_{out} - T_{in})_h < 0$, and dT_c is positive, $(T_{out} - T_{in})_c > 0$.

The equality in Eq. (7-16) does not provide the information necessary to determine the size of a heat exchanger. Empirical results for heat transfer rates must also be used. Using Eq. (7-11) and neglecting scale resistance give

$$q_c = C_c \, dT_c = U \, dA_s \, (T_h - T_c) \qquad (7\text{-}17)$$

where dA_s is the inside surface area of the control volume in contact with the tube. Since the product UA is the reciprocal of the total resistance between T_h and T_c, it is useful to write Eq. (7-17) as

$$C_c\, dT_c = UA(T_h - T_c)\frac{dx}{L} \qquad (7\text{-}18)$$

where A is the total inside surface area of the tube. Solving for dT_c gives

$$dT_c = \frac{UA(T_h - T_c)\, dx}{C_c L} \qquad (7\text{-}19)$$

Similarly, but realizing that dx is negative for the annulus control volume in counterflow, we have

$$dT_h = \frac{UA(T_h - T_c)\, dx}{C_h L} \qquad (7\text{-}20)$$

Subtracting Eq. (7-19) from Eq. (7-20) gives

$$\frac{d(T_h - T_c)}{T_h - T_c} = \frac{UA}{L}\left(\frac{1}{C_h} - \frac{1}{C_c}\right) dx \qquad (7\text{-}21)$$

Assuming constant values for $\dot{m}$, c_p, and U and integrating give

$$\ln \Delta T = \frac{UA}{L}\left(\frac{1}{C_h} - \frac{1}{C_c}\right)x + C_1 \qquad (7\text{-}22)$$

where $\Delta T = T_h - T_c$.

Using the boundary condition $\Delta T = \Delta T_1$ at $x = 0$ gives

$$\frac{\Delta T}{\Delta T_1} = \exp\left[\frac{UA}{C_c}\left(\frac{C_c}{C_h} - 1\right)\frac{x}{L}\right] \qquad (7\text{-}23)$$

for a counterflow heat exchanger.

For a parallel-flow heat exchanger, dx is positive for the annulus control volume, and Eq. (7-20) is written

$$dT_h = \frac{-UA(T_h - T_c)\, dx}{C_h L} \qquad (7\text{-}24)$$

This expression leads to

$$\frac{\Delta T}{\Delta T_1} = \exp\left[\frac{UA}{C_c}\left(-\frac{C_c}{C_h} - 1\right)\frac{x}{L}\right] \qquad (7\text{-}25)$$

for a parallel-flow heat exchanger.

Two important nondimensional parameters appear in Eqs. (7-23) and (7-25). The ratio UA/C is called the *number of transfer units* and is designated by NTU. The ratio $C_c/C_h = (\dot{m}c_p)_c/(\dot{m}c_p)_h$ is a ratio of fluid heat capacities. These two parameters are used in the design and selection of heat exchangers. Many graphs which give empirically determined relationships between NTU and C_c/C_h for various types of heat exchanger configurations are available in the literature, along with sample design calculations based upon their use. References 7-6 and 7-7 are especially useful for heat exchanger calculations.

Specifying the condition that $\Delta T = \Delta T_2 = T_{h,\text{in}} - T_{c,\text{out}}$ at $x = L$ for a counter-flow heat exchanger allows Eq. (7-23) to be written

$$\ln \frac{\Delta T_2}{\Delta T_1} = \frac{UA}{C_c}\left(\frac{C_c}{C_h} - 1\right) \qquad (7\text{-}26)$$

Combining Eqs. (7-23) and (7-26) gives

$$\frac{\Delta T}{\Delta T_1} = \exp\left(\frac{x}{L}\ln\frac{\Delta T_2}{\Delta T_1}\right) = \left(\frac{\Delta T_2}{\Delta T_1}\right)^{x/L} \qquad (7\text{-}27)$$

This same equation is also obtained for a parallel-flow heat exchanger when $\Delta T_1 = T_{h,\text{in}} - T_{c,\text{in}}$ and $\Delta T_2 = T_{h,\text{out}} - T_{c,\text{out}}$.

A mean temperature difference between T_h and T_c is defined by use of the mean value theorem of calculus. This value is given by

$$\overline{\Delta T} = \frac{1}{L}\int_0^L \Delta T\, dx \qquad (7\text{-}28)$$

Using Eq. (7-27) gives

$$\overline{\Delta T} = \frac{\Delta T_1}{L}\int_0^L \left(\frac{\Delta T_2}{\Delta T_1}\right)^{x/L} dx \qquad (7\text{-}29)$$

Equation (7-29) is integrated by use of a standard integration formula (Ref. 7-8) given by

$$\int a^x \ln a\, dx = a^x \qquad (7\text{-}30)$$

To obtain the form of Eq. (7-30), we write Eq. (7-29) as

$$\overline{\Delta T} = \frac{\Delta T_1}{\ln(\Delta T_2/\Delta T_1)}\int_0^L \left(\frac{\Delta T_2}{\Delta T_1}\right)^{x/L} \ln\frac{\Delta T_2}{\Delta T_1}\frac{dx}{L} \qquad (7\text{-}31)$$

The results of the integration can then be written as

$$\overline{\Delta T} = \frac{\Delta T_1}{\ln(\Delta T_2/\Delta T_1)}\left[\left(\frac{\Delta T_2}{\Delta T_1}\right)^{x/L}\right]_0^L \qquad (7\text{-}32)$$

or

$$\overline{\Delta T} = \frac{\Delta T_2 - \Delta T_1}{\ln(\Delta T_2/\Delta T_1)} = \frac{\Delta T_1 - \Delta T_2}{\ln(\Delta T_1/\Delta T_2)} \qquad (7\text{-}33)$$

If $\Delta T_1 = \Delta T_G$, the greatest temperature difference between T_h and T_c, and $\Delta T_2 = \Delta T_L$, the least temperature difference, then the expression for $\overline{\Delta T}$ can be written for either a parallel-flow or a counterflow heat exchanger as

$$\overline{\Delta T} = \frac{\Delta T_G - \Delta T_L}{\ln (\Delta T_G/\Delta T_L)} \qquad (7\text{-}34)$$

This expression is called the *log-mean temperature difference*. It is based upon the assumptions of constant U, constant flow rates, and constant fluid properties. The total heat transfer between the fluids in the heat exchanger is then written

$$q = UA \, \overline{\Delta T} \qquad (7\text{-}35)$$

In the process of selecting and designing heat exchangers and supporting equipment, it is desirable to optimize the selection from a cost viewpoint as well as from a physical viewpoint. The digital computer has made it possible to apply many optimizing search routines to help make optimum decisions. These techniques are discussed in Ref. 7-21.

7-5   Characteristics of Heat Exchangers

For the purpose of comparison, heat exchangers can also be classified as gas-gas, liquid-liquid, liquid-gas, and two-phase heat exchangers. The purpose of the following discussion is to point out the characteristics of these types of heat exchangers. A more detailed discussion is given in Ref. 7-6.

The magnitude of the heat transfer coefficients for gases is 10 to 100 times lower than that occurring in liquids. Thus, heat exchangers with large surface areas are required to transmit the required heat transfer rates. For this reason it is common to see fins used to increase the surface area. The heat transfer matrix often consists of many closely packed corrugated plates to separate the flow of the two gas streams. Since the heat capacity $\dot{m}c_p$ of the matrix is much greater than the heat capacity of the gases, a regenerator may be used to allow the matrix to act as an energy storage device during the cooling of the hot gas. Regenerators are being used on some automobile gas turbines to preheat the incoming air using energy from the exhaust gases. This preheating produces a substantial increase in the thermodynamic cycle efficiency. Intercooling between the air compressor stages of a gas turbine to reduce the work of compression also requires a gas-gas heat exchanger.

Liquid-liquid heat exchangers are found in many diverse applications such as controlling blood temperature by a constant temperature water bath and cooling oil to lubricate marine machinery by using sea water. The chemical and petroleum industries use many types of liquid-liquid heat exchangers. A liquid-liquid heat exchanger is characterized by a shell-and-tube-type construction rather than by fins and compact matrix design. Tube banks are often used with one liquid flowing inside the tube and the other flowing across the tube bank. Tube banks which consist of in-line tubes give a lower pressure drop but also lower values of heat transfer coefficients than staggered tube banks. Tube spacing and tube size must be considered in each design, along with tube alignment and baffle placement.

Many warm-blooded animals have an efficient liquid-liquid heat exchanger which nature has provided. Warm blood from the heart flows through a matrix of blood vessels and is cooled by the blood returning to the heart. Since the warm blood is cooled to near ambient temperature before flowing through the legs and other parts exposed to the environment, very little thermal energy is lost from the body due to heat transfer to the surroundings.

Liquid metal heat exchangers are often used to transport thermal energy at high temperatures. The operating temperature of liquid metal heat exchangers is 250–850°C. The correlations given in Table 7-4 show that high values of the convection heat transfer coefficient can be obtained with liquid metals such as sodium, bismuth, sodium-potassium eutectics, and certain chlorides, fluorides, and hydroxides.

Gas-liquid heat exchangers are perhaps the most common type. Examples include radiators for cooling water and oil, recuperators for preheating liquid fuel with hot exhaust gases, and cooling towers for reducing thermal pollution. Since the heat transfer coefficients on the gas side are orders of magnitude lower than on the liquid side, fin surfaces on the gas side are common. Also, flattened tubes, rather than circular tubes, are commonly used to carry the liquid to give a larger gas-flow area and reduce the wake behind the tubes.

The most common forms of two-phase heat exchangers are boilers and condensers. These devices are discussed in the next section. Recently, a new heat exchanger concept called a *heat pipe* has found many applications. A heat pipe is a two-phase heat exchanger which can have an effective thermal conductivity several thousand times that of metals such as copper. A sketch of a simple heat pipe is shown in Fig. 7-7. Evaporation occurs at one end of the heat pipe at the required operating temperature. This phase change develops a substantial vapor pressure. The vapor condenses at the other end, and the condensate is returned to the evaporator end by capillary action through a wick material. The wick lies along the inside surface of the pipe. The capillary driving pressure must overcome the viscous pressure loss within the wick, the loss due to the opposing vapor flow, and perhaps a pressure head due to gravity. However, the heat pipe is unique in that it can operate in a gravity-free environment.

7-6 Two-phase Heat Transfer

Heat transfer with two-phase flow occurs in boilers and condensers. Evaporation normally occurs inside boiler tubes, and condensation normally occurs in the outside of condenser tubes. As long as the tube wall is wetted with liquid, the heat exchanger surfaces within which or over which a change of phase occurs can usually be considered

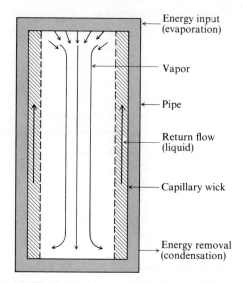

FIGURE 7-7
Heat pipe.

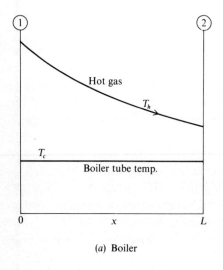

(a) Boiler

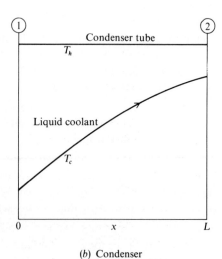

(b) Condenser

FIGURE 7-8
Two-phase heat exchangers.

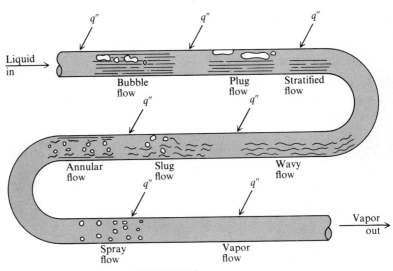

FIGURE 7-9
Horizontal two-phase flow.   (Adapted from Ref. 7-9.)

isothermal by neglecting the effect of pressure drop on the saturation temperature. The temperature profiles of the single-phase fluid and tube wall in a boiler and condenser can be represented as shown in Fig. 7-8.

The analysis of two-phase flow is currently a popular research area. An introduction to the subject was written by Gouse (Ref. 7-9). Because of the large density differences between the vapor and liquid, the orientation of the tube with respect to the gravity field strongly affects the type of two-phase flow. The flow is also a strong function of the *void fraction*, which is the ratio of the vapor volume to the liquid volume. Two-phase flow patterns that can occur within a horizontal tube with a constant heat transfer rate at the surface are indicated in Fig. 7-9. When evaporation begins, the vapor rises to the top of the pipe. This action is called *bubble flow*. The bubbles then form plugs of vapor as additional evaporation occurs. When a smooth interface exists between the two phases, the flow is called *stratified*. Stratified flow is an idealized condition. Waves normally form at the liquid-vapor interface, and these waves can cause slugs of liquid which intermittently fill the tube. For high void fractions an annular-type flow with a vapor core can form. The final two-phase flow pattern is a spray or mist flow which changes into a single-phase vapor flow.

The local convection heat transfer coefficient at the inner wall of the tube varies with both time and space coordinates. Hence, one must rely on average values based upon experimental measurements. The same type of difficulties arise when calculating the pressure drop in two-phase flow.

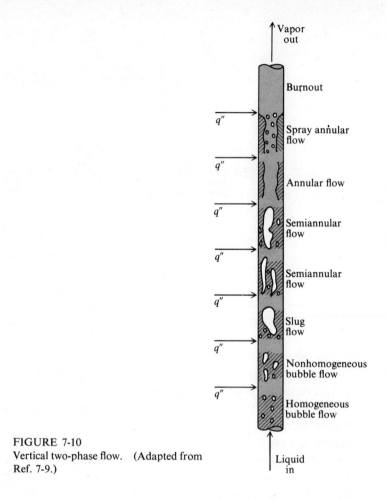

FIGURE 7-10
Vertical two-phase flow.   (Adapted from
Ref. 7-9.)

When two-phase flow occurs in a vertical tube, different flow patterns result due
to the gravity field orientation. The flow patterns tend to be symmetrical, and the
walls remain wetted with fluid at higher void fractions. For this reason, most forced-
convection boilers are designed with vertical pipe sections. Two-phase flow patterns
in a vertical tube are represented in Fig. 7-10. As heat transfer occurs from the hot
tube wall to the fluid, localized bubbles form on the surface when the local tube surface
temperature is above the liquid saturation temperature. If the bulk liquid is in the
subcooled state, the vapor bubbles lose energy to the liquid, and condensation causes
the bubbles to collapse. The period of the boiling process can be as small as 0.001 s
during subcooled boiling. Under these conditions, heat transfer rates can be estimated
by using the single-phase correlations for internal forced convection as presented in
Sec. 7-1.

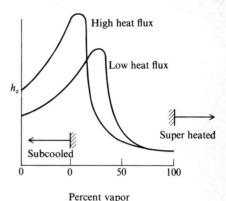

FIGURE 7-11
h_z for forced-convection boiling.

When the bulk temperature of the liquid reaches the saturation temperature, the bubbles of vapor remain in the flow. The agitation caused by the vapor-liquid inter- action greatly increases the local heat transfer coefficients. As the quality of the vapor- liquid mixture increases, the surface tension causes the vapor to form an annular flow region with a large vapor core. The vapor core is surrounded by liquid which wets the interior surface of the tube. When the vapor breaks through the liquid to the wall, the local heat transfer coefficient sharply decreases because of the insulating effect of the vapor as compared to the liquid. If the heat flux is constant along the tube length, a sharp increase in the local temperature of the tube occurs. This effect is referred to as the *burnout point*, and it is an important consideration in the design of boilers.

Due to the large viscosity difference between the liquid and vapor, the liquid velocity is only 5 to 15 percent as high as the vapor velocity. However, both the liquid velocity and vapor velocity increase as the quality of steam is increased. This increase in velocity coupled with a thinning of the liquid film at the tube wall can cause the burnout point to be reached very rapidly.

Typical axial variations of the local heat transfer coefficient are indicated in Fig. 7-11. The variation is a strong function of the rate of heat transfer from the tube to the two-phase flow. This is due to coupling between the heat transfer and pressure drop in two-phase flow. The pressure and temperature are dependent properties in the two-phase region. Thus, the pressure drop due to friction lowers the saturation temperature of the bulk fluid. The initial increase in h_z shown in Fig. 7-11 is primarily due to the higher vapor velocities caused by expansion during the evaporation. For a high heat flux at the tube surface, a vapor film rapidly forms on the tube surface, and a sharp decrease in h_z occurs at low void-fraction values. For a lower heat flux, the walls remain wetted longer until insufficient wetting at a high void fraction causes

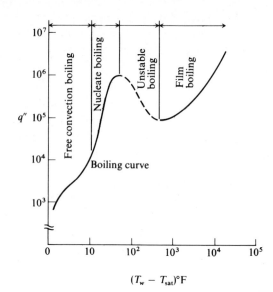

FIGURE 7-12
Pool boiling.

a sharp decrease in h_z. This behavior places a strict limitation on the heat transfer rate that can be allowed between the boiler tubes and the evaporating liquid.

The characteristics of boiling can best be explained by considering simple pool boiling since the boiling occurs in the absence of forced convection. Pool boiling occurs when a stationary pool of water is heated to its saturation temperature. Bubbles form at nucleation sites on the heated surface. Local surface temperatures at the nucleation sites are a few degrees above the saturation temperature. The actual superheat required depends on the smoothness of the surface and the purity of the water. The temperature difference that provides the potential for boiling heat transfer is the difference between the surface temperature and the saturation temperature of the boiling liquid.

The regions of pool boiling are classified as shown in Fig. 7-12. There is a similarity between the boiling heat transfer behavior in forced-convection boiling as shown in Fig. 7-11 and pool boiling as shown in Fig. 7-12. In the free-convection boiling region the bubbles have negligible effect, and the single-phase correlations for free-convective heat transfer given in Sec. 7-1 can be used to predict heat transfer rates. In the nucleate boiling region, a liquid film wets the heated surface, but agitation due to boiling produces a greatly increased value of the convection heat transfer coefficient. Heat transfer coefficients in the order of 10^5 W/m²-°C can be obtained for temperature differences $T_w - T_{sat} < 5°C$. Film boiling is much less efficient since a vapor film covers the heated surface. Thus, much higher temperature differences are needed to produce a given heat transfer rate.

The phenomenon of burnout in pool boiling is similar to that in forced convection of two-phase flow. A sudden change from nucleate boiling to film boiling can occur when a vapor film displaces the liquid film in contact with the heated surface. This leads to a sudden decrease in the convection heat transfer coefficient and is referred to as burnout. The region identified as the unstable boiling region is difficult to maintain since the heat flux decreases with an increasing temperature difference. This unstable condition is due to the rapid formation of an unstable vapor film on the heated surface, which in turn causes a rapid increase in T_w. A thorough discussion of boiling heat transfer and two-phase flow appears in Ref. 7-10.

7-7   Correlations for Boiling Heat Transfer

Because of the complexity of the velocity and temperature profiles which occur in a multiphase environment, it is not surprising that the empirical correlations found in the literature have a limited range of applicability. A few of the more widely accepted correlations are given below. It is necessary for most designers of two-phase equipment to build prototypes and make their own heat transfer measurements.

For nucleate pool boiling, Rohsenow (Ref. 7-11) gave the following correlation for the convective heat flux:

$$\frac{c_{p,\,l}(T_w - T_{\text{sat}})}{h_{fg}\,\text{Pr}^{1.7}} = C_{sf}\left[\frac{q''}{\mu_l\, h_{fg}}\sqrt{\frac{\sigma_s}{g(\rho_l - \rho_v)}}\right]^{0.33} \tag{7-36}$$

Here the subscript l refers to the liquid and v to the vapor. Latent heat of evaporation is $h_{fg} = h_g - h_f$, the difference between the enthalpy of saturated vapor and saturated liquid. Values of the surface tension σ_s of a liquid in contact with its vapor are tabulated for various substances in Ref. 7-12. The empirical constant C_{sf} depends upon the surface-fluid combination. For water-copper it is 0.013 (Ref. 7-13), and for water-brass it is 0.006 (Ref. 7-14). Values for other combinations can be found in Ref. 7-15.

For film pool boiling, Bromley (Ref. 7-16) published a correlation for the convection heat transfer coefficient given by

$$\bar{h} = 0.62\,\frac{k_v^{\,3}\rho_v(\rho_l - \rho_v)gh_{fg} + 0.4c_{p,\,v}(T_w - T_{\text{sat}})^{1/4}}{D\mu_v(T_w - T_{\text{sat}})} \tag{7-37}$$

Since boiling normally occurs at high temperatures, radiation heat transfer is also important. A radiation coefficient was defined in Chap. 1, and for $\mathscr{F} = \varepsilon$, it is given by

$$\bar{h}_r = \frac{\varepsilon\sigma(T_w^{\,4} - T_{\text{sat}}^4)}{T_w - T_{\text{sat}}}$$

There is a coupling between $\bar{h}$ and $\bar{h}_r$ in film boiling since the vapor in the boundary-layer vapor film is not entirely transparent to thermal radiation. Bromley suggested that the total heat transfer coefficient in film boiling be expressed in terms of $\bar{h}$ and $\bar{h}_r$ as

$$\bar{h}_T = \bar{h}\left(\frac{\bar{h}}{\bar{h}_T}\right)^{1/3} + \bar{h}_r \qquad (7\text{-}38)$$

The total heat transfer per unit area is then given by Newton's law of cooling written in terms of $T_w - T_{sat}$ and $\bar{h}_T$. Thus,

$$q'' = \bar{h}_T(T_w - T_{sat}) \qquad (7\text{-}39)$$

Initial efforts to find correlations for heat transfer in forced-convection boiling were based upon combining the effects of nucleate pool boiling and single-phase forced convection. (See Refs. 7-17 and 7-18.) An equation for calculating the heat flux in nucleate boiling with forced convection was suggested by Foster and Zuber, Ref. 7-19. This correlation has the form

$$q'' = \frac{K_1 C_1 {\rho_l}^3 h_{fg}(T_w - T_{sat})^3}{\sigma T_{sat}(\rho_l - \rho_v)} \qquad (7\text{-}40)$$

where K_1 and C_1 are constants which depend upon the fluid and the surface. Equation (7-40) is valid for clean commercial surfaces with two-phase bubble flow. It is relatively independent of the boiling liquid or the operating pressure and temperature. This correlation loses its validity for fully developed boiling.

A simple, approximate correlation for the convective heat flux in water undergoing forced convection with a fully developed boiling state was suggested by McAdams in Ref. 7-20. This correlation is given by

$$q'' = 0.074(T_w - T_{sat})^{3.86} \qquad 30 \text{ psia} \le P \le 100 \text{ psia} \qquad (7\text{-}41)$$

For higher pressures, Levy published (Ref. 7-22)

$$q'' = \frac{P^{4/3}}{495}(T_w - T_{sat})^3 \qquad 100 \text{ psia} \le P \le 2{,}000 \text{ psia} \qquad (7\text{-}42)$$

The most difficult problem in forced-convection boiling is to estimate when a transition from nucleate to film boiling might occur. This is much more critical than estimating heat transfer coefficients for nucleate boiling in forced convection. This transition is primarily a function of the wall heat flux and vapor quality. Experimental data of this transitional behavior can be found in Ref. 7-6.

A greater knowledge of the two-phase phenomena that occur during evaporation and condensation has become critical because of the high heat flux values that are being encountered in modern power plants. New theories and new techniques are under constant development. In spite of the great advances made in heat transfer theory during this century, there is still a lack of basic understanding in certain areas such as two-phase flow. When all the appropriate nondimensional parameters which affect the problem are understood, mathematical models and computer solutions, coupled with experimental verification, will allow a heat transfer analysis on which to base more knowledgeable decisions.

References

7-1   GIEDT, W. H.: Investigation of Variation of Point Unit-heat-transfer Coefficient around a Cylinder Normal to an Air Stream, *Trans. ASME*, vol. 71, pp. 375–381, 1949.

7-2   REYNOLDS, W. C.: "Thermodynamics," 2d ed., McGraw-Hill, New York, 1968.

7-3   KERN, D. Q.: Heat Exchanger Design for Fouling Service, *Third International Heat Transfer Conference*, Chicago, 1966.

7-4   Standards of the Tubular Exchanger Manufacturers Association, New York, 1959.

7-5   KAYS, W. M.: "Convective Heat and Mass Transfer," McGraw-Hill, New York, 1966.

7-6   FRAAS, A. P., and M. N. OZISIK: "Heat Exchange Design," Wiley, New York, 1965.

7-7   KAYS, W. M., and A. L. LONDON: "Compact Heat Exchangers," 2d ed., McGraw-Hill, New York, 1967.

7-8   HODGMAN, C. D. (ed.): "C.R.C. Standard Mathematical Tables," 10th ed., Chemical Rubber Publishing Co., Cleveland, 1955.

7-9   GOUSE, S. W., JR.: An Introduction to Two-phase Gas Liquid Flow, *M.I.T. Rept.* 8734 to ONR.

7-10   TONG, L. T.: "Boiling Heat Transfer and Two-phase Flow," Wiley, New York, 1965.

7-11   ROHSENOW, W. M.: A Method of Correlating Heat Transfer Data for Surface Boiling of Liquids, *Trans. ASME*, vol. 74, pp. 969–975, 1952.

7-12   WEAST, R. C. (ed.): "The Handbook of Chemistry and Physics," 49th ed., Chemical Rubber Publishing Co., Cleveland, 1968.

7-13   PIRET, E. L., and H. S. ISBIN: Natural Circulation Evaporation Two-phase Heat Transfer, *Chem. Eng. Progr.*, vol. 50, p. 305, 1954.

7-14   CRYDER, D. S., and A. C. FINALBARGO: Heat Transmission from Metal Surfaces to Boiling Liquids, *Trans. AIChE*, vol. 33, p. 346, 1937.

7-15   CICHELLI, M. T., and C. F. BONILLA: Heat Transfer to Liquids Boiling under Pressure, *Trans. AIChE*, vol. 41, p. 755, 1945.

7-16   BROMLEY, L. A., *et al.: Ind. Eng. Chem.*, vol. 45, p. 2639, 1953.

7-17   ROHSENOW, W. M.: "Heat Transfer," Heat Transfer with Evaporation, The University of Michigan Press, Ann Arbor, 1953.

7-18   CHEN, J. C.: A Correlation for Boiling Heat Transfer to Saturated Fluids in Convective Flow, *ASME Paper* 63-HT-34, 1963.

7-19   FOSTER, H. K., and N. ZUBER: Dynamics of Vapor Bubbles and Boiling Heat Transfer, *AIChE J.*, vol. 1, p. 531, 1955.

7-20   MC ADAMS, W. H., *et al.:* Heat Transfer at High Rates to Water with Surface Boiling, *Ind. Eng. Chem.*, vol. 41, pp. 1945–1955, 1949.

7-21   STOEKER, W. F.: "Design of Thermal Systems," McGraw-Hill, New York, 1971.

7-22   LEVY, S.: Generalized Correlation of Boiling Heat Transfer, *J. Heat Transfer*, ser. C, vol. 81, pp. 37–42, 1959.

7-23   WALLIS, G. B.: "One-dimensional Two-phase Flow," McGraw-Hill, New York, 1969.

7-24   JAKOB, M.: "Heat Transfer," vol. 1, Wiley, New York, 1949.

7-25   HILPERT, R.: Warmeabgabe von geheizen Drahten und Rohren, *Forsch. Gebiete Ingenieurw.*, vol. 4, p. 220, 1933.

7-26   FAND, R. M.: Heat Transfer by Forced Convection from a Cylinder to Water in Cross-flow, *Intern. J. Heat Mass Transfer*, vol. 8, p. 995, 1965.

7-27   KRAMERS, H.: Heat Transfer from Spheres to Flowing Media, *Physica*, vol. 12, p. 61, 1946.

7-28   YUGE, T.: Experiments on Heat Transfer from Spheres Including Combined Natural and Forced Convection, *J. Heat Transfer*, sec. C, vol. 82, p. 214, 1960.

7-29   HYMAN, S. C., C. F. BONILLA, and S. W. EHRLICH: Natural Convection Transfer Processes: I. Heat Transfer to Liquid Metals and Non-metals at Horizontal Cylinders, *Chem. Eng. Progr. Symposium Ser.*, vol. 49, no. 5, p. 21, 1953.

7-30   EMERY, A., and W. C. CHU: Heat Transfer across Vertical Layers, *J. Heat Transfer*, vol. 87, ser. C, p. 110, 1965.

7-31   GLOBE, S., and D. DROPKIN: Natural Convection Heat Transfer in Liquids Confined by Two Horizontal Plates and Heated from Below, *J. Heat Transfer*, ser. C, vol. 81, p. 24, 1959.

7-32   DITTUS, F. W., and L. M. K. BOELTER: *University of California Publications in Engineering*, vol. 2, p. 443, 1930.

7-33   SIEDER, E. N., and G. E. TATE: Heat Transfer and Pressure Drop of Liquids in Tubes, *Ind. Eng. Chem.*, vol. 8, p. 1429, 1936.

7-34   NUSSELT, W.: Der Warmeaustausch Zwischen Wand und Wasser im Rohr, *Forsch. Gebiete Ingenieurw.*, vol. 2, p. 309, 1931.

7-35   LUBARSKY, B., and S. J. KAUFMAN: Review of Experimental Investigations of Liquid Metal Heat Transfer, NACA TN 3336, 1955.

7-36   SEBAN, R. A., and T. T. SHIMAZAKE: Heat Transfer to a Fluid Flowing Turbulently in a Smooth Pipe with Walls at Constant Temperature, *Trans. ASME*, vol. 73, p. 803, 1951.

7-37   HAUSEN, H.: Darstellung des Warmeuberganges in Rohren durch verallgemeinerte Potenzbeziehunger, *VDIZ*, no. 4, p. 91, 1943.

Problems[1]

7-1   Water flows through a 3.0-m-length pipe. The inner surface temperature of the pipe is maintained at 90°C, and the environmental temperature T_∞ is 22°C. The thermal conductivity of the thick pipe is $k = 35$ W/m-K, with dimensions $r_o = 8.0$ cm and $r_i = 5.0$ cm.

---

[1] Problems whose numbers are followed by a superscript italic *c* should be analyzed by obtaining solutions on a digital computer.

(*a*) What value of the average external convection coefficient $\bar{h}$ must exist so that the outer surface temperature of the pipe is equal to 50°C, for steady-state, one-dimensional, radial conduction? (Review Sec. 4-10.)

(*b*) What is the ratio of the convective resistance to conductive resistance?

(*c*) By using empirical correlations for external cross flow over a cylinder, specify values of Re and Pr, as well as the type of convection which could physically produce the value of $\bar{h}$ required in part (*a*).

7-2   Consider a solid, horizontal copper cylinder with uniform internal generation q'''. The radius is 3 in and the length is 10 ft. A thermocouple attached to the solid surface indicates a uniform, steady-state surface temperature T_0 of 800°F. Two radiation shields 10 ft long are placed 1.0 ft from the solid surface as shown, and they are maintained at a temperature $T_s = 200°F$ by water cooling. The emissivity of all surfaces is 0.60. Air is blown over the cylinder by forced convection as shown. The free-stream air velocity is 30 ft/s, and the temperature $T_\infty = 200°F$. The cylinder and shields are located in a large chamber with a wall temperature of 200°F.

(*a*) Calculate the total heat transfer from the cylinder by forced convection.

(*b*) Show that free-convection heat transfer is negligible.

(*c*) Calculate the total net heat transfer by radiation between the cylinder and the two radiation shields using $\varepsilon = 0.6$ for all surfaces. Neglect end losses.

(*d*) Draw a radiation network and label all resistances.

(*e*) Calculate the total net heat transfer by radiation between the cylinder and the large chamber using $\varepsilon = 0.6$ for all surfaces.

(*f*) Calculate total net heat transfer from cylinder by both convection and radiation.

(*g*) For steady-state conditions, how much energy must be conducted to the surface from within the solid to maintain the surface temperature at 800°F? Hint: Apply the first law of thermodynamics to a differential system at the interface.

(*h*) Determine the value of q''' within the solid cylinder necessary to satisfy the conditions given in this problem. For copper, $k = 200$ Btu/h-ft-°F.

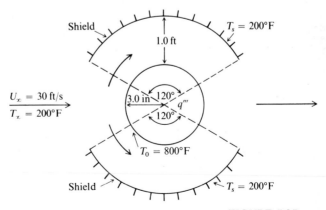

FIGURE 7-2P
Shielding assembly.

7-3 Consider a horizontal steam pipe ($k = 52$ W/m-°C) in a large room. The pipe is 6.0-cm outside diameter and 5.24-cm inside diameter. The pipe carries saturated steam at 345 kN/m², and $T_0 = 130$°C. The properties of the steam are $k = 0.03$ W/m-°C, Pr $= 0.951$, and $v = 2.8 \times 10^{-5}$ m²/s. The velocity of steam through the pipe is 3 m/s. The pipe is covered by 2.5 cm of insulation ($k = 0.15$ W/m-°C). The environmental temperature $T_\infty = 20$°C, and the pipe is 30 m long. The emittance of the outer surface of the insulation is 0.9. An equivalent resistance network is shown below, where $R_1 = $ internal convection resistance, $R_2 = $ conduction resistance in pipe, $R_3 = $ conduction resistance in the insulation, $R_4 = $ radiation resistance, and $R_5 = $ external convection resistance.

(a) Calculate the values of the five resistances.

(b) Calculate the total heat transfer for steady-state conditions. Hint: Since the external value of $\bar{h}$ is a function of $T_3 - T_\infty$ for free convection, the equation $q = (T_0 - T_\infty)/\sum_i R_i$ will have two unknowns q and R_5. A trial-and-error solution is suggested.

(c) If the quality of the steam is 100 percent at the pipe entrance, calculate the quality of steam at the exit. Assume a constant steam density of 0.48 kg/m³.

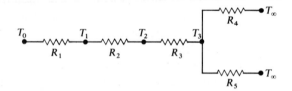

FIGURE 7-3P
Resistance network.

7-4^c Liquid bismuth flows through an annulus (2.0-in inside diameter, 2.5-in outside diameter) at a velocity of 15 ft/s. The bulk temperature is 600°F, and the average wall temperature is 800°F. A constant heat flux exists at the walls of the annulus. The properties of the bismuth are $c_p = 0.035$ Btu/lb-°F, $v = 1.62 \times 10^{-6}$ ft²/s, $k = 9.25$ Btu/h-ft-°F, Pr $= 0.0135$, and $\rho = 625$ lb/ft³.

(a) Calculate the average heat transfer coefficient.

(b) Calculate the temperature rise of the bismuth per unit length of the annulus.

7-5^c Consider laminar forced convection of air over a trapezoidal fin. Use a local variation of h_x over the fin surface as given by Eq. (6-90), with $x_0 = 0$ and x measured from the end of the fin. Review Sec. 2-5, and obtain solutions for the one-dimensional temperature distribution in the fin with a variable h_x over the surface. Compare the results with the corresponding solution using $\bar{h} = 2h_{x=L}$ over the entire fin surface. Let $U_\infty = 3$ m/s, $v = 0.019 \times 10^{-3}$ m²/s, $k = 0.026$ W/m-°C, $L = 0.15$ m, and Pr $= 0.7$. Choose your own values for N1, N2, and N3 as defined in Chap. 2.

7-6^c Review the discussion in Chap. 4 related to the transient temperature response in a fin. Obtain the steady-state solution for a fin of your choice, using $\bar{h}$ for laminar flow with Re$_L = 10^5$, $k = 0.026$ W/m-K, and Pr $= 0.8$. Suddenly the velocity of the air in the free stream is increased by a factor of 10, and thus $Re_L = 10^6$. This change causes transition to turbulent flow over most of the fin surface. Obtain the transient temperature response of the fin using $\bar{h}$ for turbulent boundary-layer flow.

7-7^c Review the solutions for steady-state, two-dimensional heat conduction with convective boundary conditions given in Sec. 3-10. The two by one rectangular region shown below has isothermal boundary conditions at each end. A laminar boundary layer exists on the top surface, and a turbulent boundary layer exists on the bottom surface. Specify appropriate local values of h_x on each surface, and calculate the steady-state, two-dimensional temperature distribution when $T_2 - T_\infty = 0.5(T_1 - T_\infty)$. The free-stream fluid is water with $Pr = 8.0$, $k = 0.34$ Btu/h-ft-°F, $\rho = 62.3$ lb$_m$/ft^3, $c_p = 1.0$ Btu/lb$_m$-°F, and $\nu = 1.2 \times 10^{-5}$ ft^2/s.

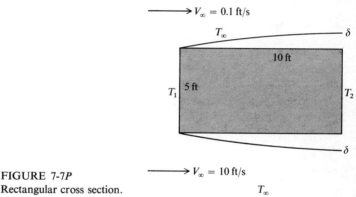

FIGURE 7-7P
Rectangular cross section.

7-8^c The 1.0-ft-square cross section shown has isothermal boundary conditions on the top and bottom surfaces maintained by a coolant. Laminar flow of liquid bismuth occurs on the left surface as shown with $T_\infty = 500$°F. The Reynolds number is 1×10^5. A laminar film-condensate boundary layer occurs on the right face with $T_\infty = 300$°F. Properties of the bismuth are $\rho = 625$ lb/ft^3, $c_p = 0.035$ Btu/lb-°F, $\mu = 1.1 \times 10^{-3}$

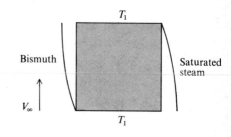

FIGURE 7-8P
Square cross section.

lb/ft-s, $v = 1.75 \times 10^{-6}$ ft^2/s, $k = 9.5$ Btu/h-ft-°F, and Pr $= 0.014$. Properties of the steam are $\rho = 0.033$ lb/ft^3, $c_p = 0.45$ Btu/lb-°F, $\mu = 1 \times 10^{-5}$ lb/ft-s, $v = 0.30 \times 10^{-3}$ ft^2/s, $k = 0.017$ Btu/h-ft-°F, and Pr $= 0.95$.

(a) Calculate the steady-state, two-dimensional temperature distribution in the solid. (Review Chap. 3.)

(b) Calculate the local heat flux vector in the center of the solid. The solid is stainless steel ($k = 9.3$ Btu/h-ft-°F).

7-9^c Consider radiation falling on a two-plate system as shown in Fig. 1-11. However, consider that the air between the plates is trapped by closing the ends of the plates. The Nusselt number for the enclosed air is given in Table 7-3. The Nusselt number for the forced convection over the top surface is given by Eq. (6-91) or (6-111). Obtain the transient response of the two plates and enclosed air by writing three coupled differential equations based upon a lumped thermal analysis, and obtain numerical solutions. Study the problem for various values of Re, Gr, Pr, and incoming irradiation.

7-10 Design an annulus with counterflow to act as a heat exchanger between two fluids. Use the following procedure:

(a) Specify the geometry and conductivity of the annulus.

(b) Specify the type of fluids and flow rates of the two energy-exchanging fluids.

(c) Using empirical correlations, calculate the resistances that give the value of the overall conductance U.

(d) Specify the inlet and outlet temperatures of both fluids.

(e) Calculate the required heat exchanger length.

7-11 Repeat Prob. 7-10 for a parallel-flow heat exchanger.

7-12 A spherical, cryogenic container is to be located on the ocean floor. The temperature of the liquid in the container is 100 K. The temperature of the surrounding sea water is 280 K. The container properties are $k = 18$ W/m-K and thickness $= 6.0$ cm. Can insulation be placed around the container to limit the thermal energy leaking into the container to less than 60 W/m^2? If so, what thermal conductivity and thickness should the insulation have?

7-13 A steel pipe ($k = 30$ W/m-K, outside diameter 15 cm) is covered with 5 cm of insulation ($k = 0.2$ W/m-K, $\alpha = \varepsilon = 0.1$). The environmental temperature is 30°C. Liquid bismuth, flowing through the pipe (inside diameter $= 13$ cm) at 1.5 m/s, enters the pipe at $T = 600$°C. If the pipe loses energy by free convection and radiation to its surroundings, estimate the outlet temperature of the bismuth if the pipe length is 15 m.

7-14 It is required to control the steady-state heat transfer through a wall which has a conduction resistance of $R = 2.0$ h-ft^2-°F/Btu. The convective conditions in the air on each side of the wall can be controlled to produce desired values of $\bar{h}_i$ and $\bar{h}_o$ and limit the steady-state heat transfer to 12 Btu/h-ft^2.

(a) Choose suitable values of $\bar{h}_i$ and $\bar{h}_o$ for $T_{\infty,L} - T_{\infty,R} = 40$°F.

(b) Specify the type of convection required on each side of the wall and the required value of the Reynolds or Grashof number to produce the convection coefficients specified in part (a).

7-15 Evaporation of water at $P = 200$ kN/m^2 occurs in a vertical boiler tube. The latent heat of evaporation $h_{fg} = 2.22$ MJ/kg, and the flow rate is 450 kg/h. Hot air at $T = 500°$C is available to provide the required energy for evaporation. Specify the gas flow rate and surface area required in a counterflow heat exchanger. What temperature drop occurs in the air?

7-16 Steady-state conduction per unit area through a composite wall is represented below. Conduction occurs through two resistances in series.
  (*a*) Define *conduction* heat transfer coefficients h_1 and h_2, and write the expression for the overall heat transfer coefficient $U = f(h_1, h_2)$.
  (*b*) If $h_1 = ah_2$, where $a > 1$, show that a change in the small coefficient h_2 produces a^2 times the change in U compared to that produced by an equal change in the larger coefficient h_1.

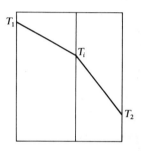

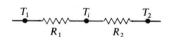

FIGURE 7-16*P*
Composite wall.   (Ref. 7-1 by permission.)

7-17 Consider the surface of a thin condenser tube. Vapor condenses on the outer surface, and convection to a stream of cooling water from the sea occurs on the inner surface. The conduction resistance across the tube wall is 3.52×10^{-4} m^2-°C/W, and the resistance caused by fouling due to mineral deposits on the inner surface is 1.2×10^{-3} m^2-°C/W. The average internal convection coefficient is 1.25×10^3 W/m^2-°C, and for film condensation the convection coefficient is 17×10^3 W/m^2-°C.
  (*a*) Calculate the overall heat transfer coefficient U.
  (*b*) A method for maintaining dropwise condensation is proposed which will increase the condensation coefficient by a factor of 5.0. What percentage increase will be gained in the overall heat transfer coefficient?
  (*c*) It is proposed to use a coolant other than salt water which will produce the same average internal convection coefficient but will slightly reduce the resistance due to mineral scale deposits from 1.2×10^{-3} to 8×10^{-4} m^2-°C/W. What percentage increase will be gained in the overall heat transfer coefficient?

7-18ᶜ An electrical switch has a metal filament which is heated and cooled to open and close the switch. During heating, thermal energy (I^2R) is generated within the filament and the surfaces of the filament lose energy by convection and radiation. During cooling, no current flows through the filament. During heating, the equation for the temperature change is given by

$$I^2R - \bar{h}A_s(T - T_\infty) - \varepsilon A_s(T^4 - T_\infty^4) = \rho Vc(dT/dtc)$$

(a) Nondimensionalize this equation and identify the important parameters. Explain the physical meaning of each parameter.

(b) Obtain solutions for the transient response of the filament temperature as a function of the governing parameters. Values of $\bar{h}$ must be estimated by the student, based upon the proposed design and environment of operation.

RUNGE-KUTTA
INTEGRATION SCHEME

A-1   Derivation of Equivalent First-order Equations

The Runge-Kutta method may be used to determine the solution to an nth-order ordinary differential equation of the form

$$\frac{d^n y}{dx^n} = f\left(x, y, \frac{dy}{dx}, \ldots, \frac{d^{n-1}y}{dx^{n-1}}\right) \qquad \text{(A-1)}$$

It is usually more convenient to work with the equivalent set of first-order differential equations which can be obtained from Eq. (A-1). The mathematical transformation of variables which leads to this equivalent set of equations is obtained by letting

$$Y_1 = y$$

$$Y_2 = \frac{dy}{dx}$$

$$\vdots$$

$$Y_n = \frac{d^{n-1}y}{dx^{n-1}}$$

$$\text{(A-2)}$$

and then defining

$$F_1 = \frac{dY_1}{dx} = \frac{dy}{dx} = Y_2$$

$$F_2 = \frac{dY_2}{dx} = \frac{d^2y}{dx^2} = Y_3 \qquad \text{(A-3)}$$

$$\vdots$$

$$F_n = \frac{dY_n}{dx} = \frac{d^ny}{dx^n} = f(x, Y_1, Y_2, \ldots, Y_n)$$

For example, consider a third-order differential equation given by

$$\frac{d^3y}{dx^3} = \left(\frac{dy}{dx}\right)^2 - \frac{dy}{dx}\frac{d^2y}{dx^2} - 1 \qquad \text{(A-4)}$$

with initial conditions at $x = 0$ specified by

$$\frac{d^2y}{dx^2} = C \qquad \frac{dy}{dx} = 0 \qquad y = 0 \qquad \text{(A-5)}$$

We now let

$$Y_1 = y$$

$$Y_2 = \frac{dy}{dx} \qquad \text{(A-6)}$$

$$Y_3 = \frac{d^2y}{dx^2}$$

and then define

$$F_1 = \frac{dY_1}{dx} = Y_2$$

$$F_2 = \frac{dY_2}{dx} = Y_3 \qquad \text{(A-7)}$$

$$F_3 = \frac{dY_3}{dx} = Y_2^2 - Y_2 Y_3 - 1$$

with initial conditions at $x = 0$ given by

$$Y_1 = 0 \qquad Y_2 = 0 \qquad Y_3 = C \qquad \text{(A-8)}$$

The above three equations for F_1, F_2, and F_3 are three first-order differential equations which are equivalent to Eq. (A-4). There are now three dependent variables Y_1, Y_2, and Y_3 instead of one.

A-2   Derivation of Recursion Formula

We now focus our attention on obtaining a numerical solution to a first-order differential equation of the form

$$F = \frac{dy}{dx} = f(x, y) \qquad \text{(A-9)}$$

The objective of a numerical solution is to approximate the solution $y(x)$ at many selected values of the independent variable x over a specified interval $a \leq x \leq b$. When the value of $y(x)$ is known at some starting point, say $y = y_i$ at $x = x_i$, the value of y_{i+1} at $x_i + \Delta x$ can be calculated by use of the Runge-Kutta recurrence formula given by

$$y_{i+1} = y_i + W_1 k_1 + W_2 k_2 + \cdots + W_n k_n \qquad \text{(A-10)}$$

The weighting functions $W_1, W_2, \ldots, W_n$ are determined by matching the terms in the recurrence formula to an nth-order Taylor series expansion of y about y_i. The k values depend upon derivatives dy/dx given by the differential equation (A-9) evaluated within the increment Δx. For an nth-order approximation the k's are given by

$$
\begin{aligned}
k_1 &= f(x_i, y_i)\, \Delta x \\
k_2 &= f(x_i + p_1 \Delta x, y_i + q_{11} k_1)\, \Delta x \\
k_3 &= f(x_i + p_2 \Delta x, y_i + q_{21} k_1 + q_{22} k_2)\, \Delta x \\
&\ \vdots \\
k_n &= f(x_i + p_{n-1} \Delta x, y_i + q_{n-1,\,1} k_1 + q_{n-1,\,2} k_2 + q_{n-1,\,3} k_3 \\
&\qquad + \cdots + q_{n-1,\,n-1} k_{n-1})\, \Delta x
\end{aligned}
\qquad \text{(A-11)}
$$

As will be illustrated below, the p's and q's are also determined by matching Eqs. (A-11) with terms of a Taylor series expansion.

The recurrence formula for the second-order Runge-Kutta technique requires an equation of the form

$$y_{i+1} = y_i + W_1 k_1 + W_2 k_2 \qquad \text{(A-12)}$$

with

$$k_1 = f(x_i, y_i)\, \Delta x \qquad \text{(A-13)}$$

and

$$k_2 = f(x_i + p\, \Delta x, y_i + q k_1)\, \Delta x \qquad \text{(A-14)}$$

The motivation of the Runge-Kutta method is to evaluate y_{i+1} in terms of y_i and the first derivative dy/dx. This method has advantages over a second-order Taylor series given by

$$y_{i+1} = y_i + \frac{dy_i}{dx} \Delta x + \frac{1}{2!} \frac{d^2 y_i}{dx^2} (\Delta x)^2 \qquad \text{(A-15)}$$

since the second derivative may not be known or well-behaved. The Taylor series expansion leaves room for modification since the terms on the right side of Eq. (A-15) are all evaluated at the beginning of the ith increment of x.

Using Eq. (A-9) we can write $dy_i/dx = f(x_i, y_i)$ and $d^2 y_i/dx^2 = dF_i/dx$. Since F is a point function of x and y, the chain rule of calculus gives

$$dF = \frac{\partial f}{\partial x} dx + \frac{\partial f}{\partial y} dy \qquad \text{(A-16)}$$

It follows that

$$\frac{dF}{dx} = \frac{\partial f}{\partial x} (1) + \frac{\partial f}{\partial y} \frac{dy}{dx} \qquad \text{(A-17)}$$

Combining the above relationships with Eq. (A-15) and using Eq. (A-9) gives

$$y_{i+1} = y_i + f(x_i, y_i) \, \Delta x$$
$$+ \frac{1}{2!} \left[\frac{\partial f(x_i, y_i)}{\partial x} + \frac{\partial f(x_i, y_i)}{\partial y} f(x_i, y_i) \right] (\Delta x)^2 \qquad \text{(A-18)}$$

This expression of a second-order Taylor series expansion serves as the standard of accuracy that must be met.

By defining k_1 and k_2 in terms of the function $f(x, y)$ in Eqs. (A-13) and (A-14), we force these coefficients to be a function of the first derivative dy/dx. The value of k_1 is based upon the derivative dy_i/dx at the beginning of each interval in x. The value of k_2 is based upon the derivative at some unknown location within the increment Δx since $0 < p < 1$.

We now expand the derivative $dy/dx = f(x, y)$ about $f(x_i, y_i)$ by use of a Taylor series expansion for two variables. This expansion is given by

$$f(x_i + \Delta x, y_i + \Delta y) = f(x_i, y_i) + \frac{\partial f(x_i, y_i)}{\partial x} \Delta x + \frac{\partial f(x_i, y_i)}{\partial y} \Delta y$$
$$+ \frac{\partial^2 f(x_i, y_i)}{\partial x^2} \frac{(\Delta x)^2}{2} + \frac{\partial^2 f(x_i, y_i)}{\partial x \, \partial y} \Delta x \, \Delta y + \frac{\partial^2 f(x_i, y_i)}{\partial y^2} \frac{(\Delta y)^2}{2} + \cdots \qquad \text{(A-19)}$$

The second-order terms can be neglected along with the higher-order terms. However, to be consistent with the definition of k_2, given by Eq. (A-14), we want an expansion for $f(x_i + p \, \Delta x, y_i + qk_1)$. This expression is

$$f(x_i + p \, \Delta x, y_i + qk_1) = f(x_i, y_i) + \frac{\partial f(x_i, y_i)}{\partial x} p \, \Delta x + \frac{\partial f(x_i, y_i)}{\partial y} qk_1$$

Multiplying each side of this equation by Δx and using the definition of k_1, given by Eq. (A-13), allows one to write

$$k_2 = f(x_i + p \, \Delta x, y_i + qk_1) \, \Delta x$$
$$= f(x_i, y_i) \, \Delta x + \frac{\partial f(x_i, y_i)}{\partial x} p(\Delta x)^2 + \frac{\partial f(x_i, y_i)}{\partial y} qf(x_i, y_i)(\Delta x)^2 \qquad \text{(A-20)}$$

Using the derived expressions for k_1 and k_2 in Eq. (A-12) gives

$$y_{i+1} = y_i + W_1 f(x_i, y_i) \, \Delta x + W_2 f(x_i, y_i) \, \Delta x$$
$$+ W_2 \left[\frac{\partial f(x_i, y_i)}{\partial x} p(\Delta x)^2 + \frac{\partial f(x_i, y_i)}{\partial y} qf(x_i, y_i)(\Delta x)^2 \right] \qquad \text{(A-21)}$$

To obtain the accuracy of a second-order Taylor series expansion about y_i, we choose values of W_1 and W_2 so that Eq. (A-21) agrees with Eq. (A-18). Equating coefficients of like terms requires that

$$W_1 + W_2 = 1$$
$$W_2 p = \tfrac{1}{2} \qquad \text{(A-22)}$$
$$W_2 q = \tfrac{1}{2}$$

The above three equations have four unknowns. If one looks at the form of Eq. (A-14), it is seen that a choice of $p = 1$ will require that k_2 be evaluated at the convenient location of $x = x_i + \Delta x$. Letting $p = 1$, it follows from Eqs. (A-22) that $W_2 = \frac{1}{2}$, $W_1 = \frac{1}{2}$, and $q = 1$. Equations (A-12) to (A-14) can now be written

$$y_{i+1} = y_i + \tfrac{1}{2}(k_1 + k_2) \tag{A-23}$$

where

$$k_1 = f(x_i, y_i)\,\Delta x \tag{A-24}$$

and

$$k_2 = f(x_i + \Delta x, y_i + k_1)\,\Delta x \tag{A-25}$$

The second-order Runge-Kutta method may be used as follows: First the initial value of y_i at $x = x_i$ must be known. Then $f(x_i, y_i)$ gives the initial value of k_1 from Eq. (A-24) when the value of Δx is specified. The value of $y_i + k_1$ can then be calculated. Note that $k_1 = (dy/dx)_i\,\Delta x = \Delta y$, the increment in y assuming a linear variation across Δx and a slope of k_1. Next, $f(x, y)$ can be evaluated at $x_i + \Delta x$ and $y_i + k_1$ to obtain k_2. Then Eq. (A-23) is used to calculate y_{i+1}. Finally, this value of y and the value of $x = x_i + \Delta x$ are used in Eq. (A-24) to repeat the process for the next increment in x.

When the same procedure is followed for a third-order Runge-Kutta technique, the following results are obtained:

$$y_{i+1} = y_i + \frac{k_1 + 4k_2 + k_3}{6} \tag{A-26}$$

$$k_1 = f(x_i, y_i)\,\Delta x$$

$$k_2 = f\left(x_i + \frac{\Delta x}{2}, y_i + \frac{k_1}{2}\right)\Delta x \tag{A-27}$$

$$k_3 = f(x_i + \Delta x, y_i - k_1 + 2k_2)\,\Delta x$$

The results for a fourth-order technique suggested by Runge are

$$y_{i+1} = y_i + \frac{k_1 + 2k_2 + 2k_3 + k_4}{6} \tag{A-28}$$

$$k_1 = f(x_i, y_i)\,\Delta x$$

$$k_2 = f\left(x_i + \frac{\Delta x}{2}, y_i + \frac{k_1}{2}\right)\Delta x$$

$$k_3 = f\left(x_i + \frac{\Delta x}{2}, y_i + \frac{k_2}{2}\right)\Delta x \tag{A-29}$$

$$k_4 = f(x_i + \Delta x, y_i + k_3)\,\Delta x$$

This set of equations forms the algorithm that is used throughout the book to obtain numerical solutions to ordinary differential equations. The fourth-order Runge-Kutta algorithm can also be explained from a graphical point of view. This is done in Sec. 1-10.

A-3 Simultaneous First-order Differential Equations

As previously indicated, higher-order differential equations may be recast into an equivalent set of first-order differential equations. One must then solve a set of simultaneous first-order equations.

Consider a second-order equation of the form

$$\frac{d^2y}{dx^2} = f\left(x, y, \frac{dy}{dx}\right) \qquad \text{(A-30)}$$

Letting

$$Y_1 = y$$

$$Y_2 = \frac{dy}{dx}$$

and defining

$$F_1 = Y_2 \qquad \text{(A-31)}$$

$$F_2 = F(x, Y_1, Y_2) \qquad \text{(A-32)}$$

gives two ordinary differential equations, where $F_1 = dY_1/dx$ and $F_2 = dY_2/dx$.

In general we consider two coupled ordinary differential equations given by

$$\frac{dY_1}{dx} = f[x, Y_1(x), Y_2(x)] \qquad \text{(A-33)}$$

$$\frac{dY_2}{dx} = g[x, Y_1(x), Y_2(x)] \qquad \text{(A-34)}$$

with initial conditions at $x = x_0$ given by $Y_1 = Y_1(0)$ and $Y_2 = Y_2(0)$. The fourth-order Runge-Kutta equations are then

$$Y_{1,\,i+1} = Y_{1,\,i} + \frac{k_1 + 2k_2 + 2k_3 + k_4}{6}$$

$$k_1 = f(x_i, Y_{1,\,i}, Y_{2,\,i})\,\Delta x$$

$$k_2 = f\left(x_i + \frac{\Delta x}{2}, Y_{1,\,i} + \frac{k_1}{2}, Y_{2,\,i} + \frac{l_1}{2}\right)\Delta x$$

$$k_3 = f\left(x_i + \frac{\Delta x}{2}, Y_{1,\,i} + \frac{k_2}{2}, Y_{2,\,i} + \frac{l_2}{2}\right)\Delta x$$

$$k_4 = f(x_i + \Delta x, Y_{1,\,i} + k_3, Y_{2,\,i} + l_3)\,\Delta x \qquad \text{(A-35)}$$

$$Y_{2,\,i+1} = Y_{2,\,i} + \frac{l_1 + 2l_2 + 2l_3 + l_4}{6}$$

$$l_1 = g(x_i, Y_{1,\,i}, Y_{2,\,i})\,\Delta x$$

$$l_2 = g\left(x_i + \frac{\Delta x}{2}, Y_{1,\,i} + \frac{k_1}{2}, Y_{2,\,i} + \frac{l_1}{2}\right)\Delta x$$

$$l_3 = g\left(x_i + \frac{\Delta x}{2}, Y_{1,\,i} + \frac{k_2}{2}, Y_{2,\,i} + \frac{l_2}{2}\right)\Delta x$$

$$l_4 = g(x_i + \Delta x, Y_{1,\,i} + k_3, Y_{2,\,i} + l_3)\,\Delta x$$

This procedure may be used for any number of simultaneous first-order differential equations. A set of equations is used for each dependent variable.

A-4   Example

Consider the first-order equation given by

$$\frac{dx}{dt} = xt \qquad \text{(A-36)}$$

with an initial condition $t_0 = 0$ at $x_0 = 1.0$. An analytical solution to this equation is given by

$$x = \exp \tfrac{1}{2}t^2 \qquad \text{(A-37)}$$

We now compare a numerical solution obtained by using a second-order Runge-Kutta scheme to the known analytical solution. We arbitrarily choose the independent-variable increment $\Delta t = h = 0.1$. The recurrence equations are

$$x(t_i + h) = x(t_i) + \frac{k_1 + k_2}{2} \qquad \text{(A-38)}$$

$$k_1 = f(t_i, x_i)\,\Delta t \qquad \text{(A-39)}$$
$$k_2 = f(t_i + \Delta t, x_i + k_1)\,\Delta t \qquad \text{(A-40)}$$

Using the initial condition $t_0 = 0$, at $x_0 = 1.0$, we find that $k_1 = 0$ and $k_2 = f(0 + 0.1, 1.0 + 0)\,\Delta t = (0.1)(0.1) = 0.01$. Then by Eq. (A-38)

$$x(t_0 + h) = x(t_0) + \frac{k_1 + k_2}{2} = 1.0 + \frac{0 + 0.01}{2} = 1.0050$$

This value of $x(t) = 1.005$ at $t = 0.1$ is now used to calculate the value of $x(t)$ at $t = 0.2$:

$$k_1 = f(0.1, 1.005)\,\Delta t = (0.1005)(0.1) = 0.01005$$
$$k_2 = f(0.2, 1.015)\,\Delta t = (0.2030)(0.1) = 0.02030$$

Then by Eq. (A-38)

$$x(0.2) = 1.0050 + \frac{0.01005 + 0.02030}{2} = 1.0201$$

This procedure is repeated over the independent-variable interval of interest. The analytical solution is compared with the numerical solution for $0 \le t \le 1.0$ in Table A-1. Higher-order Runge-Kutta schemes give an even closer comparison between the analytical and numerical results. Higher-order methods also allow larger increments, $\Delta t = h$, to be used without sacrificing the accuracy of the numerical solution (see Table A-2). In general, for a given integration scheme the accuracy decreases as the increment in the independent variable increases. Table A-1 also shows the relative effect of increasing h when using a second-order Runge-Kutta technique on this example problem. The value of $x(t)$ at $t = 1.0$ is calculated using intervals of 0.1, 0.2, 0.5, and 1.0. As can be seen, the error is a strong function of the increment size.

A-5 Conclusion

Many other algorithms may be used to obtain solutions to ordinary differential equations. These are discussed in books on numerical analysis. The Runge-Kutta method was chosen as the integration scheme since it is self-starting when the required initial conditions and/or boundary conditions are specified. This characteristic leads to a simple computer program which has wide applicability for solving engineering problems. Since the function $f(x,y)$ must be evaluated several times between each independent-variable increment, it is not necessarily the most economical method with respect to computing time. However, the slight increase in computing time is usually compensated for by the ease in programming. The accuracy obtained by using the fourth-order Runge-Kutta recurrence formula can be kept well within engineering requirements by using appropriate incremental step sizes in the independent variable.

Table A-1   SECOND-ORDER ACCURACY SOLUTION OF THE DIFFERENTIAL EQUATION $dx/dt = xt$ **USING DIFFERENT SIZE INTERVALS OF INTEGRATION**

t	Numerical Solution, $x(t)$ $h = 0.1$	Numerical Solution, $x(t)$ $h = 0.2$	Numerical Solution, $x(t)$ $h = 0.5$	Numerical Solution, $x(t)$ $h = 1.0$	Analytical Solution, $x(t) = \exp \frac{1}{2} t^2$
0	1.0000	1.0000	1.0000	1.0000	1.0000
0.1	1.0050				
0.2	1.0202	1.0200			1.0202
0.3	1.0460				
0.4	1.0832	1.0828			1.0833
0.5	1.1330		1.1250		
0.6	1.1970	1.1963			1.1972
0.7	1.2773				
0.8	1.3767	1.3753			1.3771
0.9	1.4987				
1.0	1.6478	1.6449	1.6172	1.5000	1.6487

Table A-2   HIGHER-ORDER ACCURACY

Order accuracy	$x(1.0)$, with $h = 1.0$
Second	1.5000
Third	1.6667
Fourth	1.6458

SATISFACTION OF ASYMPTOTIC BOUNDARY CONDITIONS

B-1 Introduction

The boundary conditions associated with ordinary differential equations of the boundary-layer type are of the two-point asymptotic class. Two-point boundary conditions have values of the dependent variable specified at two different values of the independent variable. Specification of an asymptotic boundary condition implies that the first derivative (and in the context of the boundary-layer equations all higher derivatives) of the dependent variable approaches zero as the outer specified value of the independent variable is approached.

One method of numerically integrating a two-point asymptotic boundary-value problem of the boundary-layer type, the initial-value method, requires that it be recast as an initial-value problem. Thus it is necessary to estimate as many boundary conditions at the surface as were previously given at infinity. The governing differential equations are then integrated with these assumed surface boundary conditions. If the required outer boundary condition is satisfied, a solution has been achieved. However, this is not generally the case. Hence, a method must be devised to logically estimate the new surface boundary conditions for the next trial integration. Asymptotic boundary-value problems such as those governing the boundary-layer equations are further complicated by the fact that the outer boundary condition is specified at infinity. In the trial integrations infinity is numerically approximated by some large value of the independent variable. There is no *a priori* general method of estimating this value. Selecting too small a maximum value for the independent variable may not allow the solution to asymptotically converge to the required accuracy. Selecting too large a value may result in divergence of the trial integrations or in slow convergence to the surface boundary conditions required to satisfy the asymptotic outer boundary conditions. Selecting too large a value of the independent variable is expensive in terms of computer time.

In order to effectively illustrate the method used to overcome these problems, it is applied to the Falkner-Skan equation. For comparison purposes the Newton-Raphson iteration method for estimating the surface boundary conditions for the Falkner-Skan equation is developed. Subsequently the Nachtsheim-Swigert (cf. Ref. B-1) iteration method is discussed.

B-2   Newton-Raphson Iteration Technique: Falkner-Skan Equation

Consider the Newton-Raphson technique specifically for the boundary-value problem associated with the Falkner-Skan equation. This boundary-value problem is given by

$$f''' + ff'' + \beta(1 - f'^2) = 0$$
$$f(0) = f'(0) = 0$$
$$f'(\eta \to \infty) \to 1$$

To recast this as an initial-value problem, an estimate of $f''(0)$ is needed to perform the required trial numerical integration. Further, the maximum value of the independent variable for the numerical integration must be specified. This value is designated as η_{max}, and it numerically represents infinity. Thus the boundary conditions may be restated as

$$f(0) = f'(0) = 0 \qquad f''(0) = k_1 \qquad (B-1)$$

with the resulting numerical solution required to satisfy the asymptotic outer boundary condition

$$f'(\eta_{max}) = 1 + \delta_1 \qquad (B-2)$$

where δ_1 is some small number.

All iteration methods that are used to estimate the required values of unknown surface boundary conditions required to satisfy the asymptotic boundary condition at η_{max}, Eq. (B-2), depend on the integrals of the boundary-layer equations at large values of the independent variable being functions of the unknown surface boundary conditions. For the Falkner-Skan equation this may be stated mathematically as

$$f'(\eta_{max}) = f'[f''(0)] = 1 + \delta_1 \qquad (B-3)$$

In order to obtain a correction equation for the value of $f''(0)$ required to satisfy the outer boundary condition at η_{max}, Eq. (B-3), $f'(\eta_{max})$ is expanded in a first-order Taylor series,

$$f'(\eta_{max}) = f'_c(\eta_{max}) + \frac{\partial f'(\eta_{max})}{\partial f''(0)} \, df''(0) + \cdots$$

where $f'(\eta_{max})$ = the required boundary value of f' at η_{max}

$f'_c(\eta_{max})$ = the value of f' at η_{max} calculated with the assumed value of $f''(0)$

$\partial f'/\partial f''(0)$ = the change in f' at η_{max} with a change in the assumed boundary condition at the surface $f''(0)$

$df''(0)$ = the required change in $f''(0)$ necessary to satisfy the required boundary condition at η_{max}

Using a finite difference representation of the differentials, letting $f''(0) = x$, and solving for Δx gives

$$\Delta x = \left(\frac{1 - f'_c}{f'_x} \right)_{\eta = \eta_{max}} \qquad \text{(B-4)}$$

where

$$f'_x = \frac{\partial f'(\eta_{max})}{\partial f''(0)}$$

This algebraic equation can be solved for any given trial integration of the governing equation, provided the derivative f'_x can be evaluated at η_{max}. A method of determining this derivative (cf. Ref. B-1) is to numerically integrate the perturbation equation obtained by differentiating the Falkner-Skan equation with respect to x, that is $f''(0)$. The resulting boundary-value problem for the perturbation equation is an initial value problem,

$$f'''_x = -(ff''_x + f''f_x + 2\beta f' f'_x)$$

with $$f_x(0) = f'_x(0) = 0 \qquad f''_x(0) = 1.0$$

For an assumed value of $f''(0)$ the perturbation equation can be integrated along with the Falkner-Skan equation, the derivative f'_x evaluated, and the subsequent values of $f''(0)$ determined from Eq. (B-4). Since the perturbation equation is third-order and f'_x is required, integration of the perturbation equation, along with the Falkner-Skan equation, represents integration of a fifth-order system of equations. Considerable simplification results if a finite difference representation of the derivative f'_x is used. In particular we assume that

$$f'_x = \frac{\Delta f'(\eta_{max})}{\Delta f''(0)} = \frac{f'_2(\eta_{max}) - f'_1(\eta_{max})}{f''_2(0) - f''_1(0)} \qquad \text{(B-5)}$$

where $f''_1(0) = k_1$, some assumed value

$f''_2(0) = k_1 + \varepsilon$, where ε is a small quantity, say between 10^{-5} and 10^{-2}

$f'_1(\eta_{max})$ = the value of f' at η_{max} obtained by numerical integration of the Falkner-Skan equation using $f''_1(0)$

$f'_2(\eta_{max})$ = the value of f' at η_{max} obtained by numerical integration of the Falkner-Skan equation using $f''_2(0)$

By performing two trial integrations with the assumed values of $f''_1(0)$ and $f''_2(0)$ the required convergence derivative f'_x may be approximately determined using Eq. (B-5). Equation (B-4) then yields the required correction to $f''(0)$. For a simple one-parameter representation such as given in Eq. (B-3), the approximate value of f'_x may be updated by using the current estimate of $f''(0)$ and the current value of $f'(\eta_{max})$ to determine f'_x. However, this technique does not solve the problem of choosing η_{max} nor does the Newton-Raphson method assure asymptotic convergence to the required outer boundary condition. Figure B-1 is

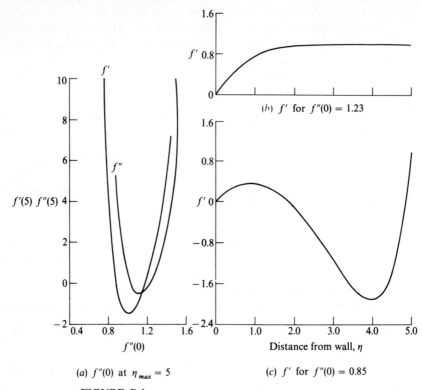

(a) $f''(0)$ at $\eta_{max} = 5$ (c) f' for $f''(0) = 0.85$

FIGURE B-1
Asymptotic convergence requirement for Falkner-Skan equation $\beta = 1.0$

taken from Ref. B-1. Examination of Fig. B-1a shows that the boundary condition $f'(\eta_{max}) = 1.0$ is approximately satisfied at two values of $f''(0)$, namely $f''(0) = 0.85$ and $f''(0) = 1.23$. However, further examination of Fig. B-1a shows that $f''(\eta_{max})$ is very large for $f''(0) = 0.85$, whereas it is quite small for $f''(0) = 1.23$. These observations are confirmed by Fig. B-1b and c. However, if a Newton-Raphson iteration scheme is used, convergence to $f'(\eta_{max}) = 1.0 \pm \delta_1$ will occur for two values of $f''(0)$, i.e., approximately 0.85 and 1.23 with $\eta_{max} = 5.0$. Here is an illustration of the fact that the Newton-Raphson iteration scheme does not yield asymptotic convergence; i.e., it does not assure that $f''(\eta_{max}) \to 0$ but only that $f'(\eta_{max}) = 1.0 \pm \delta_1$ regardless of $f''(\eta_{max})$. Further, a Newton-Raphson iteration scheme will not yield a unique solution for $f''(0)$.

B-3   Nachtsheim-Swigert Iteration Scheme: Falkner-Skan Equation

Nachtsheim and Swigert (cf. Ref. B-1) developed an iteration method which overcomes these difficulties. It is presented below in a somewhat modified form. To satisfy the asymptotic boundary condition at infinity, that is, $f'(\eta \to \infty)$ asymptotically, requires that both $f'(\eta \to \infty)$

$\rightarrow 1$ and $f''(\eta \rightarrow \infty) \rightarrow 0$. However, since there is only one adjustable parameter, namely $f''(0)$, it is not possible to simultaneously satisfy both conditions. We thus consider that the required asymptotic boundary condition will be satisfied at η_{max} if

$$f'(\eta_{max}) = 1 + \delta_1 \qquad \text{(B-2)}$$

and

$$f''(\eta_{max}) = \delta_2 \qquad \text{(B-6)}$$

where δ_1 and δ_2 are small numbers. Again performing a first-order Taylor series expansion gives

$$f'(\eta_{max}) = f'_c(\eta_{max}) + \frac{\partial f'(\eta_{max})}{\partial f''(0)} df''(0)$$

and

$$f''(\eta_{max}) = f''_c(\eta_{max}) + \frac{\partial f''(\eta_{max})}{\partial f''(0)} df''(0)$$

Using a finite difference representation of the differentials yields

$$\delta_1 = f'_c + f'_x \Delta x - 1 \qquad \text{(B-7)}$$
$$\delta_2 = f''_c + f''_x \Delta x \qquad \text{(B-8)}$$

where the required values for boundary conditions at η_{max} [cf. Eqs. (B-2) and (B-6)] were used to obtain these results. Nachtsheim and Swigert (Ref. B-1) obtained a least-squares solution to Eqs. (B-7) and (B-8). The procedure to minimize $\delta_1{}^2 + \delta_2{}^2$ with respect to x is to differentiate with respect to x and set the result equal to zero. Solving the result for $\Delta x = \Delta f''(0)$ yields

$$\Delta x = \frac{f'_x(1 - f'_c) - f''_x f''_c}{f'_x{}^2 + f''_x{}^2} \qquad \text{at } \eta_{max} \qquad \text{(B-9)}$$

Here again Nachtsheim and Swigert used a perturbation equation to determine the required convergence derivatives, that is, f'_x and f''_x. However, the finite difference representation of the convergence derivatives described above is successful. Examination of Eq. (B-9) gives the fundamental reason. If the finite difference representation of f'_x and f''_x yields values which are too large but with $f'_c(\eta_{max})$ and $f''_c(\eta_{max}) = 0(1)$, then the denominator will be proportionately larger than the numerator and Δx will be smaller. The next iteration will yield improved values of $f''(0)$ and $f'_c(\eta_{max})$, and hence with continuous upgrading of f'_x and f''_x, convergence to the required value of $f''(0)$ will occur. However, when using *either* the appropriate perturbation equations or a finite difference method to determine the convergence derivatives, it must be possible to determine these derivatives in some meaningful manner. This procedure is not possible if the initial estimate of $f''(0)$ is so poor that the solution diverges. Such a case is shown in Fig. B-2. Figure B-2 shows that for $\beta = 1.0$, trial solutions of the Falkner-Skan equation for $f''(0) = 1.23$ behave in approximately the same manner as the actual solution. Determination of the required convergence derivatives using trial solutions with $f''(0) = 1.23$ will yield acceptable values. Convergence to $f''(0) = 1.2325878$, the value required to satisfy the asymptotic outer boundary condition, occurs rapidly. However, trial solutions with $f''(0) = 1.0$ or 1.5 diverge or oscillate. Attempts to determine the required convergence

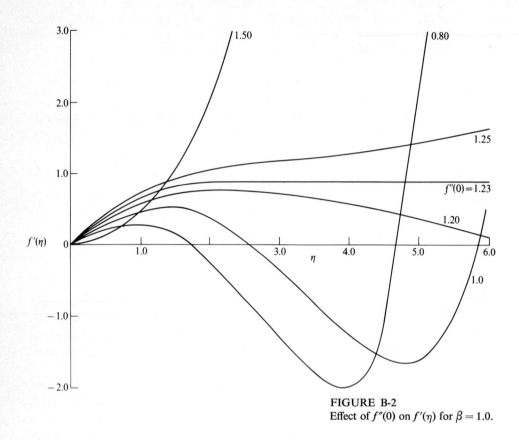

FIGURE B-2
Effect of $f''(0)$ on $f'(\eta)$ for $\beta = 1.0$.

derivatives from these trial solutions do not yield meaningful results. Convergence to the required value of $f''(0)$ does not occur. Nachtsheim and Swigert (Ref. B-1) present a technique for overcoming this difficulty. From Fig. B-2 observe that for small values of η the trial solutions for even poor estimates or $f''(0)$ behave approximately in the same manner as the correct solution. Thus by initially using a small value for η_{max}, say 2.0, the required convergence derivatives for the Nachtsheim-Swigert iteration technique can be approximately determined. This allows determination of a nearly correct value of $f''(0)$. Using this nearly correct value of $f''(0)$ and a larger value of η_{max}, say 5.0 or 6.0, a Nachtsheim-Swigert iteration scheme will rapidly yield the value of $f''(0)$ which will satisfy the required outer boundary conditions. Nachtsheim and Swigert studied the effect of η_{max} on convergence. Their results for $\beta = 1.0$ are shown in Table B-1.

Along with the discussion of the effect of η_{max} on convergence to the required solution, it is appropriate to consider the effect of η_{max} on acceptable errors in satisfying the asymptotic boundary conditions, that is, δ_1 and δ_2. Accepting the sum of the squares of the deviations δ_1 and δ_2 as a measure of the error in the solution, we write

$$E(\text{error}) = \delta_1{}^2 + \delta_2{}^2 = (1 - f_c')^2 \qquad \text{at} \quad \eta_{max}$$

Fig. B-3 may be determined. Figure B-3 shows that, unlike the Newton-Raphson method, the Nachtsheim-Swigert least-squares method yields a unique solution for $f''(0)$.

B-4   Nachtsheim-Swigert Iteration Scheme: Falkner-Skan Equation—Zero Shearing Stress Solutions $\sim \beta_0$

In determining the solutions for the Falkner-Skan equation it may be necessary to determine the *separation solution;* i.e., the value of β for which $f''(0) = 0$ and the required asymptotic boundary conditions are satisfied. For this case consider

$$f'(\eta_{max}) = f'(\beta) = 1 + \delta_1 \qquad \text{(B-10)}$$

and

$$f''(\eta_{max}) = f''(\beta) = \delta_2 \qquad \text{(B-11)}$$

where

$$f''(0) = 0.$$

Again expanding in a first-order Taylor series and seeking a least-squares solution gives

$$\Delta\beta = \frac{f'_\beta(1 - f'_c) + f''_\beta f''_c}{f'^2_\beta + f''^2} \qquad \text{(B-12)}$$

where $f'_\beta = \partial f'/\partial\beta$ and $f''_\beta = \partial f''/\partial\beta$ at η_{max}.

Table B-1   EFFECT OF η_{max} ON CONVERGENCE RATE

$f''(0)$ (initial guess)	$\eta_{max} = 2$		$\eta_{max} = 5$	
	$f''(0)$ (after two trials)	Percent difference from required value	$f''(0)$ (after two trials)	Percent difference from required value
0.25	1.2799034	3.83	0.26706180	−78.33
0.50	1.2449846	1.00	0.69205364	−43.85
0.75	1.2292604	−0.26	0.88486032	−28.21
1.00	1.2266764	−0.47	1.0413712	−15.51
1.25	1.2266282	−0.48	1.2325888	0.00
1.50	1.2266765	−0.47	1.2444813	0.96
1.75	1.2277231	−0.39	1.2890732	4.58
2.00	1.2312423	−0.10	1.3564389	10.04
2.25	1.2383960	0.47	1.8834930	52.80
2.50	1.2498889	1.40	2.4946776	102.39
2.75	1.2660459	2.71	· · ·	· · ·
3.00	1.2869264	4.40	· · ·	· · ·

Trial	Correction	η_{max}	$f''(0)$
1	· · ·	2.0	1.0
2	1	2.0	1.2463981
3	2	5.0	1.2266764
4	3	5.0	1.2326729
5	4	5.0	1.2325878
· · ·	5	· · ·	1.2325878

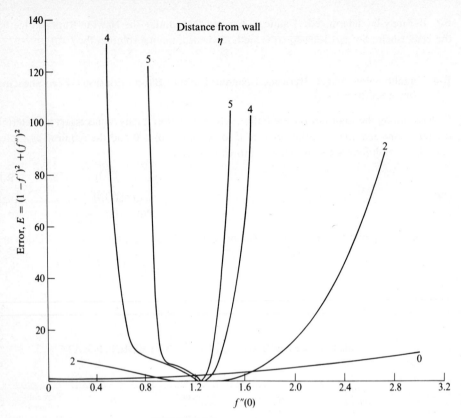

FIGURE B-3
Error as a function of $f''(0)$ for different values of η_{max}, $\beta = 1.0$.

B-5   Nachtsheim-Swigert Iteration Scheme: Energy Equation

The boundary conditions for the incompressible energy equation (5-45) may be stated as

$$\theta(0) = 1.0 \qquad \text{(B-13a)}$$
$$\theta(\eta \to \infty) \to 0 \qquad \text{(B-13b)}$$

The asymptotic outer boundary condition may functionally be stated as

$$\theta(\eta_{max}) = \theta[\theta'(0)] = \delta_1 \qquad \text{(B-13c)}$$
$$\theta'(\eta_{max}) = \theta[\theta'(0)] = \delta_2 \qquad \text{(B-13d)}$$

Expanding in a first-order Taylor series and seeking a least-squares solution for Δy yields

$$\Delta\theta'(0) = \Delta y = \frac{\theta_y\,\theta_c - \theta'_y\,\theta'_c}{\theta_y{}^2 + \theta'_y{}^2} \qquad \text{at} \qquad \eta_{max} \qquad \text{(B-14)}$$

where
$$\theta_y = \frac{\partial\theta(\eta_{max})}{\partial\theta'(0)} \qquad \text{and} \qquad \theta'_y = \frac{\partial\theta'(\eta_{max})}{\partial\theta'(0)}$$

The energy equation can be stated in an alternate form with $\bar{\theta} = (T - T_w)/(T_\infty - T_w)$ and with boundary conditions given by

$$\bar{\theta}(0) = 0 \qquad \text{(B-15a)}$$
$$\bar{\theta}(\eta \to \infty) \to 1.0 \qquad \text{(B-15b)}$$

The required correction is then

$$\Delta y = \frac{\bar{\theta}_y(1 - \bar{\theta}_c) - \bar{\theta}_y' \bar{\theta}_c'}{\bar{\theta}_y{}^2 + \bar{\theta}_y'{}^2} \qquad \text{(B-16)}$$

B-6 Nachtsheim-Swigert Iteration Scheme: Energy Equation Adiabatic Wall Solution, θ_{aw}

When there is no heat transfer at the surface, the solution obtained for the energy equation is the adiabatic wall solution. For the adiabatic wall solution the appropriate boundary conditions for the energy equation are

$$\bar{\theta}'(0) = 0 \qquad \bar{\theta}(\eta \to \infty) \to 1 \qquad \text{(B-17)}$$

The asymptotic outer boundary conditions are then functionally expressed as

$$\bar{\theta}(\eta_{\max}) = \bar{\theta}(\bar{\theta}_{aw}) = 1 + \delta_1 \qquad \text{(B-18a)}$$
$$\bar{\theta}'(\eta_{\max}) = \bar{\theta}'(\bar{\theta}_{aw}) = \delta_2 \qquad \text{(B-18b)}$$

The Nachtsheim-Swigert correction to $\bar{\theta}_{aw}$ is then

$$\Delta \bar{\theta}_a = \frac{\bar{\theta}_a(1 - \bar{\theta}_c) - \bar{\theta}_a' \bar{\theta}_c'}{\bar{\theta}_a{}^2 + \bar{\theta}_a'{}^2} \qquad \text{(B-19)}$$

where $\bar{\theta}_a = \partial\theta(\eta_{\max})/\partial\theta_{aw}$ and $\bar{\theta}_a' = \partial\theta'(\eta_{\max})/\partial\theta_{aw}$.

B-7 Nachtsheim-Swigert Iteration Scheme: Systems of Equations—The Free-Convection Boundary Layer—$f''(0)$, $\theta'(0)$

Extension of the Nachtsheim-Swigert iteration scheme to systems of differential equations is straightforward, as was shown in Ref. B-1. This extension is illustrated by considering the boundary-value problem associated with free convection from a vertical isothermal flat plate

$$f''' + 3ff'' - 2f'^2 + \theta = 0$$
$$\theta'' + 3\mathrm{Pr}f\theta' = 0$$

with boundary conditions

$$f(0) = f'(0) = 0 \qquad \theta(0) = 1$$

and

$$f'(\eta \to \infty) \to 0 \qquad \theta(\eta \to \infty) \to 0$$

Here there are two asymptotic boundary conditions and hence two unknown surface conditions $f''(0)$ and $\theta'(0)$. Within the context of the initial-value method and the Nachtsheim-Swigert iteration technique the outer boundary conditions may be functionally represented as

$$f'(\eta_{\max}) = f'[f''(0),\theta'(0)] = \delta_1 \qquad \text{(B-20)}$$
$$\theta(\eta_{\max}) = \theta[f''(0),\theta'(0)] = \delta_2 \qquad \text{(B-21)}$$

with the asymptotic convergence criteria given by

$$f''(\eta_{max}) = f''[f''(0), \theta'(0)] = \delta_3 \qquad \text{(B-22)}$$

$$\theta'(\eta_{max}) = \theta'[f''(0), \theta'(0)] = \delta_4 \qquad \text{(B-23)}$$

Expanding in a first-order Taylor series after using Eqs. (B-20) to (B-23) yields

$$f'(\eta_{max}) = f_c'(\eta_{max}) + f_x' \, \Delta x + f_y' \, \Delta y = \delta_1 \qquad \text{(B-24)}$$

$$\theta(\eta_{max}) = \theta_c(\eta_{max}) + \theta_x \, \Delta x + \theta_y \, \Delta y = \delta_2 \qquad \text{(B-25)}$$

$$f''(\eta_{max}) = f_c''(\eta_{max}) + f_x'' \, \Delta x + f_y'' \, \Delta y = \delta_3 \qquad \text{(B-26)}$$

$$\theta'(\eta_{max}) = \theta_c'(\eta_{max}) + \theta_x' \, \Delta x + \theta_y' \, \Delta y = \delta_4 \qquad \text{(B-27)}$$

where $x = f''(0)$, $y = \theta'(0)$, and the x and y subscripts indicate partial differentiation, e.g., $f_y' = \partial f'(\eta_{max})/\partial \theta'(0)$. The c subscript indicates the value of the function at η_{max} determined from the trial integration.

Solution of these equations in a least-squares sense requires determining the minimum value of $E = \delta_1^2 + \delta_2^2 + \delta_3^2 + \delta_4^2$ with respect to x and y. Differentiating E with respect to x yields

$$(f_x'^2 + \theta_x^2 + f_x''^2 + \theta_x'^2) \, \Delta x + (f_x' f_y' + \theta_x \theta_y + f_x'' f_y'' + \theta_x' \theta_y') \, \Delta y$$
$$= -(f_c' f_x' + \theta_c \theta_x + f_c'' f_x'' + \theta' \theta_x)$$

or
$$Q_{xx} \, \Delta x + Q_{xy} \, \Delta y = +Q_{cx}$$

Similarly differentiating E with respect to y yields

$$(f_x' f_y' + \theta_x \theta_y + f_x'' f_y'' + \theta_x' \theta_y') \, \Delta x + (f_y'^2 + \theta_y^2 + f_y''^2 + \theta_y'^2) \, \Delta y$$
$$= -(f_c' f_y' + \theta_c \theta_y + f_c'' f_y'' + \theta_c' \theta_y')$$

or
$$Q_{xy} \, \Delta x + Q_{yy} \, \Delta y = +Q_{cy}$$

The solutions for Δx and Δy, the corrections to $f''(0)$ and $\theta'(0)$ which will yield asymptotic convergence to the required outer boundary conditions, are given by

$$\Delta x = \frac{Q_{cx} Q_{yy} - Q_{cy} Q_{xy}}{Q_{xx} Q_{yy} - Q_{xy}^2} \qquad \text{(B-28)}$$

and

$$\Delta y = \frac{Q_{xx} Q_{cy} - Q_{xy} Q_{cx}}{Q_{xx} Q_{yy} - Q_{xy}^2} \qquad \text{(B-29)}$$

The partial derivatives required above may be determined from the appropriate perturbation equations (cf. Ref. B-1) or by using the finite difference approximation previously discussed in Sec. B-2. The FREEBL program in Chap. 5 uses the finite difference approximation. In order to implement the finite difference approximation for a two-parameter iteration, it is necessary to perform three trial integrations with appropriate values of $f''(0)$ and $\theta'(0)$. These values are

1   $f''(0) = k_1$ $\theta'(0) = l_1$
2   $f''(0) = k_1 + \varepsilon$ $\theta'(0) = l_1$
3   $f''(0) = k_1$ $\theta'(0) = l_1 + \varepsilon$

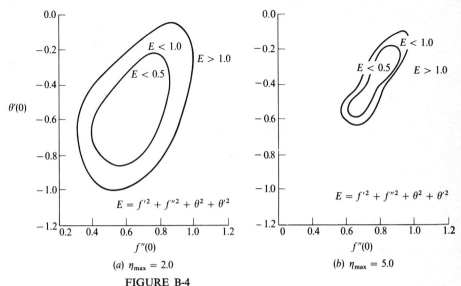

(a) $\eta_{\max} = 2.0$

(b) $\eta_{\max} = 5.0$

FIGURE B-4
Level curves of error for the free convection boundary-layer equations.

where k_1 and l_1 each is an assumed constant value, and ε is a small number, say between 10^{-5} and 10^{-2}. Runs 1 and 2 yield the change in f', f'', θ, and θ' at $\eta_{\max}$ with a change in $f''(0)$ and hence the convergence derivatives f'_x, f''_x, θ_x, and θ'_x. Runs 1 and 3 yield the change in f', f'', θ, and θ' at $\eta_{\max}$ with a change in $\theta'(0)$ and hence the convergence derivatives f'_y, f''_y, θ_y, and θ'_y. It should be noted that for a two-parameter iteration as described above, continuous upgrading of the convergence derivatives is not possible when a finite difference approximation is used to estimate the derivatives. Since the arguments used in Sec. B-3 also apply in this case, this restriction is not as important as it appears.

The error in satisfying the asymptotic boundary conditions is

$$E = \delta_1{}^2 + \delta_2{}^2 + \delta_3{}^2 + \delta_4{}^2 = f_c'{}^2 + f_c''{}^2 + \theta_c{}^2 + \theta_c'{}^2 \qquad \text{at} \qquad \eta_{\max}$$

The resulting level curves of error for the free-convection problem are shown in Fig. B-4. Here we see that increasing $\eta_{\max}$ decreases the possible range of initial estimates of $f''(0)$ and $\theta'(0)$ which will eventually yield convergence to the required accuracy. In fact for very large values of $\eta_{\max}$ or very small values of E the level curve of error reduces to a single point; i.e., one must guess the exact solution.

Reference

B-1   NACHTSHEIM, P. R., and P. SWIGERT: Satisfaction of Asymptotic Boundary Conditions in Numerical Solution of Systems of Non-linear Equations of the Boundary Layer Type, NASA TN D-3004, October 1965.

APPENDIX C

SYSTEMS OF UNITS

Length: 1 ft $= 0.3048$ m (exactly).

Mass: 1 lb $= 0.45359237$ kg (exactly).

Density, concentration: 1 kg/m^3 $= 10^{-3}$ g/cm^3 $=$ g/l.
Density of water is about 10^3 kg/m^3.
Density of air is about 1 kg/m^3 at NTP.
1 lb/ft^3 $= 16.0185$ kg/m^3.

Acceleration: Standard gravity $= 9.80665$ m/s^2.

Force: 1 N $= 10^5$ dyn. The weight of an apple is about 1 newton!
1 kg$_f$ $= 9.80665$ N (exactly). 1 lb$_f$ $= 4.44822$ N.
1 pdl $= 0.138255$ N.

Pressure: 1 N/m^2 $= 10$ dyn/cm^2. 10^5 N/m^2 is called the *bar* and is roughly
equal to atmospheric pressure.
1 lb$_f$/in^2 $= 6{,}894.76$ N/m^2 ≈ 7 kN/m^2.
1 kg$_f$/cm^2 (*technical atmosphere*) $= 98{,}066.5$ N/m^2 (exactly).
1 atm (*normal* or *standard atmosphere*) $= 101{,}325$ N/m^2 (exactly).
1 mm Hg $= 133.322$ N/m^2. 1 in H$_2$O $= 249.089$ N/m^2.
1 torr $= 133.322$ N/m^2.

Dynamic viscosity:	1 N-s/m^2 (poiseuille) = 10 poise.
	Dynamic viscosity of water is about 10^{-3} poiseuille, or 1 mPl.
	1 lb$_f$ s/ft^2 = 47.8803 Pl. 1 lb/ft-h = 4.13377×10^{-4} Pl.
Kinematic viscosity:	1 m^2/s = 10^4 St. 1 ft^2/s = 0.0929030 m^2/s.
	1 ft^2/h = 2.58064×10^{-5} m^2/s.
Energy, work, heat, etc.:	1 J = 10^7 erg.
	1 ft-lb$_f$ = 1.35582 J. 1 kg$_f$ m = 9.80665 J.
	1 cal. (intern. table) = 4.1868 J (exactly).
	1 Btu (intern. table) = 1,055.06 J (exactly).
	1 kW-h = 3.6×10^6 J.
Power, heat flux, etc.:	1 W = 10^7 erg/s.
	1 kg$_f$-m/s = 9.80665 W. 1 ft-lb$_f$/s = 1.35582 W.
	1 hp (British) = 745.700 W.
	1 hp (metric) = 735.499 W.
	1 kcal$_{IT}$/h = 1.163 W (exactly).
	1 Btu/h = 0.293071 W.
Heat flux density:	1 kcal$_{IT}$/m^2-h = 1.163 W/m^2 (exactly).
	1 Btu/ft^2-h = 3.15459 W/m^2.
Heat transfer coefficient:	1 kcal$_{IT}$/m^2-h-°C = 1.163 W/m^2-°C.
	1 Btu/ft^2-h-°F = 5.67826 W/m^2-°C.
Enthalpy, latent heat (specific):	1 kcal$_{IT}$/kg = 4.1868 kJ/kg (exactly).
	1 Btu/lb = 2,326 J/kg (exactly).
	Latent heat of boiling water is about 2 MJ/kg.
Heat capacity (specific):	1 kcal$_{IT}$/kg°C = 1 Btu/lb-°F = 4.1868 kJ/kg°C (exactly).
	Specific heat capacity of water is about 4 kJ/kg-°C; of air, 1 kJ/kg-°C at NTP.
Thermal conductivity:	1 kcal/m-h-°C = 1.163 W/m-°C.
	1 Btu/ft-h-°F = 1.7307 W/m-°C.
	Thermal conductivity of water is about 0.6 W/m-°C.

The SI System

The six basic units of the SI System are defined as follows:

Length—METER (m): the length equal to 1,650,763.73 wavelengths in vacuum of the radiation corresponding to the transition between two specified levels of the krypton-86 atom.

Mass—KILOGRAM (kg): the mass of the international prototype held at the Bureau International des Poids et Mesures in Paris. Practically, it is the weight of 1,000 cm^3 of distilled water at 0°C.

Time—SECOND (s): the fraction 1/31,556,925.9747 of the tropical year for 1900 January 1 at 12 h ephemeris time.

Electrical Current—AMPERE (A): the constant current which, if maintained in two parallel conductors of infinite length of negligible circular cross-section and placed one meter apart in vacuum, would produce a force equal to 2×10^{-7} newton per meter length between the conductors.

Thermodynamic Temperature—DEGREE KELVIN (K): the degree interval of the thermodynamic scale on which the temperature of the triple point of water is 273.16 degrees.

Luminous Intensity—CANDELA (cd): the luminance of a full radiator, at a temperature equal to the solidification temperature of platinum, which emits 60 units of luminous intensity per square centimeter.

Five derived SI units of special interest in this text are:

Force—NEWTON (N): that force which, when applied to a body having a mass of one kilogram, gives the mass an acceleration of one meter per second per second. $N \equiv kg\text{-}m/s^2$.

Work, Energy, Heat—JOULE (J): the work done when the point of application of a force equal to one newton is displaced through a distance of one meter in the direction of the force. $J \equiv N\text{-}m$.

Power—WATT (W): a rate of energy transference equivalent to one joule per second. $W \equiv J/s$.

Dynamic Viscosity—POISEUILLE (Pl): a measure of viscosity equal to one newton-second per square meter. $Pl \equiv N\text{-}s/m^2$.

Kinematic Viscosity—STOKE (St): a measure of kinematic viscosity equal to 10^{-4} square meters per second. $St \equiv (m^2/s) \times 10^{-4}$.

APPENDIX D

PROPERTY DATA

Many handbooks and indexes contain tabulated values of the transport and thermodynamic properties for modern engineering materials. Tabulations of required property data for materials in the solid, liquid, and gas phases, under various environmental conditions, are extensive. Most materials find use over wide ranges of temperature and pressure. When making a final selection of material for a certain application, one should consult standard references for accurate property data for a specific material within the appropriate ranges of temperature and pressure. Some references are suggested at the end of this appendix.

This text encourages a generalized, nondimensional approach to heat transfer analysis. Many of the governing nondimensional parameters contain thermodynamic and transport properties, and approximate values of these properties must be known to help estimate reasonable orders of magnitude for the parameters. The limited property data given here are for the convenience of the reader and allows a realistic parametric analysis to be carried out. Most of the suggested problems in each chapter give typical property data for the material suggested in the problem. The data presented in this appendix can be used to obtain an indication of the effect on heat transfer if various other materials are used for a given application.

Appendix D is arranged as follows:

Table D-1 PROPERTY DATA FOR METALS

Metal type	Temperature range T °C	Density ρ g/cm³	Specific heat c kJ/kg-°C	Thermal conductivity k W/m-°C	Total emittance ε
Aluminum	0–400	2.72	0.895	204–250	0.04–0.06 (polished) 0.07–0.09 (commercial) 0.2–0.3 (oxidized)
Antimony	0–300	6.90	0.38	18–15	0.28–0.31 (polished)
Brass (70% Cu, 30% Zn)	100–300	8.5	0.37	104–139	0.03–0.07 (polished) 0.2–0.25 (commercial) 0.45–0.55 (oxidized)
Bronze	0–100	8.5	0.38	58	0.03–0.07 (polished) 0.4–0.5 (oxidized)
Constantan (60% Cu, 40% Ni)	0–100	8.9	0.42	22–26	0.03–0.06 (polished) 0.2–0.4 (oxidized)
Copper	0–600	9.0	0.38	385–350	0.02–0.04 (polished) 0.1–0.2 (commercial)
Inconel (cast)	0–500	8.3	0.46	15	0.2 (polished) 0.5–0.8 (oxidized)
Iron ($C \approx 4\%$, cast)	0–1,000	7.26	0.419	52–35	0.2–0.25 (polished) 0.55–0.65 (oxidized) 0.6–0.8 (rusted)

Table D-1 PROPERTY DATA FOR METALS (*continued*)

Metal type	Temperature range T °C	Density ρ g/cm³	Specific heat c kJ/kg-°C	Thermal conductivity k W/m-°C	Total emittance ε
Iron (C ≈ 0.5%, wrought)	0–1,000	7.85	0.460	59–35	0.3–0.35 (polished) 0.9–0.95 (oxidized)
Lead	0–300	11.4	0.13	35–29	0.05–0.08 (polished) 0.3–0.6 (oxidized)
Magnesium	0–300	1.75	1.01	170–155	0.07–0.13 (polished)
Mercury	0–300	13.4	0.125	8–10	0.1–0.12
Molybdenum	0–1,000	1.02	0.251	125–99	0.06–0.10 (polished)
Monel metal	0–500	8.85	0.545	27	0.17 (polished) 0.4–0.46 (oxidized)
Nickel	0–400	8.9	0.46	90–59	0.05–0.07 (polished) 0.35–0.49 (oxidized)
Platinum	0–1,000	21.4	0.24	70–75	0.05–0.10 (polished) 0.07–0.11 (oxidized)
Silver	0–400	10.5	0.23	410–360	0.01–0.03 (polished) 0.02–0.04 (oxidized)
Steel (C ≈ 1%)	0–1,000	7.80	0.47	43–28	0.07–0.17 (polished)
Steel (Cr ≈ 1%)	0–1,000	7.86	0.46	62–33	0.07–0.17 (polished)
Steel (Cr ≈ 18%, Ni ≈ 8%)	0–1,000	7.81	0.46	16–26	0.07–0.17 (polished)
Tin	0–200	7.3	0.23	66–57	0.04–0.06 (polished)
Tungsten	0–1,000	19.3	0.13	166–76	0.04–0.08 (polished) 0.1–0.2 (filament)
Zinc	0–400	7.2	0.38	110–93	0.02–0.03 (polished) 0.10–0.11 (oxidized) 0.2–0.3 (galvanized)

Table D-2   PROPERTY DATA FOR NONMETALS

Material	Temperature range T °C	Density ρ g/cm³	Specific heat c kJ/kg-°C	Thermal conductivity k W/m-°C	Total emittance ε
Asbestos	100–1,000	0.58	0.82	0.15–0.22	0.93–0.97
Brick	100–1,000	1.76	0.84	0.38–0.43	0.90–0.95 (rough red) 0.80–0.85 (silica) 0.75 (fireclay)
Clay	0–200	1.46	0.88	1.3	0.91
Concrete	0–200	2.1	0.88	0.81–1.4	0.94
Glass (quartz)	0–600	2.2	0.84	0.78–1.0	0.94–0.66
Ice	0	0.91	1.9	2.2	0.97–0.99
Limestone	100–400	2.5	0.92	1.3	0.95–0.80
Marble	0–100	2.56	0.79	2.8	0.93–0.95
Plasterboard	0–100	1.25	0.84	0.43	0.92
Rubber (hard)	0–100	1.2	1.42	0.15	0.94
Sandstone	0–300	2.24	0.8	1.7 (moist) 1.3 (dry)	0.83–0.9
Wood (oak)	0–100	0.6–0.8	2.4	0.17–0.21	0.90

Table D-3   TOTAL SURFACE EMITTANCE FOR COATINGS

Coating	Temperature range T °C	Total surface emittance ε
White enamel	0–100	0.9
Lampblack	0–400	0.96–0.95
Gypsum	0–100	0.9
Lubricating oil	0–100	0.3 (0.001-in-thick film) 0.5 (0.002-in-thick film) 0.7 (0.005-in-thick film)
Oil paints	0–100	0.92–0.96
Black lacquer	0–100	0.80–0.95
Black shellac	0–100	0.80–0.83
Aluminum paints (10% Al, 22% lacquer)	0–100	0.5
Aluminum paints (26% Al, 27% lacquer)	0–100	0.3
Paper	0–50	0.92
Plaster	0–100	0.94
Water	0–100	0.95–0.97

Table D-4   ABSORPTANCES FOR SOLAR IRRADIATION*

Material surface	Total absorptance α
Aluminum (polished)	0.26
Asphalt	0.92
Brick (red)	0.7–0.75
Chromium	0.5
Concrete	0.87
Copper (polished)	0.26
Iron (rusted)	0.74
Iron (galvanized)	0.89
Nickel (electrolytically deposited)	0.60
Paper (roofing)	0.88
Water (open sea)	0.96

* As compiled by H. SCHENCK: "Heat Transfer Engineering," Prentice-Hall, Englewood Cliffs, N.J., 1959.

Table D-5   RANGE OF PROPERTY DATA FOR SATURATED LIQUIDS

Liquid	Temperature T °C	Density ρ g/cm³	Specific heat c kJ/kg-°C	Thermal conductivity k W/m-°C	Kinematic viscosity ν cm²/s	Isobaric compressibility $\bar{\beta}$ 1/K	Prandtl number
Ammonia	−50	0.71	4.6	0.55	0.44×10^{-2}		2.6
	0	0.64	4.6	0.54	0.37×10^{-2}	2.45×10^{-3}	2.1
	50	0.58	5.0	0.49	0.28×10^{-2}		2.0
Carbon dioxide	−50	1.15	1.84	0.085	0.119×10^{-2}		2.96
	0	0.92	2.5	0.01	0.11×10^{-2}	6.6×10^{-3}	2.38
	30	0.60	3.6	0.07	0.08×10^{-2}		28.7
Ethylene glycol	0	1.13	2.3	0.24	57.5×10^{-2}	0.65×10^{-3}	615
	100	1.06	2.76	0.26	2.05×10^{-2}		22
Freon	−50	1.54	0.88	0.067	0.31×10^{-2}		6.2
	0	1.39	0.92	0.073	0.215×10^{-2}	2.52×10^{-4}	3.8
	50	1.22	1.0	0.067	0.185×10^{-2}		3.5
Fuel JP4	0	0.82	2.0	0.142	2.02×10^{-2}		22
	100	0.72	2.34	0.135	0.57×10^{-2}		8.0
Glycerin	0	1.28	2.26	0.28	8.26×10^{2}	0.51×10^{-3}	85×10^{3}
	50	1.25	2.6	0.29	1.49		1.6×10^{3}
Hydraulic fluid (MIL-H-5606)	0	0.88	1.75	0.135	0.94		1×10^{3}
	100	0.80	2.22	0.10	0.103		1×10^{2}
Hydrogen	−260	0.074	6.7	0.095	0.336×10^{-2}		1.85
	−242	0.051	20.8	0.14	0.194×10^{-2}		0.8
Oil (engine)	0	0.89	1.8	0.147	4.28×10	0.7×10^{-3}	47.1×10^{3}
	150	0.80	2.5	0.132	0.056		84
Oxygen	−212	1.28	1.67	0.19	0.45×10^{-2}		5.0
	−130	0.90	1.67	0.087	0.092×10^{-2}		1.5
Sulfur dioxide	−50	1.56	1.34	0.242	0.48×10^{-2}		4.24
	0	1.44	1.34	0.208	0.29×10^{-2}	1.95×10^{-3}	2.38
	50	1.30	1.38	0.173	0.16×10^{-2}		1.6
Water	0	1.0	4.18	0.55	1.77×10^{-2}		13.6
	100	0.96	4.18	0.69	0.30×10^{-2}	0.18×10^{-3}	6.5
	300	0.72	4.35	0.54	0.14×10^{-2}		1.0

Table D-6 RANGE OF PROPERTY DATA FOR LIQUID METALS

Liquid metal	Temperature T °C	Density ρ g/cm^3	Specific heat c kJ/kg-°C	Thermal conductivity k W/m-°C	Kinematic viscosity ν cm^2/s	Prandtl number
Bismuth	320 760	10 9.5	0.146 0.163	16.5 15.6	0.162×10^{-2} 0.094×10^{-2}	0.0142 0.0083
Lead	370 700	10.6 10.2	0.159 0.159	16.1 14.9	0.228×10^{-2} 0.135×10^{-2}	0.024 0.017
Lead (44.5%), bismuth (55.5%)	150 650	10.5 9.9	0.146 0.146	9.0 12.0	0.167×10^{-2} 0.179×10^{-2}	0.024 0.024
Lithium	200 650	5.12 4.65	4.18 4.18	38 38	1.16×10^{-2} 0.84×10^{-2}	0.065 0.065
Mercury	10 150	13.6 13.3	0.138 0.138	8.1 11.6	1.17×10^{-2} 0.81×10^{-2}	0.027 0.012
Potassium	150 700	0.8 0.68	0.8 0.75	45 33	0.46×10^{-2} 0.19×10^{-2}	0.0066 0.0031
Sodium	94 700	0.93 0.77	1.38 1.25	86 60	0.75×10^{-2} 0.23×10^{-2}	0.011 0.0038
Sodium (56%), potassium (44%)	94 700	0.88 0.74	1.13 1.05	25.6 29	0.66×10^{-2} 0.22×10^{-2}	0.026 0.0058
Sodium (22%), potassium (78%)	94 760	0.85 0.69	0.94 0.88	24.4 26.6	0.58×10^{-2} 0.21×10^{-2}	0.019 0.0068

Table D-7 RANGE OF PROPERTY DATA FOR GASES AND VAPORS

Gas or vapor	Temperature T °C	Density ρ kg/m³	Specific heat c kJ/kg·°C	Thermal conductivity k W/m·°C	Kinematic viscosity ν cm²/s	Prandtl number
Air ($P = 1$ atm)	−170	3.61	1.02	0.0092	1.92×10^{-2}	0.77
	400	0.513	1.10	0.050	62.5×10^{-2}	0.68
	1,400	0.208	1.27	0.095	272×10^{-2}	0.75
Ammonia ($P = 1$ atm)	−50	0.385	2.2	0.017	19×10^{-2}	0.93
	200	0.44	2.4	0.047	37.5×10^{-2}	0.84
Carbon dioxide ($P = 1$ atm)	−50	2.4	0.78	0.0107	4.5×10^{-2}	0.82
	300	0.96	1.05	0.0415	30×10^{-2}	0.67
Carbon monoxide ($P = 1$ atm)	−50	1.60	1.04	0.019	8.9×10^{-2}	0.76
	300	0.58	1.08	0.043	52×10^{-2}	0.72
Helium ($P = 1$ atm)	−240	1.48	5.2	0.035	3.42×10^{-2}	0.74
	0	0.192	5.2	0.0135	96×10^{-2}	0.70
	600	0.056	5.2	0.288	710×10^{-2}	0.72
Hydrogen ($P = 1$ atm)	−240	0.85	10.9	0.0225	1.9×10^{-2}	0.76
	0	0.088	14.2	0.165	94×10^{-2}	0.71
	1,000	0.019	15.5	0.50	$1,230 \times 10^{-2}$	0.73
Nitrogen ($P = 1$ atm)	−170	3.53	1.05	0.0095	1.97×10^{-2}	0.78
	230	0.64	1.05	0.040	38×10^{-2}	0.68
	970	0.29	1.21	0.073	156×10^{-2}	0.75
Oxygen ($P = 1$ atm)	−170	3.84	0.92	0.0097	1.95×10^{-2}	0.80
	300	0.67	1.0	0.047	49×10^{-2}	0.70
Steam ($P = 1$ atm)	100	0.59	1.88	0.0242	21.4×10^{-2}	0.98
	500	0.284	2.1	0.069	95×10^{-2}	0.89
Steam ($P = 68.04$ atm)	300	33.1	5.0	0.057	0.6×10^{-2}	1.8
	500	20.5	2.4	0.078	1.42×10^{-2}	9.8

References

D-1 PERRY, J. H.: "Chemical Engineers Handbook," 4th ed., McGraw-Hill, New York, 1963.

D-2 MARKS, L. S.: "Mechanical Engineers Handbook," 6th ed., McGraw-Hill, New York, 1964.

D-3 "International Critical Tables," McGraw-Hill, New York, 1926–1930.

D-4 KOWALCZYK, L. S.: Thermal Conductivity and Its Variability with Temperature and Pressure, *Trans. ASME*, vol. 77, pp. 1021–1036, October 1955.

D-5 HILSENRATH, J., *et al.*: Tables of Thermal Properties of Gases, *Natl. Bur. Std. Circ.* 564, November 1955.

D-6 SCOTT, R. B.: "Cryogenic Engineering," Van Nostrand, Princeton, N.J., 1959.

D-7 "Aerospace Applied Thermodynamics Manual," Society of Automotive Engineers, Inc., February 1960.

D-8 ECKERT, E. R. G., and R. M. DRAKE: "Heat and Mass Transfer," McGraw-Hill, New York, 1959.

D-9 "Handbook of Chemistry and Physics," 44th ed., C. D. Hodgman (ed.), Chemical Rubber Publishing Co., Cleveland, 1956–1957.

APPENDIX E

NOMENCLATURE

To be consistent with the accepted nomenclature, it is sometimes necessary to use the same symbol for more than one meaning. To avoid confusion, each symbol is defined when it is used in the textual material. Also, separate tables exist in the text to define the computer nomenclature used in documented programs.

Symbol	Meaning	Symbol	Meaning
A	Area	D	Diameter
a	Geometrical parameter	D_H	Hydraulic diameter
B	Constant of integration	E	Emissive power or total energy
b	Geometrical parameter	e	Geometrical parameter
C	Constant of integration or fluid heat capacity	F_{i-j}	Radiation shape factor between black surfaces
C_f	Skin friction	$\mathscr{F}_{i-j}$	Radiation shape factor between gray surfaces
c	Specific heat or parameter		
c_n	Mathematical constants	f	Nondimensional dependent variable or Fanning friction factor
c_p	Specific heat at constant pressure		

G	Irradiation or mass flow rate per unit area	r	Radial distance or geometrical parameter
g	Acceleration due to gravity	$\bar{r}$	Nondimensional independent variable
g_i	Orthogonal functions		
H	Mathematical function	s	Constant
h	Static enthalpy	T	Temperature
$\bar{h}$	Average convection heat transfer coefficient	t	Time
		$\bar{t}$	Nondimensional time
h_x	Local convection heat transfer coefficient	U	Total internal energy or reference velocity or overall conductance
h_{fg}	Latent heat of fusion		
I	Radiation intensity	u	Specific internal energy or x direction velocity or radial velocity
I_n	Modified Bessel function		
$\hat{\imath}$	Unit vector		
J	Radiosity	V	Velocity or volume
J_0	Bessel function of first kind	v	y direction velocity or circumferential velocity
$\hat{\jmath}$	Unit vector		
k	Thermal conductivity	δW	Net work
L	Characteristic length	w	z direction velocity or axial velocity
l	Length		
m	Mass or constant exponent	X	Function of x
$\dot{m}$	Mass flow rate	x	Dimensional independent variable
n	Normal direction or constant		
P	Pressure	$\bar{x}$	Nondimensional independent variable
p	Perimeter		
δQ	Net heat transfer	Y	Function of y
q	Heat transfer rate	Y_0	Bessel function of second kind
q''	Heat transfer rate per unit area	y	Dimensional independent variable
q'''	Internal energy generation per unit volume	$\bar{y}$	Nondimensional independent variable
R	Radius or resistance to heat transfer	Z	Elevation
		z	Dimensional independent variable
R_T	Recovery factor		

Greek letter	Meaning		
α	Absorptance or thermal diffusivity	δ^*	Displacement thickness
		ε	Emittance
β	Pressure gradient parameter	ε_H	Eddy diffusivity of momentum
$\bar{\beta}$	Isobaric compressibility	ε_M	Eddy diffusivity of heat
Γ	Energy thickness	ζ	Ratio of thermal boundary-layer thickness to velocity boundary-layer thickness
δ	Incremental radial distance or boundary-layer thickness		

η	Nondimensional independent variable	ν	Kinematic viscosity
Θ	Momentum thickness	ξ	Nondimensional distance
θ	Nondimensional temperature or angular measurement	ρ	Reflectance or density
κ	Thermal boundary-layer parameter	σ	Stefan-Boltzmann constant
		σ_s	Surface tension
λ	Wavelength or eigenvalue or velocity boundary-layer parameter	T	Function of time
		τ	Transmittance or nondimensional time or shearing stress
μ	Dynamic viscosity	ϕ	Angular measurement
N	Energy thickness parameter	ψ	Angular measurement or stream function
		ω	Solid angle

Nondimensional parameters		Definition
Bi	(Biot number)	$\dfrac{\hbar l}{k_s}$
E	(Eckert number)	$\dfrac{U^2}{c_p \Delta T}$
F	(Fourier number increment)	$\dfrac{\alpha \Delta t}{(\Delta x)^2}$
Fo	(Fourier number)	$\dfrac{\alpha t}{l^2}$
fo	(Fourier modulus)	$\dfrac{\alpha \Delta t}{l^2}$
Gr	(Grashof number)	$\dfrac{g\bar{\beta} \Delta T L^3}{\nu^2}$
Gz	(Graetz number)	$RePr\dfrac{D}{L}$
L1	(length to diameter ratio)	$\dfrac{L}{D}$
N1, N2, N3, N4, N5, N6, N7, N8 (parameters)		defined in text

NTU	(number of transfer units)	$\dfrac{UA}{C}$
$\overline{Nu}$	(average Nusselt number)	$\dfrac{\hbar L}{k_f}$
Nu_x	(local Nusselt number)	$\dfrac{h_x x}{k_f}$
Pe	(Peclet number)	$RePr$
Pr	(Prandtl number)	$\dfrac{\mu c_p}{k_f}$
Re	(Reynolds number)	$\dfrac{UL}{\nu}$
Ro	(radiation parameter)	ml
St	(Stanton number)	$\dfrac{\hbar}{\rho U c}$
$\overline{\Delta T}$	(log-mean temperature difference)	$\dfrac{\Delta T_G - \Delta T_L}{\ln(\Delta T_G/\Delta T_L)}$